Nuclear Weapons Databook

Volume I
U.S. Nuclear Forces and Capabilities

Nuclear Weapons Databook

Volume I
U.S. Nuclear Forces and Capabilities

Thomas B. Cochran, William M. Arkin, and Milton M. Hoenig

A book by the
Natural Resources Defense Council, Inc.

BALLINGER PUBLISHING COMPANY
Cambridge, Massachusetts
A Subsidiary of Harper & Row, Publishers, Inc.

Copyright © 1984 by the Natural Resources Defense Council, Inc. All rights reserved.
No part of this publication may be reproduced, stored in a retrieval system,
or transmitted in any form or by any means, electronic, mechanical, photocopy,
recording or otherwise, without the prior written consent of the publisher.

International Standard Book Number: 0-88410-172-X (C)

Library of Congress Catalog Card Number: 82-24376 (P)

Printed in the United States of America

Library of Congress Cataloging in Publication Data

Cochran, Thomas B.
  Nuclear weapons databook.

  Includes bibliographical references and photographs index
    1. Atomic weapons.   I. Arkin, William, M.
II. Hoenig, Milton M.   III. Title.
U264.C6 1983   355.8′25119   82-24376
  ISBN 0-88410-172-X (v. 1)
  ISBN 0-88410-173-8 (pbk.: v. 1)

About the Authors

Thomas B. Cochran is a Senior Staff Scientist and Director of the *Nuclear Weapons Databook* Project at the Natural Resources Defense Council, Inc. He has served as a consultant to numerous government agencies and non-government organizations on energy and nuclear nonproliferation matters. He is the author of *The Liquid Metal Fast Breeder Reactor: An Environmental and Economic Critique* (Washington, DC: Resources for the Future, 1974). He holds a Ph.D. in physics from Vanderbilt University.

William M. Arkin is Director of the Arms Race and Nuclear Weapons Research Project at the Institute for Policy Studies in Washington, DC. He has been an intelligence analyst with the U.S. Army in Berlin and a Senior Staff Member of the Center for Defense Information. He is author of *Research Guide to Current Military and Strategic Affairs* (Washington, DC: IPS, 1981) and *SIOP: The Secret U.S. Plan for Nuclear War* (with Peter Pringle) (New York: W. W. Norton, 1983).

Milton M. Hoenig is a consultant to the Natural Resources Defense Council, Inc. In the past he has been at the U.S. Arms Control and Disarmament Agency in the Nonproliferation Bureau. He holds a Ph.D. in theoretical nuclear physics from Cornell University.

Contents

List of Figures ix
List of Tables xii
Foreword xiv
Preface xvi
Acknowledgements xvii
How to Use the *Nuclear Weapons Databook* ... xviii

Chapter One: The Nuclear Weapons System: An Overview 2
Nuclear Weapons and Delivery Systems:
   Definitions 2
The Nuclear Stockpile Today 3
Nuclear Weapons Deployments 5
History of the Nuclear Weapons Stockpile .. 6
   Early Years (1945-1955) 6
   Peak Production Years (1955-1967) 11
   Stockpile Stabilization and Refinement (1967-1980) 12
   New Generation of Nuclear Warheads (1980-Present) 13
Future Nuclear Weapons Developments 14
   Strategic Weapons Developments 17
   Theater and Tactical Weapons Developments . 19

Chapter Two: Nuclear Weapons Primer ... 22
Nuclear Fission and Fusion 22
Fission Weapons 22
   Chain Reaction 22
   Fissionable Materials 23
   Critical Mass 24
   Fission Weapon Design 25
Fusion Weapons 26
   Fusion Reactions 26
   Thermonuclear Fuels 27
   Thermonuclear Boosted Fission Weapons ... 27
   Thermonuclear Weapon Design 27
   Enhanced Radiation Weapons (Neutron Bombs) . 28
Third Generation Weapons 29
Warhead Features 30
   Warhead Safety and Control 30
   Selectable Yields 31
   Weights and Yields of Nuclear Weapons ... 31

Chapter Three: U.S. Nuclear Stockpile .. 38
Nuclear Warheads in the Stockpile 41
   W25 41
   B28 42
   W31 45
   W33 47
   B43 49
   W44 51
   W45 52
   W48 54
   W50 56
   B53 58
   W53 59
   W54 60
   W55 61
   W56 62
   B57 63
   B61 65
   W62 68
   W68 69
   W69 71
   W70 72
   W76 74
   W78 75
   W79 77
   W80 79

Chapter Four: The Role of Nuclear Weapons in U.S. and Allied Military Forces 82
Air Force Nuclear Weapons Roles 84
Army Nuclear Weapons Roles 86
Marine Corps Nuclear Weapons Roles 89
Navy Nuclear Weapons Roles 91
Allied Nuclear Weapons Roles 94

Contents

Chapter Five: Strategic Forces 100
Land-Based Missiles 100
Sea-Based Systems 103
Strategic Bomber Force 105
Ballistic Missile Reentry Vehicles 106
   Mk-21/Mk-5 (Advanced Ballistic Reentry
      Vehicle) (ABRV) 109
   Advanced Maneuverable Reentry Vehicle
      (AMaRV) 109
   Mk-500 EVADER Maneuverable RV (MaRV) .. 110
   Precision Guided Reentry Vehicle (PGRV) 110
Land-Based Missile Systems 111
   TITAN II (LGM-25C) 111
   MINUTEMAN II (LGM-30F) 113
   MINUTEMAN III (LGM-30G) 116
   PEACEKEEPER/MX Missile System 120
   MX Missile 121
   MX Warhead and Reentry Vehicle 125
   W87 126
   MX Basing 128
      Multiple Protective Structure (MPS) Basing 130
      Deep Basing 130
      Continuous Patrol Aircraft 130
      Ballistic Missile Defense 131
      Closely Spaced Basing/Dense Pack 131
Small Missile 132
Sea-Based Missile Systems 134
   POSEIDON Submarine 134
   POSEIDON C3 Missile System (UGM-73A) ... 136
   TRIDENT Submarine 138
   TRIDENT I C4 Missile System 142
   TRIDENT II D5 Missile 144
   Shallow Underwater Missile System (SUM) ... 147
   SSBN-X 147
Strategic Bomber Force 148
   B-52 STRATOFORTRESS 148
   FB-111 152
   Short-Range Attack Missile (SRAM) (AGM-69A) 154
   New Bombers 156
   B-1B 158
   Advanced Technology Bomber (ATB) ("Stealth") 162
Strategic Defensive Systems 163
   Ballistic Missile Defense 163
   SENTRY 166
   Bomber Interception 168
      GENIE (AIR-2A) 168

Chapter Six: Cruise Missiles 172
Air-Launched Cruise Missile (ALCM) (AGM-86B) 174
Ground-Launched Cruise Missile (GLCM)
   (BGM-109) 179
   W84 182
TOMAHAWK Sea-Launched Cruise Missile
   (SLCM) (BGM-109) 184
HARPOON Missile (AGM-84A/RGM-84A/
   UGM-84A) 188
Advanced Technology Cruise Missiles 191
Advanced Strategic Air-Launched Missile
   (ASALM) 193

Chapter Seven: Nuclear Capable Aircraft and Bombs 198
Nuclear Bombs 198
   B83 200
Tactical Nuclear Air-Launched Missiles 202
Future Nuclear Capable Delivery Aircraft 202
   Tactical Fighter Derivative 203
   Advanced Tactical Fighter 203
Nuclear Capable Aircraft 205
   A-4 SKYHAWK 205
   A-6 INTRUDER 207
   A-7 CORSAIR II 209
   AV-8B HARRIER II 211
   CF-101B VOODOO 213
   F-4 PHANTOM II 215
   F-15 EAGLE 217
   F-16 FIGHTING FALCON 220
   F-18/A-18 HORNET 223
   F-100 SUPERSABRE 226
   F-104 STARFIGHTER 228
   F-106 DELTA DART 230
   F-111 232
   P-3 ORION 234
   S-3 VIKING 236
   SH-3 SEA KING 238
   SH-60 SEAHAWK 239
   TORNADO 241

Contents

Chapter Eight: Naval Nuclear Weapons   244
The Reagan Naval Program 246
Attack Submarines 247
    LOS ANGELES Class Submarines (SSN-688) . . 249
Aircraft Carriers. 251
    NIMITZ Class Aircraft Carriers (CVN-68) 253
Surface Combatant Ships 255
    Battleship Reactivation 256
    TICONDEROGA Class Cruisers (CG-47) 257
    BURKE Class Destroyers (DDG-51) 261
    OLIVER HAZARD PERRY Class Frigates
    (FFG-7) 263
Naval Nuclear Missile Systems 264
    Vertical Launching System 264
    Anti-Submarine Nuclear Weapon Systems 264
        ASROC (RUR-5A) 267
        SUBROC (UUM-44A) 269
        ASW Standoff Weapon 271
    Anti-Air Nuclear Weapon Systems 272
        TERRIER (RIM-2) 273
        STANDARD-2 (SM-2) Missile (RIM-67B) 275
        W81 277

Chapter Nine: Army Nuclear Weapons . . 280
Missiles and Rockets 282
    HONEST JOHN (MGR-1B) 282
    LANCE (MGM-52C) 284
    NIKE-HERCULES (MiM-14) 287
    PERSHING 1a (MGM-31A/B) 289
    PERSHING II 292
        W85 297
    Corps Support Weapon System 298
Nuclear Artillery 300
    M109 155mm Gun 302
    M198 155mm Gun 304
    M110 8-inch (203mm) Gun 306
    Future Nuclear Artillery 308
        W82 309
Atomic Demolition Munitions and Earth
    Penetration Weapons 311

Glossary 313
Glossary of Terms 314
Glossary of Abbreviations and Acronyms 324

Index 329

List of Figures

Figure 1.1 MINUTEMAN III (LGM-30G) missile ... 4

Figure 1.2 B28RI nuclear bomb ... 13

Figure 1.3 U.S. nuclear stockpile, 1949 to present (with future projections) ... 14

Figure 1.4 Advanced Strategic Air-Launched Missile (ASALM) ... 17

Figure 1.5 MX missile ... 19

Figure 2.1 Nuclear fission and fusion reactions ... 23

Figure 2.2 Interior components of *Fat Man* type bomb ... 25

Figure 2.3 *Mike* device, the first successful thermonuclear test ... 26

Figure 2.4 The Mk-17, first droppable thermonuclear bomb ... 27

Figure 2.5 B-52 atop the TRESTLE electromagnetic pulse simulator ... 29

Figure 2.6 The 100 foot tower at ground zero of the Trinity Site at Alamogordo, New Mexico ... 31

Figure 2.7 The "Gadget"—prototype for the *Fat Man* weapon design ... 31

Figure 2.8 Plutonium fission weapon of the *Fat Man* type ... 32

Figure 2.9 Nuclear weapon of the *Little Boy* type ... 32

Figure 2.10 *Davy Crockett*, the smallest nuclear weapon ever deployed ... 33

Figure 2.11 The Mk-5 nuclear weapon, first lightweight tactical gravity bomb ... 33

Figure 2.12 Mk-23, 16-inch projectile for Navy battleship guns ... 33

Figure 2.13 Army's 280mm atomic cannon ... 34

Figure 2.14 W19 projectile for the Army's 280mm howitzer ... 35

Figure 2.15 Yield-to-Weight Ratios vs. Yield and Weight for Current Stockpile ... 36

Figure 3.1 Genie (AIR-2A) rocket ... 41

Figure 3.2 B28IN bomb ... 42

Figure 3.3 Four B28FI bombs ... 43

Figure 3.4 Building Block Concept for W28 ... 44

Figure 3.5 HONEST JOHN (MGR-1B), left, and NIKE-HERCULES (MiM-14), right ... 45

Figure 3.6 8-inch conventional artillery projectiles ... 47

Figure 3.7 B43 bomb ... 49

Figure 3.8 ASROC (RUR-5A) ... 51

Figure 3.9 TERRIER (RIM-2D) ... 52

Figure 3.10 Medium Atomic Demolition Munition (MADM) ... 52

Figure 3.11 155mm nuclear artillery projectile ... 54

Figure 3.12 PERSHING 1a (MGM-31A/B) ... 56

Figure 3.13 PERSHING 1a missile ... 57

Figure 3.14 B53 bomb ... 58

Figure 3.15 W53 warhead ... 59

Figure 3.16 Special Atomic Demolition Munition (SADM) ... 60

Figure 3.17 SUBROC (UUM-44A) ... 61

Figure 3.18 MINUTEMAN II (LGM-30F) ... 62

Figure 3.19 B57 bomb ... 63

Figure 3.20 B61 bomb ... 65

Figure 3.21 Mock-up of Mk-12 reentry vehicle ... 68

Figure 3.22 POLARIS A3 (UGM-27C), left, and POSEIDON C3 (UGM-73A), right ... 69

Figure 3.23 Short-Range Attack Missiles (SRAM) (AGM-69) ... 71

Figure 3.24 LANCE (MGM-52) ... 72

Figure 3.25 TRIDENT I C4 (UGM-93A) ... 74

Figure 3.26 Mk-12A reentry vehicles ... 75

Figure 3.27 M753 8-inch nuclear artillery projectile and M110 gun ... 77

Figure 3.28 Air-launched Cruise Missiles (AGM-86B) ... 79

List of Figures

Figure 3.29 TOMAHAWK (BGM-109) Sea-launched Cruise Missile ... 79

Figure 4.1 Nuclear Weapons Locations in the United States ... 82

Figure 4.2 F-16 FALCON ... 85

Figure 4.3 MINUTEMAN missile launch site ... 87

Figure 4.4 Two GENIE (AIR-2A) rockets mounted under Air Force F-101 ... 88

Figure 4.5 PERSHING 1a (MGM-31A/B) platoon ... 89

Figure 4.6 M109 155mm howitzer ... 90

Figure 4.7 AV-8B HARRIER ... 91

Figure 4.8 U.S.S. *Virginia* (CGN-38) ... 93

Figure 4.9 U.S.S. *Ohio* (SSBN-726), first TRIDENT submarine ... 93

Figure 4.10 Deployment of U.S./NATO nuclear weapons in Europe ... 96

Figure 4.11 Deployment of U.S. nuclear weapons in Asia ... 97

Figure 5.1 TITAN II (LGM-25C) missile ... 111

Figure 5.2 MINUTEMAN II (LGM-30F) missile ... 113

Figure 5.3 Comparison of MINUTEMAN II and MINUTEMAN III ... 114

Figure 5.4 MINUTEMAN III (LGM-30G) missile in silo ... 116

Figure 5.5 MINUTEMAN III launch sequence ... 117

Figure 5.6 Full scale mock-up of MX missile ... 121

Figure 5.7 MX prototype vehicle launch validation test ... 123

Figure 5.8 MX bus with four Mk-21 reentry vehicles ... 126

Figure 5.9 One-eighth scale MX silo test model ... 129

Figure 5.10 Reinforcing steel bar skeleton used in the MX silo model test ... 129

Figure 5.11 MINUTEMAN I (LGM-30B) missile ... 130

Figure 5.12 Closely Spaced Basing ... 131

Figure 5.13 Small missile prototype ... 132

Figure 5.14 U.S.S. *Sam Rayburn* (SSBN-635) ... 134

Figure 5.15 POSEIDON C3 (UGM-73A) missile ... 136

Figure 5.16 U.S.S. *Ohio* (SSBN-726), the first TRIDENT submarine ... 138

Figure 5.17 Missile hatches open on TRIDENT submarine ... 139

Figure 5.18 Interior of missile compartment of TRIDENT submarine ... 139

Figure 5.19 TRIDENT I C4 (UGM-93A) missile ... 142

Figure 5.20 TRIDENT I C4 missile being loaded into U.S.S. *Ohio* ... 143

Figure 5.21 B-52G bomber ... 148

Figure 5.22 B-52G with SRAMs (AGM-69A) ... 149

Figure 5.23 FB-111 with SRAMs ... 152

Figure 5.24 Short-Range Attack Missile (SRAM) (AGM-69A) ... 154

Figure 5.25 B-1 bomber ... 158

Figure 5.26 B-1 bomber ... 159

Figure 5.27 B-1 bomber being refueled by KC-135 tanker aircraft ... 160

Figure 5.28 SAFEGUARD complex ... 163

Figure 5.29 SPARTAN missile test ... 163

Figure 5.30 SPRINT missile ... 166

Figure 5.31 GENIE rocket, loaded into missile bay of F-106 ... 168

Figure 6.1 Air-Launched Cruise Missile (AGM-86B) ... 174

Figure 6.2 ALCM soon after drop from B-52 bomber ... 175

Figure 6.3 Ground-Launched Cruise Missile (BGM-109) test firing ... 179

Figure 6.4 GLCM convoy ... 181

Figure 6.5 GLCM, with missile launcher erect ... 181

Figure 6.6 BGM-109 cutaway diagram ... 182

Figure 6.7 TOMAHAWK Sea-Launched Cruise Missile (BGM-109) ... 184

Figure 6.8 First launch of TOMAHAWK missile ... 185

Figure 6.9 TOMAHAWK missile with inert warhead scores direct hit on Navy ship ... 186

Figure 6.10 Air-launched HARPOON (AGM-84A) missile ... 188

List of Figures

Figure 6.11 Ship-launched HARPOON (RGM-84A) missile 189

Figure 6.12 Advanced Strategic Air-Launched Missile (ASALM). 194

Figure 7.1 F-111 delivering B83 bomb prototype 200

Figure 7.2 A-4M SKYHAWK. 205

Figure 7.3 A-6 INTRUDER. 207

Figure 7.4 A-7 CORSAIR 209

Figure 7.5 AV-8B HARRIER 211

Figure 7.6 F-101B VOODOO 213

Figure 7.7 F-4D PHANTOM II 215

Figure 7.8 F-15 EAGLE. 217

Figure 7.9 F-15 EAGLE 218

Figure 7.10 F-16 FALCON 220

Figure 7.11 F-16 FALCON 221

Figure 7.12 F-18 HORNET 223

Figure 7.13 F-18 HORNET. 224

Figure 7.14 F-100 SUPERSABRE 226

Figure 7.15 F-104G STARFIGHTER 228

Figure 7.16 F-106 DELTA DART 230

Figure 7.17 F-111 232

Figure 7.18 P-3 ORION 234

Figure 7.19 S-3 VIKING 236

Figure 7.20 SH-3H SEA KING 238

Figure 7.21 SH-60 SEAHAWK 239

Figure 7.22 TORNADO 241

Figure 8.1 U.S.S. *Los Angeles* (SSN-688) attack submarine. 249

Figure 8.2 U.S.S. *Nimitz* (CVN-68) aircraft carrier 253

Figure 8.3 U.S.S. *New Jersey* (BB-62) battleship . . 256

Figure 8.4 U.S.S. *Ticonderoga* (CG-47) cruiser . . . 257

Figure 8.5 U.S.S. *Ticonderoga* (CG-47) cruiser . . . 258

Figure 8.6 U.S.S. *Spruance* (DD-963) destroyer . . 262

Figure 8.7 Exercise Swordfish, 11 May 1962, with effects of underwater nuclear blast 264

Figure 8.8 ASROC (RUR-5A) missile 267

Figure 8.9 Mk-16 eight tube box launcher aboard Naval ship 267

Figure 8.10 SUBROC (UUM-44A) missile 269

Figure 8.11 Anti-Air Nuclear Weapon Systems . . 272

Figure 8.12 TERRIER (RIM-2) missiles 273

Figure 8.13 STANDARD-2 (RIM-67B) missile. . . . 275

Figure 8.14 STANDARD-2MR missiles. 276

Figure 8.15 STANDARD-2 (RIM-67B) missile . . . 277

Figure 9.1 HONEST JOHN (MGR-1B) missile . . . 282

Figure 9.2 LANCE (MGM-52C) missile 284

Figure 9.3 The accuracy of a nuclear armed LANCE missile 285

Figure 9.4 NIKE-HERCULES (MiM-14) missile . . 287

Figure 9.5 NIKE-HERCULES missile launch 287

Figure 9.6 PERSHING 1a (MGM-31A/B) missile launch 289

Figure 9.7 Launch sequence of PERSHING 1a . . . 290

Figure 9.8 PERSHING II missile launch 293

Figure 9.9 Flight sequence of PERSHING II missile 294

Figure 9.10 Comparison of PERSHING 1a and PERSHING II missiles 294

Figure 9.11 Assault Breaker prototype missile . . . 298

Figure 9.12 First live nuclear artillery test, Shot *Grable*, 25 May 1953 300

Figure 9.13 M109 155mm self-propelled gun. . . . 302

Figure 9.14 M198 155mm towed gun 304

Figure 9.15 M110 8-inch (203mm) self-propelled howitzer. 306

Figure 9.16 Medium Atomic Demolition Munition (MADM) mock-up 311

List of Tables

Table 1.1 Nuclear Warheads in the Stockpile (1983) 3

Table 1.2 Strategic Nuclear Weapons (1983) 4

Table 1.3 Theater and Tactical Nuclear Weapons (1983) 5

Table 1.4 U.S. Nuclear Warheads (1945-Present) .. 7

Table 1.5 Inactive Nuclear Delivery Systems (1945-Present) 10

Table 1.6 U.S. Nuclear Weapons Stockpile 15

Table 1.7 Projected Nuclear Warhead Production, 1983 to mid-1990s 16

Table 1.8 Nuclear Weapons Development Phases . 17

Table 1.9 Nuclear Weapons Research and Development Programs (1983) 18

Table 2.1 Warhead Safeguard Features 30

Table 3.1 U.S. Nuclear Weapons Stockpile (1983) . 39

Table 3.2 B28 Modifications 43

Table 4.1 Allocation of Nuclear Warheads in the Service Branches (1983) 83

Table 4.2 Personnel with Nuclear Weapons Duties 84

Table 4.3 Air Force Nuclear Weapons Units 86

Table 4.4 Strategic Bomber Force Basing (1983) .. 86

Table 4.5 ICBM Deployments 87

Table 4.6 Strategic Interception Forces (1983) ... 88

Table 4.7 Army Nuclear Weapons Units 90

Table 4.8 Allocation of Nuclear Weapons in Army Units 91

Table 4.9 Navy Nuclear Weapons Units 92

Table 4.10 Naval Bases for Nuclear Armed Ships and Submarines (1983) 94

Table 4.11 Allied Nuclear Capabilities (1983) ... 95

Table 5.1 Features of "Legs" of the Strategic TRIAD 101

Table 5.2 Strategic Nuclear Forces (1971-1981) ... 101

Table 5.3 Strategic Nuclear Forces (1983) 102

Table 5.4 Strategic Missile Submarines (1983) ... 103

Table 5.5 Strategic Submarine Chronology 104

Table 5.6 Strategic Submarine Forces (1979-1990) 105

Table 5.7 Strategic Bomber Force Loadings 106

Table 5.8 Bomber Forces Funding (1970-1980) ... 106

Table 5.9 U.S. Ballistic Missile Reentry Vehicles . 107

Table 5.10 RV Developments 108

Table 5.11 RV Chronology 108

Table 5.12 ABRES/ASMS Costs 109

Table 5.13 MINUTEMAN Chronology 118

Table 5.14 MX Missile System Costs 122

Table 5.15 Major MX Contractors 124

Table 5.16 MX Chronology 125

Table 5.17 MX Basing Options 128

Table 5.18 Major POSEIDON Subcontractors ... 136

Table 5.19 TRIDENT Submarine Construction .. 140

Table 5.20 Submarine-Launched Ballistic Missile Options 144

Table 5.21 B-52 Bomber Force 150

Table 5.22 B-52 Modifications 151

Table 5.23 Candidate Systems for B-52 Bomber Replacement: Long-Range Combat Aircraft/ Multi Role Bomber 156

Table 5.24 New Bomber Funding 157

Table 5.25 Nuclear Weapons Loads for B-1B Bomber 160

Table 5.26 Major B-1B Subcontractors 161

Table 5.27 BMD Funding 164

Table 5.28 Major Ballistic Missile Defense Contractors 165

List of Tables

Table 6.1 Major TOMAHAWK Cruise Missile Contractors 173

Table 6.2 ALCM Chronology 175

Table 6.3 Major ALCM Subcontractors 176

Table 6.4 ALCM Program Schedules 177

Table 6.5 TOMAHAWK SLCM Types 185

Table 6.6 SLCM Deployments 186

Table 6.7 SLCM Funding and Procurement 187

Table 7.1 Nuclear Bombs 199

Table 7.2 Nuclear Capable Tactical Aircraft 199

Table 7.3 Future Tactical Fighter Aircraft Programs 204

Table 7.4 F/A-18 Deployments 224

Table 8.1 Nuclear Capable Ships and Submarines 244

Table 8.2 U.S. Naval Forces (1983) 245

Table 8.3 Nuclear Capable Attack Submarines . . 247

Table 8.4 Nuclear Capable Aircraft Carriers 251

Table 8.5 Nuclear Weapons and Systems on Aircraft Carriers 252

Table 8.6 Nuclear Capable Surface Ships (1983) 255

Table 8.7 Nuclear Capable Cruisers 260

Table 8.8 Nuclear Capable Destroyers 262

Table 8.9 Nuclear Capable Frigates 263

Table 8.10 Nuclear Capable Shipboard Missile Launchers 265

Table 8.11 Vertical Launching System Platforms 266

Table 9.1 Nuclear Artillery Guns 301

Table 9.2 Comparison of Old and New 155mm Nuclear Artillery Shells 308

Foreword

Generation Upon Generation

The *Nuclear Weapons Databook* is for those who want to understand the nuclear arms race and are not frightened by numbers. This first volume is the most authoritative and complete reference work available on U.S. forces and capabilities. I expect it to join the classic government publication, *The Effects of Nuclear Weapons,* on the bookshelves of those who are interested in the technical issues relating to the nuclear arms race.[1]

The comprehensive material on U.S. nuclear weapons and weapons-development programs presented in this volume will be used by professionals in Congress, academia, public interest groups, and the media in assessing nuclear weapons policy alternatives. They will find that the *Databook* will make them less dependent for their information upon the generosity of Executive Branch officials.

The *Databook* will also be extremely helpful to the increasingly large group of citizen-activists who wish to challenge, on a technical level, the arguments which are used to rationalize the continuation of the nuclear arms race.

Just leafing through Volume I of the *Databook* teaches one some important facts about the arms race. Many readers will be surprised to learn, for example, just how "nuclearized" the U.S. military establishment is. They will learn that, in addition to the relatively familiar "strategic" nuclear weapons systems which give the U.S. the capability to destroy the Soviet Union—or any other nation for that matter—from thousands of miles away, virtually every unit of the U.S. Armed Forces has the capability to deliver nuclear destruction at shorter ranges.

For example, the Army has nuclear weapons ranging from man-portable atomic demolition mines and nuclear artillery shells to nuclear-tipped surface missiles able to attack Moscow from West Germany. The Army, Navy, and Air Force all have anti-aircraft missiles with nuclear warheads. The Navy has nuclear depth charges which can be dropped from aircraft or shot by rockets from surface ships or submarines. And there are thousands of "tactical" bombs with yields ranging from one third to one hundred times that of the Hiroshima bomb.

Another message implicit in the *Databook* which struck this reader with particular force is the fact that there is always another generation of nuclear weapons under development.

Consider modern long-range cruise missiles. Breakthroughs in the development of tiny efficient jet engines and in terrain-recognizing microprocessors have finally made these miniature pilotless aircraft such effective nuclear weapons delivery vehicles that all the armed services are spending billions on them.

The Air Force is equipping its B-52G/H bombers to carry a total of over 3000 cruise missiles. Each of these precision-guided drones can carry a nuclear warhead with more than ten times the explosive power of the Hiroshima bomb to a target 1500 miles away. The Navy plans to deploy hundreds of nuclear-armed cruise missiles on its surface ships and on its attack submarines. (The latter are to be kept as an "enduring reserve" to strengthen the position of the U.S. in a post-nuclear war world.) The Air Force is also, amid furious controversy, attempting to deploy 464 cruise missiles in Western Europe.

Meanwhile, follow-on "advanced" cruise missiles are under development. In the short term, there will be evolutionary improvements to increase the range and reduce the radar reflectivity of the current generation of cruise missiles. And, in the longer term, an "advanced strategic air-launched missile" is planned which will travel at four times the speed of sound. We learn from the *Databook* that the program to develop a supersonic cruise missile was initiated in June 1974—less than one year after development work began on the current generation of cruise missiles!

We can also learn from the *Databook* that the many cycles of "modernization" of the U.S. nuclear arsenal have not increased its destructive power over the past 20 years. Overkill was achieved within a decade of Hiroshima. Since that time, the designers of strategic nuclear weapons systems have been concentrating on other areas—perhaps most ominously on first-strike capabilities.

---

1 Edited by Samuel Glasstone and Phillip J. Dolan, 3rd edition jointly published by the Departments of Defense and Energy, 1977.

Foreword

As a result of these development programs, most missiles in the next generation of U.S. nuclear weapons will have the capabilities to destroy "hardened" military targets and most will be able to destroy many. The MX with its ten accurate warheads is designed to destroy up to ten Soviet missiles in their silos, for example. The submarine-launched TRIDENT II missile is to have similar capabilities.

Strikes with these new "war-fighting" nuclear weapon systems could hardly be described as "surgical," however. Paradoxically, as the accuracy of U.S. missiles is being dramatically increased, the power of their warheads is also being increased. For example, the TRIDENT II is to carry warheads with ten times the explosive power of those on the POSEIDON missile which it is to replace. The total explosive power carried by TRIDENT II would be hundreds of times greater than that of the bomb which destroyed Hiroshima. Tens of millions of innocent civilians would therefore certainly die if the U.S. were to attack the Soviet Union's "hard" military targets.

Perhaps the most important message of the *Databook* is implicit in the fact that it could be written. That fact proves that systematic research, using public sources such as the "sanitized" transcripts of Congressional hearings, can glean enough information to lay the basis for a fully informed public debate over U.S. nuclear weapons policy. In the past, when the public was willing to leave policy-making to the "experts," this fact was irrelevant. Now, when a large fraction of the public has concluded that the nuclear arms race is too important to be left to unsupervised experts, the availability of the information in the *Databook* will make a significant difference.

Frank von Hippel
June 1983

Dr. Frank von Hippel, a theoretical physicist, is a Professor of Public Policy and a faculty associate of the Center for Energy and Environmental Studies at Princeton University. He is also currently the elected chairman of the Federation of American Scientists.

Preface

The *Nuclear Weapons Databook* is meant to be a current and accurate encyclopedia of information about nuclear weapons. It is intended to assist the many people who are today actively working on the problems of the nuclear arms race. In our society today, there is no greater threat to the human environment than a nuclear holocaust. Because of the obvious and terrifying consequences of the use of nuclear weapons, the Natural Resources Defense Council (NRDC) has followed every aspect of nuclear development, including nuclear weapons development, for over a decade. NRDC has long believed that accurate information is critical in understanding the imperative for and implications of arms control. Information about nuclear weapons, policy, plans, and implications remains shrouded in secrecy. Informed public decisions on nuclear arms questions can only occur if better and more information on the subject is available. The purpose of this *Databook* is to help overcome this barrier.

Since 1980, NRDC has sponsored the research required to produce this first of several volumes on all aspects of the production and deployment of nuclear weapons worldwide. As now planned the *Nuclear Weapons Databook* will consist of at least eight volumes:

I. U.S. Nuclear Forces and Capabilities
II. U.S. Nuclear Weapons Production Complex
III. Soviet Nuclear Weapons
IV. Other Foreign Nuclear Weapons
V. Environment, Health, and Safety
VI. Command and Control of Nuclear Weapons and Nuclear Strategy
VII. Arms Control
VIII. The History of Nuclear Weapons.

Volume I of the *Nuclear Weapons Databook* is based as much as possible on original documentation, and the source of information is indicated in the extensive footnotes accompanying the text and fact sheets. The *Databook*, however, is only as useful as the accuracy of the information presented. We therefore strongly encourage the reader to contribute to this effort—to advise us of errors and new information. We also wish to be advised of additional subject areas that should be included in future editions and recommended changes in the format of the data presented. Experts who are willing to serve as contributors or reviewers of the various sections of the *Databook*, particularly subject areas not now covered, are also desired.

Please address all correspondence to the authors at the Natural Resources Defense Council, 1725 I Street, N.W., Suite 600, Washington, DC 20006 (202/223-8210).

The publication of the first volume of the *Databook* may appear to be imbalanced because of a lack of comparison with Soviet nuclear weapon systems. This "omission" simply reflects our view that publication of the U.S. material should not be held up pending work on foreign nuclear arsenals. Even upon publication of the third volume (now in preparation), this appearance of imbalance may continue due to the much more limited availability of data on the Soviet nuclear weapon system in open literature. Furthermore, the *Databook* is not intended to be another document on the assessment of U.S.-Soviet military balance. The basic material to be presented on both the U.S. and Soviet weapons systems is meant to serve as a step toward a more sophisticated understanding of the dynamics of the two systems.

Thomas B. Cochran
William M. Arkin
Milton M. Hoenig

Acknowledgments

This first Volume of the *Nuclear Weapons Databook* could not have been compiled without the invaluable assistance of many other institutions and individuals. We are grateful to the U.S. Departments of Defense and Energy for their responsiveness to our numerous requests for information. The Department of Energy's National Atomic Museum was particularly helpful. The Arms Control Association, Federation of American Scientists, and the Center for Defense Information made available to us extensive data from their files.

Frank von Hippel contributed valuable information and insights. Others who we want to thank for reviewing information in the manuscript include Stan Norris of the Center for Defense Information, Christopher Paine of the Federation of American Scientists, Bill Hartung of the Council on Economic Priorities, Chuck Hansen, Milton Leitenberg of the Swedish Institute of International Affairs, Paul Rusman of Leiden University, the Netherlands, and Duncan Campbell. Valuable research assistance was also provided by David Leech, Flora Montealegre, and Richard Fieldhouse of the Institute for Policy Studies and the Arms Race and Nuclear Weapons Research Project, of which William Arkin is Director. Nevertheless, responsibility for all facts and analyses in the *Databook* remains solely that of the authors.

Information and photographs from manufacturers were valuable in the research for the *Databook* and we would like to thank Boeing, Grumman Aerospace, General Dynamics, McDonnell Douglas, Lockheed, Martin Marietta, and Goodyear for their assistance. Photographs were also generously provided by Bettie Sprigg of the Department of Defense; Bob Carlisle, Department of the Navy; the National Atomic Museum; Department of the Army; Department of the Air Force; Sandia National Laboratories; and Tom Greenberg of the Center for Defense Information.

The Natural Resources Defense Council and the authors wish to acknowledge gratefully the support and encouragement given to the *Nuclear Weapons Databook* by the Bydale Foundation; the Columbia Foundation; the W. Alton Jones Foundation; the Joyce Mertz-Gilmore Foundation; the New Hope Foundation; the Ploughshares Fund; the Rockefeller Family Fund; the Samuel Rubin Foundation; the Wallace Genetic Foundation; David B. Arnold, Jr.; Gloria and Jeff Coolidge; John F. Hogan; and Philip S. Weld. We must especially recognize Joan K. Davidson and James Marshall, both of whom are Trustees of NRDC, for their enthusiasm and commitment which have been a continuous source of inspiration to this project.

We appreciate the continuing encouragement and support of the entire Board of Trustees and Staff of the Natural Resources Defense Council, particularly Adrian W. DeWind, Chairman of the Board, John H. Adams, Executive Director, and S. Jacob Scherr, Senior Staff Attorney. We are deeply indebted to Barbara J. Pratt, who typed initial drafts of the text, and to John Mercer, who edited the final draft. Wayne E. Nail developed a word processor format which simplified revision and typesetting, and used his graphic skills to design this volume. We are grateful to the people at Alexander Typesetting, Inc., who worked diligently with the many complex tables and fact sheets. Finally, we would like to thank Carol Franco for her unfailing support of the *Databook* project, and Barbara Roth, Steve Kramer, and Gerry Galvin of Ballinger, for their assistance and patience during the difficult production of this book.

How to Use the *Databook*

How to Use the Nuclear Weapons Databook

The *Databook* is designed primarily for those who need to find basic facts about nuclear weapons. It is not designed to replace any existing reference books, but to supplement and hopefully contribute to the already existing volumes. Since the *Databook* is both factual and comprehensive, it will provide a more easily available and accessible source than either the numerous specialized publications which are known to the experts or the less authoritative and secondary sources of information which are commonly available.

Three chapters provide an overview and explain how nuclear weapons work. Six subsequent chapters contain an overview, fact sheets, and descriptions of nuclear warheads, delivery systems, and research programs. In these chapters, the development of the nuclear arsenal, from the oldest weapons to the newest to the future, is presented. It is hoped that in this way an understanding of the continual exploitation of technology for nuclear weapons can be clearly seen.

The *Databook* is not meant to be read straight through, although reading the first two and fourth chapters, along with the introduction to each of the remaining six chapters, can provide valuable background for using the Databook as a reference work. The table of Contents, page headings, and index should enable any user to quickly find any information needed. A detailed glossary and list of abbreviations and acronyms used in the book is provided; the abbreviation and acronym list is particularly important as the key to deciphering the shorthand source citations. Numerous tables and figures are used throughout the book to help illustrate the difficult technical material, and each fact sheet to the extent possible contains common information and characteristics.

A sample fact sheet will contain numerous categories of information, some of which are well known and some of which are not. In each case, with a weapon or warhead, we have tried to provide information on the manufacturer, evolution, cost, characteristics, and use of the system. Many gaps in the data reflect the fact that we have been unable to get all details for every system.

Do not let the details frighten you. You do not have to be a physicist or defense expert to use this book. There is an abundance of data that should be useful regardless of one's level of expertise.

The Dangerous Decade Ahead

For more than 30 years, the United States has conducted foreign relations in the shadow of nuclear arms; now the nuclear umbra is darker and more extensive than ever. There are many more weapons than ever before, but most significantly, the range, accuracy, targeting flexibility, and payload of intercontinental nuclear weapon systems have been markedly improved:

- During the past decade, the warhead count of intercontinental nuclear weapons went up 200 percent.
- Their estimated explosive power (equivalent megatonnage) grew some 30 percent.
- Their pin-point targeting (hard target kill) potential increased 200 percent.

U.S. Joint Chiefs of Staff,
U.S. Military Posture for FY 1982

1
Definitions

Chapter One
The Nuclear Weapons System: An Overview[1]

The dominant factor in East-West relations is the nuclear weapon. Since the first explosion of a nuclear device over the New Mexico desert in July 1945, nuclear weapons have gained a preeminent position in U.S. and Soviet military and foreign policies. This has led to the creation of large military infrastructures to support nuclear weapons.

Today, 38 years after the first atom bomb was exploded, there are approximately 26,000 nuclear warheads in the United States arsenal. Well over 200,000 people and an annual budget of over $35 billion are involved in U.S. development and production of new warheads, the care for those already in the so-called "stockpile," and the planning for their use. This volume presents a detailed picture of the present and future nuclear weapons capabilities in the U.S., including the nuclear weapons arsenal, the military structure which exists to support and eventually use those weapons, and the state of current and future nuclear weapons technology.

U.S. policy governing the control and possible use of nuclear weapons has gone through significant changes over the past 38 years. The use of the new and powerful atomic weapon was not initially treated as a fundamental break from previous "conventional" military requirements, particularly strategic bombing. U.S. nuclear strategy then evolved to a position of "deterrence," where the maintenance of large nuclear arsenals and the mutual consequences of U.S. and Soviet nuclear warfare were thought to "assure" that nuclear weapons would never be used. Today, policy is based on the belief that the limited use of nuclear weapons is possible. Indeed, a "war fighting" strategy involving nuclear weapons is seen as the only credible deterrent.

However one interprets policy, the vast arsenal of weapons and trends in its technological development provide insight into the dynamics of the nuclear arms race and evidence of its increasing dangers.

Nuclear Weapons and Delivery Systems: Definitions

The terms "nuclear device," "nuclear warhead," and "nuclear weapon" are often used interchangeably, but the distinctions between them are noteworthy. A nuclear explosive device (or simply "nuclear device") is an assembly of nuclear and other materials and fuzes which could be used in a test, but generally cannot be reliably delivered as part of a weapon. A nuclear warhead implies further refinement in design and manufacture resulting in a mass produced, reliable, predictable nuclear device capable of being carried by missiles, aircraft, or other means. A nuclear weapon is a fully integrated nuclear warhead with its delivery system.

Although definitions are often subject to transient political considerations, nuclear weapons are generally categorized according to their intended use, as "strategic," "theater," or "tactical."

Strategic (Nuclear) Weapons. The category of long-range weapons generally allocated for attacking the homeland of the enemy or protecting the homeland. This includes intercontinental missiles, both land based (ICBMs) and sea based (SLBMs); long-range heavy bombers and their carried weapons (bombs and air-launched missiles); long-range cruise missiles not carried on bombers; and homeland defense missiles, that are both ground and air launched.

Theater (Nuclear) Weapons.[2] All other nuclear weapons earmarked for use in regional plans and confrontations where the intent is not merely tactical surprise or advantage, but the destruction of "targets"—bases and support facilites—that provide reinforcement for a battle. Theater weapons comprise bombs and depth charges on non-strategic aircraft, cruise missiles (air, sea and land based), short-range ballistic missiles used in surface-to-surface and surface-to-air missions, artillery projectiles, and atomic demolition munitions (nuclear land mines).

---

[1] Information on the history of the U.S. nuclear weapons stockpile is contained in David Alan Rosenberg, "U.S. Nuclear Stockpile, 1945 to 1950," *The Bulletin of the Atomic Scientists*, May 1982, pp. 25-30; Milton Leitenberg, "Background Information on Tactical Nuclear Weapons," *Tactical Nuclear Weapons: European Perspectives* (SIPRI, 1978); Norman Polmar, *Strategic Weapons: An Introduction* (New York: Crane Russak, 1982 (Revised Edition)).

[2] "Theater" nuclear weapons and forces have undergone the most changes in terminology. They have been labeled both "intermediate-range" and "non-strategic" nuclear forces by the Reagan Administration due to the perceived negative connotation of the word "theater" in the European political debate which equates its use (as in "theater of war") with a postulated American policy to attempt to restrict the use of these weapons to Europe and spare U.S. territory in a nuclear war originating in Europe. In addition, "theater" is often used synonymously with "tactical," in referring to short-range weapons.

1
Stockpile Today

Table 1.1
Nuclear Warheads in the Stockpile (1983)[1]

Warhead[2] / Reentry Vehicle Model	Weapon System
STRATEGIC OFFENSE	
W53/Mk-6	TITAN II
W56/Mk-11C	MINUTEMAN II
W62/Mk-12	MINUTEMAN III
W68/Mk-3	POSEIDON C3
W69	SRAM
W76/Mk-4	TRIDENT I C4
W78/Mk-12A	MINUTEMAN III
W80-1	ALCM
STRATEGIC DEFENSE	
W25	GENIE
TACTICAL	
W31	HONEST JOHN / NIKE-HERCULES
W33	8-inch howitzer
W44	ASROC
W45-1	TERRIER
W48	155mm howitzer
W50	PERSHING 1a
W55	SUBROC
W70*	LANCE
W79*	8-inch howitzer
ATOMIC DEMOLITION MUNITIONS (ADMs)	
W45-3	Medium ADM
W54	Special ADM
BOMBS[3]	
B28	Tactical and Strategic Aircraft
B43	Tactical and Strategic Aircraft
B53	B-52 Aircraft
B61	Tactical and Strategic Aircraft
NUCLEAR DEPTH BOMB / BOMB	
B57	ASW Patrol, Tactical and Strategic Aircraft

1 Two warheads—W66 and W71—are in inactive storage and are being retired.
2 All current nuclear bombs are referred to as "B-" followed by the warhead program number, e.g., B-61 (or simply B61). If the warhead of a nuclear weapon has other applications, it is designated with a "W." Modification(s) to the major assembly design of a warhead are designated by Mod. numbers (e.g., "B-61 Mod 1" or simply "B61-1"). Mod 0 is the first version of a weapon design. Subsequent modifications of the weapon system are numbered.
3 The B28 and B61 bombs have numerous known Mods.
* Enhanced radiation yield.

Tactical (Nuclear) Weapons. Refers to those "theater" weapons, more precisely termed "short-range" and "battlefield" weapons, whose purpose is to affect directly the course of a tactical maneuver or a battle. Tactical weapons include bombs, short-range missiles, nuclear artillery, and atomic demolition munitions.

The Nuclear Stockpile Today

The U.S. nuclear weapons stockpile contains 24 warhead types (see Table 1.1). The oldest warhead is the W33, a gun assembly, low yield, fission nuclear artillery projectile, first deployed in 1956. The newest is the W80-1, a small thermonuclear warhead for the strategic Air-Launched Cruise Missile (ALCM), deployed in 1981. The stockpile of about 26,000 nuclear warheads consists of eight strategic missile types, one strategic defensive warhead, eleven tactical warheads for missiles, artillery and atomic demolition munitions, and five nuclear bomb types. The bombs are carried by both strategic and tactical aircraft.[3]

The nuclear weapons stockpile remained fairly constant throughout the 1970s, stabilizing at about 25,000; a marked increase in the rate of production and retirements of nuclear weapons which began in 1981, will significantly change the complexion of the stockpile. While the stockpile was made up predominantly of tactical

---

3 Only one of the bombs, the large, nine megaton B53, is solely carried by B-52 bombers.

1

Stockpile Today

Table 1.2
Strategic Nuclear Weapons (1983)

Delivery Platforms	Systems	Warheads
Bombers	FB-111, B-52	B28, B43, B53, B57, B61, W69, W80-1
Interceptors	F-4, F-15, F-106	W25
Land-Based Missiles	TITAN II, MINUTEMAN II, MINUTEMAN III	W53, W56, W62, W78
Submarine-Based Missiles	POSEIDON, TRIDENT I	W68, W76

warheads in the 1960s, the mix is now about evenly split between strategic and tactical weapons.[4]

In the strategic forces, there are currently 2149 warheads on more than 1000 land-based strategic missiles, another 4960 on submarine-launched missiles, and 2580 allocated to be carried on strategic bombers (see Table 1.2). These weapons are referred to as the "force loadings" and do not include maintenance spares or "weapons reserved for restrike (reserves) and weapons on inactive status."[5]

Four warhead types are deployed with land-based strategic missile forces: the W53 nine megaton TITAN II warhead, the W56 1.2 megaton MINUTEMAN II warhead, the W62 170 Kt triple warhead on the MINUTEMAN III, and the W78 335 Kt triple warhead on the MINUTEMAN III. Submarine missiles carry two warhead types: the W68 40-50 Kt warhead on the POSEIDON (each can carry 7-14 warheads) and the W76 100 Kt warhead on the TRIDENT I (each missile carries 8 warheads). Bomber forces carry five nuclear bomb types depending on the mission and targets: the B28, with yields from 70-1450 Kt; the B43, with one megaton yield; the B53, with nine megaton yield; the B57, with a low Kt yield; and the B61, with a 300-500 Kt yield. The low yield W25 warhead on the GENIE air-to-air rocket is also deployed with fighter interceptor strategic units.

Theater and tactical nuclear warheads are currently deployed on a variety of rocket and missile systems, aircraft, artillery, and land mines (see Table 1.3). Their explosive yields vary from .01 kiloton to over one megaton. While virtually all strategic systems are armed only with nuclear warheads, most theater and tactical systems are dual capable—they can be armed with conventional or nuclear warheads. Only two systems are solely nuclear capable: the Navy SUBROC (W55) anti-submarine rocket and the Army PERSHING 1a (W50) missile.

Of the rockets and missiles, one free-flight rocket—the HONEST JOHN (with W31 warhead)—remains deployed. Although retired from American forces, is still used in allied forces. Army surface-to-surface ballistic missiles include the 100+ km range LANCE (W70) and the 500+ km range PERSHING 1a (W50). Both the Army and Navy have nuclear armed surface-to-air missile systems that double as surface-to-surface systems: the Army NIKE-HERCULES (W31) and Navy ship-based

Figure 1.1 MINUTEMAN III (LGM-30G) missile.

---

[4] In the early 1960s when the stockpile had between 23,000 and 30,000 warheads, there were only about 7000 strategic warheads, most of which were bombs carried on B-47s and the new B-52s; DOD, FY 1984 Annual Report, p. 52.

[5] Force loadings are defined as "those independently-targetable weapons associated with on-line ICBMs, SLBMs and UE (unit equipment) strategic aircraft"; ACDA, FY 1979 ACIS, p. 31.

1
Deployments

TERRIER (W45). Two nuclear armed Navy anti-submarine rockets (ASROC with W44 and SUBROC with W55 warheads) are deployed on a variety of ships and submarines. Two types of low yield atomic demolition munitions (ADMs) are in use by the Army, Marine Corps, and Navy: the medium atomic demolition munition (MADM) (W45) and the man-portable special atomic demolition munition (SADM) (W54).

The Air Force, Marine Corps, and Navy all fly nuclear capable aircraft (see Chapter Seven) and their aircraft for use in theater and tactical nuclear warfare are assigned three different nuclear bombs: the B43, B57, and B61. The most commonly deployed bomb is the newest and most versatile, the 300-500 kiloton yield B61, which by virtue of its design and low weight is able to be carried by every nuclear certified aircraft type. The B57, which doubles as a light weight bomb or depth charge, is carried by tactical fighters, maritime patrol aircraft, or helicopters. A fourth bomb, the B28, is in use by the Air Force and some NATO countries.

One of the major changes in the nuclear stockpile as new, more accurate weapons have been introduced has been the reduction in gross explosive megatonnage. The peak explosive capacity of the stockpile occurred in 1960. Since 1960, as the total number of warheads in the stockpile peaked and then decreased, a significant reduction in megatonnage resulted.[6] Deployment of single warhead low yield missiles allowed a reduction in bombers with their larger yield bombs. According to one official report, "the total number of megatons was four times as high in 1960 than in 1980."[7] With the introduction of many new warheads, "the stockpile yield will not change appreciably in the foreseeable future."[8]

Nuclear Weapons Deployments

Nuclear weapons are in use in all four of the armed services for strategic contingency and regional war plans. Military equipment, units and personnel are all required to have special selection and certification before they can carry out nuclear duties. Nuclear weapons are available for different military missions—anti-aircraft, ground attack, ship attack, anti-submarine warfare—each mission providing for the warheads and delivery systems to be kept during peacetime in various states of readiness. Over 9000 strategic weapons are kept on constant alert; a smaller number of theater nuclear weapons (tactical aircraft and PERSHING 1a missiles) are also maintained at a high state of readiness (so called "quick reaction alert"). Most other weapons are kept on a lower level of readiness, either in storage sites or, in the Navy, in ammunition lockers and special ships.

Nuclear weapons are widely dispersed. They are deployed at about 200 storage sites and bases, both inside the United States and in nine foreign countries (see Chapter Four). Within the United States, they are present in 34 states at a number of central storage sites, at naval bases, at strategic bomber and fighter interceptor bases, at research and development facilities, and in over 1000 underground silos (throughout ten states) holding nuclear armed intercontinental missiles. Over-

Table 1.3
Theater and Tactical Nuclear Weapons (1983)

Delivery Platforms	Systems	Warheads
Anti-Submarine Aircraft	P-3, S-3, NIMROD (UK)	B57
Anti-Submarine Helicopters	SH-3	B57
Anti-Submarine Missiles	ASROC, SUBROC	W44, W55
Artillery	155mm and 8-inch guns	W48, W33, W79
Atomic Demolition Munitions	Medium ADM, Special ADM	W45, W54
Attack Aircraft	A-4, A-6, A-7, F/A-18	B43, B57, B61
Attack Submarines	SUBROC	W55
Fighter Aircraft	F-4, F-15, F-16, F-100, F-104, F-111, TORNADO (NATO)	B28, B43, B57, B61
Surface Ships	ASROC, TERRIER	W44, W45
Surface-to-Air Missiles	NIKE-HERCULES, TERRIER	W31, W45
Surface-to-Surface Missiles	HONEST JOHN, LANCE, PERSHING 1a	W31, W70, W50

---

6 SASC, FY 1983 DOE, p. 54; HASC, FY 1982 DOE, p. 142.
7 DOD, FY 1984 Annual Report, p. 55.

8 Senate Report No. 97-173, 30 July 1981.

1

Early History

seas, thousands of warheads are stored at over 100 locations, the majority of which are in West Germany.

United States policy in every administration since the Truman years has provided for the continued deployment of U.S. nuclear warheads abroad. The first formal agreements were concluded in 1954 with NATO allies in Europe. From 1958-1964, a large number of nuclear warheads were deployed overseas for the first time, and bilateral agreements were concluded with a number of nations covering the deployment and shared use of nuclear warheads. In Europe, there were about 7000 nuclear weapons by 1964, about the number estimated there today. In the Pacific, about 1000 nuclear weapons are estimated to be deployed at land bases.

History of the Nuclear Weapons Stockpile

During the first two decades of U.S. nuclear weapons history there was a massive scientific investment in nuclear weapons research and development. This resulted in significant advances in technology and a high level of weapons turnover as new weapons were continually deployed. Advances in nuclear warhead design, including progress in electronics miniaturization, resulted in more efficient uses of fissile materials and fabrication of small nuclear warheads (see Chapter Two). Small warheads and rapid developments in warhead delivery systems (particularly in missile technology) led to the wide adoption of nuclear weapons within U.S. military forces. Old technology was replaced with new capabilities, with new warhead designs taking advantage of the latest efficiency, control, and safety features. Each new delivery system incorporated additional "improvements": increased range, better accuracy, improved mobility, and greater lethality.

The practice has been that as new warheads are produced and enter the stockpile, old warheads are retired. Changes in the size of the stockpile thus have been and are still based upon differences in the build and retirement rates. In the last three decades, the retention period of warheads in the stockpile has tripled—the average age is now 13 years.[9] The cost of retaining warheads has also greatly increased as weapons sophistication has increased. It is important to note, however, that the development of new warheads and the diversity of delivery modes and weapon systems is not merely a technological phenomenon. Each development also has been a response to the nuclear policy and strategy of the day.

Nuclear weapons developments have progressed through four periods: an early research oriented period (1945-1955), a peak production and growth period (1955-1967), a period of numerical stabilization during which significant operational refinements were made (1967-1980), and a second growth period in which a new generation of warheads is being produced and new technologies are being widely adapted (1980-present).

Early Years (1945-1955)

Atomic bomb developments immediately after World War II focused on perfection of implosion design (due to the scarcity of fissile material)[10] and on improvements in the ballistics, efficiency, reliability, and explosive yield of the *Fat Man* design dropped on Nagasaki. (*Fat Man* became the basis for the Mk-III and Mk-IV nuclear bombs). During the first five years of the stockpile, the number of nuclear warheads grew slowly, limited by the availability of fissile materials.[11] In 1945, the stockpile had only two weapons; in 1946, it had nine; in 1947, it had 13; and in 1948, it had 50.[12]

The practice developed of conserving the fissile materials by keeping them separate from the larger number of available bomb casings (this was also a method of control), and of utilizing the fissile materials taken from retired weapons in new warheads.[13] The growth in the supply of fissile material and its more efficient use, the design of smaller and lighter warheads, the first test of a Soviet nuclear weapon in 1949, and the outbreak of the Korean War in June 1950 all influenced decisions to expand the nuclear arsenal and diversify the types of weapons. In January 1950, President Truman decided to place the thermonuclear (fusion) development program on a crash basis. Weapons research and production then proceeded quickly along two parallel courses: development of efficient, usable fission weapons (with yields up to several hundred kilotons), and development of more powerful fusion weapons (with yields from 1 Mt to 40 Mt).

---

9 DOD, FY 1984 Annual Report, p. 55.
10 The *Little Boy* bomb (later Mk-I) dropped on Hiroshima was a gun assembly weapon, while the *Fat Man* bomb dropped on Nagasaki was an implosion weapon. Implosion weapons require less fissile material (see Chapter Two).
11 Senate Report No. 97-517, 5 August 1982, p. 3.
12 In 1982, the State Department released an undated memorandum containing these early stockpile numbers as partial refutation of some reports of a nuclear weapons build-up.

13 The early generation of nuclear warheads were designed in such a way that the fissile material was separated and stored apart from the nuclear "casing," both for safety and security reasons. The Atomic Energy Commission produced and maintained custody of the fissiles cores, while the military developed and maintained the bomb casings. This practice was discontinued in the mid 1950s when "complete" weapons, the final assemblies, were first turned over to the Department of Defense and new integrated warhead types designed and produced, eliminating the old separation requirement.

1
Warheads (1945-Present)

Table 1.4
U.S. Nuclear Warheads (1945-Present)

Warhead Designator[1]	Delivery System	Service	Lab	Lab Assignment (yr)[2]	Stockpile Entry (yr)[3]	Retirement from Active Service (yr)
TRINITY	Test	—	LANL	1943	1945	1945 at Alamogordo[4]
LITTLE BOY*	Bomb	AF	LANL	1943	1945	1945 at Hiroshima[5]
FAT MAN	Bomb	AF	LANL	1943	1945	1945 at Nagasaki[6]
Mk-I*[7]	Bomb	AF	LANL	1945	1945	1945-1946
Mk-II[8]	Bomb	AF	LANL	?	(cancelled)	
Mk-III[9]	Bomb	AF	LANL	1943	1947	1950
Mk-IV[10]	Bomb	AF/N	LANL	1947	1949	1953
Mk-5†[11]	Bomb	AF/N	LANL	1949	1952	1963
	REGULUS I	N	LANL	1949	1952	1954
	MATADOR	AF				
Mk-6	Bomb	AF/N	LANL	1949	1951	1961
Mk-7†[12]	Bomb	AF/N	LANL	1949	1952	1967
	BOAR Bomb	N	LANL	1949	(never deployed)	
	CORPORAL (Mod 1)	A			1953	1967
	HONEST JOHN (Mod 2)	A			1954	1967
	ADM-B (Mod 3)	A			1954	1967
Mk-8*†	Improved LITTLE BOY (Mod 3)	N	LANL	1950	1951	1956
Mk-9*	280mm howitzer	A	LANL	1950	1952	1957
Mk-10*	Bomb	AF/N	LANL	1950	(cancelled 1952)	
Mk-11*	Bomb	AF/N	LANL	1950	1956	1960
Mk-12[13]	Bomb	AF/N	LANL	?	1954	1962
Mk-13	Bomb	AF	LANL	?	(cancelled 1954)	
Mk-14**	Bomb	AF	LANL	1952	(cancelled 1954)	
Mk-15**	Bomb	AF/N	LANL	1952	1955	1965
Mk-16**	Bomb	AF	LANL	1952	(cancelled 1953)	
Mk-17[14]**	Bomb	AF	LANL	1952	1954	1957
Mk-18	Bomb	AF	LANL	1951	1953	1957
W19*	280mm howitzer	A	LANL	1953	1956	1963
B20	Bomb	AF	LANL	?	(cancelled 1954)	
B21	Bomb	AF	LANL	?	1955	1957
B22	Bomb	?	LANL	?	(cancelled 1954)	
W23*	16-inch naval gun	N	LANL	1953	1956	1959
B24	Bomb	AF	LANL	1952	1954	1956
W25	GENIE	AF	LANL	1954	1957	(active)
B26	Bomb	AF	LANL	?	(cancelled 1955)	
T4*	ADM	A	LANL	?	1957	1963
B27	Bomb	N	LANL/LLNL	1955	(cancelled 1958-9)	
W27	REGULUS II	N	LANL/LLNL	1955	1958	1964
B28†[15]	Bomb	AF	LANL	1955	1958	(active)
W28†	HOUND DOG	AF	LANL	1955	1958	1975
	MACE	AF			1959	1969
B29	Bomb	AF	LANL	?	(cancelled 1955)	
W29	REDSTONE	A	LANL	?	(cancelled 1956)	
W30†	TALOS	N	LANL	1955	1959	1979
	TADM	A/MC			1959	1966
W31†	HONEST JOHN	A	LANL	1954	1958	(active)
	NIKE HERCULES	A			1958	(active)
	ADM	A/MC			1958	1965
W32	240mm howitzer	A	LANL	?	(cancelled 1955)	
W33*	8-inch howitzer	A/MC	LANL	1954	1956	(active)

Nuclear Weapons Databook, Volume I 7

1

Warheads (1945-Present)

Table 1.4 Continued
U.S. Nuclear Warheads (1945-Present)

Warhead Designator[1]	Delivery System	Service	Lab	Lab Assignment (yr)[2]	Stockpile Entry (yr)[3]	Retirement from Active Service (yr)
W34†	LULU (Mk-101 depth bomb)	N		1955	1958	1976
	HOTPOINT (Mk-104)	N		?	1958	(retired)
	ASTOR	N		1955	1958	1976
W35†	ATLAS	AF	LANL	1955	(cancelled 1957-58)	
	TITAN	AF			(cancelled 1957-58)	
	THOR	AF			(cancelled 1957-58)	
	JUPITER	A			(cancelled 1957-58)	
B36	Bomb	AF/N	LANL	1953	1956	1962
W37	NIKE HERCULES	A	LANL	?	(cancelled 1959)	
W38	ATLAS D/E	AF	LLNL	?	1958	1965
	TITAN I	AF			1960	1965
B39	Bomb	AF	LANL	1955	1957	1966
W39†	SNARK	AF	LANL	1956	1958	1961
	REDSTONE	A			1958	1965
W40†	BOMARC	AF	LANL	1956	1959	1972
	LA CROSSE	A/MC			1959	1964
B41	Bomb	AF	LANL/LLNL	1957	1960	1976
W42†	HAWK	A	?	?	(cancelled)	
	FALCON	AF				
	SPARROW	AF				
B43†	Bomb	AF/N/MC	LANL	1956	1961	(active)
W44	ASROC	N	LANL	1956	1961	(active)
W45†	MADM	A/MC	LLNL	1956	1965	(active)
	LITTLE JOHN	A			1962	(retired)
	TERRIER	N			1962	(active
	BULLPUP B	AF/N			?	1978
W46	Unknown	?	?	?	(cancelled 1958)	
W47	POLARIS A1/A2	N	LLNL	1957	1960	1968
W48	155mm howitzer	A/MC	LLNL	1957	1963	(active)
W49†	THOR	AF	LANL	1957	1958	1963
	JUPITER	A			1959	1963
	ATLAS E/F	AF			1960	1965
	TITAN I	AF			1960	1965
W50	PERSHING I	A	LANL	1958	1963	(active)
W51	?	?	?	?	?	?
W52	SERGEANT	A	LANL	1960	1962	1977
B53	Bomb	AF	LANL	1958	1962	(active)
W53	TITAN II	AF	LANL	1960	1962	(active)
W54†	FALCON	AF	LANL	1959	1961	1972
	DAVY CROCKETT	A			1960	1971
	Special ADM	A/MC/N			1960	1964	(active)
W55	SUBROC	N	LLNL	1959	1964	(active)
W56	MINUTEMAN II	AF	LLNL	1960	1965	(active)
B57	ASW Depth Bomb	N/MC/AF	LANL	1960	1963	(active)
W58	POLARIS A3	N	LLNL	1960	1964	1981
W59	MINUTEMAN I	AF	LANL	1960	1961	1969
W60	TYPHOON	N	?	?	(cancelled 1964)	
B61	Bomb	AF/MC/N	LANL	1963	1968	(active)
W62	MINUTEMAN III (Mk-12)	AF	LLNL	1964	1970	(active)
W64	LANCE	A	?	?	(cancelled 1964)[16]	
W66	SPRINT	A	LANL	1968	1974	(inactive storage 1976)[17]
W67	MINUTEMAN III	AF	?	?	(cancelled 1967)[18]	
	POSEIDON	N	?	?	(cancelled 1967)[19]	
W68	POSEIDON	N	LLNL	1966	1971	(active)

Warheads (1945-Present)

Table 1.4 Continued
U.S. Nuclear Warheads (1945-Present)

Warhead Designator[1]	Delivery System	Service	Lab	Lab Assignment (yr)[2]	Stockpile Entry (yr)[3]	Retirement from Active Service (yr)
W69	SRAM	AF	LANL	1967	1970	(active)
W70	LANCE (Mod-1/2)	A	LLNL	1969	1973	(active)
	LANCE (Mod-3) (ER Warhead)	A	LLNL	1976	1981	(active)
W71	SPARTAN	A	LLNL	1968	1974	(inactive storage 1976)[20]
W72	WALLEYE	N	LANL	1969	1970	1979
W73	CONDOR	AF	?	?	(cancelled 1978)[21]	
W74	155mm howitzer	A/MC	?	?	(cancelled 1973)[22]	
W75	8-inch howitzer	A/MC	?	?	(cancelled 1973)[23]	
W76	TRIDENT I	N	LANL	1973	1978	(active)
B77	Bomb	AF	LLNL	1974	(cancelled 1978)	
W78	MINUTEMAN III (Mk-12A)	AF	LANL	1974	1979	(active)
W79	8-inch howitzer (ER Warhead)	A/MC	LLNL	1975	1981	(active)
W80	ALCM (Mod-1)	AF	LANL	1976	1980	(active)
	SLCM (Mod-0)	N			(1984)	—
W81	STANDARD-2	N	LANL	1977	(1986)	—
W82	155mm howitzer (ER Warhead)	A/MC	LLNL	1978	(1986)	—
B83	Bomb	AF/N	LLNL	1979	(1984)	—
W84	GLCM	AF	LLNL	1978	(1983)	—
W85	PERSHING II (air/surface burst)	A	LANL	1979	(1983)	—
W86	PERSHING II (earth penetrator)	A	LANL	1979	(cancelled 1981)	
W87	MX	AF	LLNL	1981	(1986)	—

Sources: National Atomic Museum, Albuquerque, NM; Los Alamos National Laboratory, Los Alamos, NM; Livermore National Laboratory, Livermore, CA; Correspondence with Chuck Hansen, Mountain View, CA.

* Gun-assembly weapons. All other LANL entries are implosion weapons.
** First Thermonuclear designs.
† Warhead modified for various applications.

1 All current nuclear bombs are referred to as "B-" followed by the warhead program number, e.g., B61. If the warhead of a nuclear weapon has other applications, it is designated as "W." Prior to the 1960s nuclear weapons were assigned "Mark" ("Mk") numbers. "Mk" is now used for reentry vehicles. "B" numbers were later given to gravity bombs and one Atomic Demolition Munition (ADM), and "W" numbers to other warheads. The T4, another ADM, is the exception. In this table warheads prior to number 19 have been left with the Mk prefix; all others are designated "W" or "B."
2 The W25 was the first weapon developed under formal procedures agreed on with the Department of Defense. Start of development for earlier weapons are estimates.
3 Stockpile entry dates vary according to different sources. Prior to assembly of warheads with nuclear materials, some dates refer to casings only. "First Production Unit" date is officially used by DOD to refer to three distinct deliveries: the date the nuclear weapons trainer was delivered, the date a nuclear warhead was delivered for operational suitability testing by DOD, and the date the first production "war reserve unit" was delivered to DOD. The stockpile entry date listed is thought to correspond as closely as possible to the last date.
4 TRINITY Test, Alamogordo, NM, 16 June 1945 at 5:29 am.
5 Hiroshima, Japan, 6 August 1945, at 8:15 am.
6 Nagasaki, Japan, 9 August 1945, at 11:02 am.
7 Production model of *Little Boy*.
8 First implosion design; cancelled because of inability to achieve efficiency and reliability of design.
9 First production model of *Fat Man*.
10 First standard production model nuclear weapon. Mk-IV was developed to improve the performance of the Mk-III.
11 First light weight (3000 lb) tactical gravity bomb.
12 Versatile tactical gravity bomb (1700 lb) designed for employment on tactical aircraft, carried internally and externally; one variant, called "Betty," was a Navy depth bomb.
13 Light weight (1000 lb) bomb capable of delivery at supersonic speeds.
14 First droppable thermonuclear bomb to be tested. It weighed 21 tons.
15 Earlier versions of B28 bomb have been retired.
16 Weapons never tested; SASC, FY 1981 DOE, p. 150.
17 Dismantling of the SPRINT and SPARTAN missiles began in FY 1983.
18 Weapons never tested; SASC, FY 1981 DOE, p. 150.
19 *Ibid*.
20 Dismantling of the SPRINT and SPARTAN missiles began in FY 1983.
21 Weapons never tested; SASC, FY 1981 DOE, p. 150.
22 *Ibid*.
23 *Ibid*.

1

Inactive Delivery Systems

Table 1.5
Inactive Nuclear Delivery Systems (1945-present)

System	Warheads	Yield[1]	Active[2]
AIRCRAFT[3]			
AD-4B SKYRAIDER	Mk-7, Mk-8	?	1953-?
AD-5N SKYRAIDER	Mk-7, Mk-8	?	1951-?
AJ-1 SAVAGE	Mk-IV, Mk-5, Mk-6, Mk-7, Mk-8, Mk-15	20 Kt-Mt	1950-1961
AJ-2 SAVAGE	Mk-5	?	1949-1960
A-1 SKYRAIDER	?	?	1962-?
A-3 SKYWARRIOR	Mk-5, Mk-15, B27, B28, B43	40 Kt-Mt	1955-1970
A-5 VIGILANTE	B27, B28, B43	Mt range	1962/3-1970
B-29 SUPERFORTRESS	Mk-III, Mk-IV, Mk-5, Mk-6	40 Kt range	1945-1956
B-36 (NO NAME)	Mk-III, Mk-IV, Mk-5, Mk-6, Mk-17, B18, B24, B36, B39	40 Kt-24 Mt	1949-1958
B-45 TORNADO	Mk-5, Mk-7	?	1948-1959
B-47 STRATOJET	Mk-5, Mk-6, Mk-15, B18, B28, B36, B41, B53	40 Kt-10 Mt	1951-1966
B-50 SUPERFORTRESS	Mk-III, Mk-IV, Mk-5, Mk-6	40 Kt range	1948-1965
B-57 INTRUDER[4]	Mk-7, B43	10 Kt-1 Mt	1955-?
B-58 HUSTLER	B39, B43, B53	Mt range	1960-1970
B-66 DESTROYER	B28, B43	Mt range	1956-1965
XB-70A VALKYRIE	B41, B53	10 Mt	Test only
FJ-4B FURY	Mk-7, Mk-8, Mk-12	10-60 Kt	1954-1962
F2H-2B BANSHEE	Mk-8	20-50 Kt	1949-?
F3H-2N DEMON	?	Kt range	1956-1964
F9F-8B COUGAR	Mk-12	Kt range	1952/3-?
F-84G THUNDERJET[5]	Mk-7	10-60 Kt	1948-1970
F-86H SABRE	Mk-7, Mk-12	10-60 Kt	1952-?
F-89A/B/C/D/H/J SCORPION	FALCON, GENIE	Low Kt	1951-1968
F-100A/C/F SUPERSABRE	Mk-7, B23, B43	10 Kt-1 Mt	1954-?
F-101A/C VOODOO	Mk-7, B23, B43, FALCON, GENIE	10 Kt-1 Mt	1957-1980
F-102A DELTA DAGGER	FALCON, GENIE	Low Kt	1954-?
F-105 THUNDERCHIEF	B28, B43, B61 BULLPUP	Low Kt-1 Mt	1958-1982
P2V3C NEPTUNE	Mk-IV, W34	20 Kt range	1949-?
P2V5 NEPTUNE	W34	Low Kt	1950-?
P5M2 MARLIN	W34	Low Kt	1952-1967
S-2 TRACKER	W34	Low Kt	1952-?
ARTILLERY			
(280mm)	W9, W19	15 Kt	1952-1963
(16-inch howitzer)	W23	10-15 Kt	1956-1959
M44, M53, M59, M114 (155mm)	W48	Sub Kt	1963-?
M55, M115 (203mm)	W33	Sub-12 Kt	1956-?
NAVAL WEAPONS			
ASTOR	W34	low Kt	1963-?
REGULUS I	W5	40-50 Kt	1952-1954
REGULUS II	W27	low Mt	1958-1964
TALOS	W30	5 Kt	1959-1979
TACTICAL MISSILES			
BULLPUP B	W45	20 Kt	?-1976
CORPORAL	W7	10-60 Kt	1953-1967
DAVY CROCKETT	W54	Sub Kt	1960-1971
FALCON	W54	1.5 Kt	1961-1972
HONEST JOHN	W7	10-60 Kt	1954-1974
JUPITER	W35, W49	5 Mt	1959-1963
LACROSSE	W40	low Kt	1959-1964
LITTLE JOHN	W45	1-10 Kt	1962-?
MACE A/B	W28	1 Mt	1959-1969

1
Peak Production

Table 1.5 Continued
Inactive Nuclear Delivery Systems (1945-present)

System	Warheads	Yield[1]	Active[2]
MATADOR	W5	40-50 Kt	1951-?
REDSTONE	W29, W39	Mt range	?-1965
SERGEANT	W52	60 Kt	1962-1977
THOR	W35, W49	500 Kt range	1958-1963
WALLEYE	W72	100 Kt	1970-1979
STRATEGIC MISSILES			
ATLAS D/E/F	W35, W38, W49	500 Kt-1 Mt	1958-1965
BOMARC	W40	400-500 Kt	1958-1972
BOMARC B	W40	400-500 Kt	1959-1972
HOUND DOG	W28	1 Mt range	1958-1975
MINUTEMAN I	W59	1 Mt	1961-1969
POLARIS A1	W47	800 Kt	1960-1966
POLARIS A2	W47	800 Kt	1960-1968
POLARIS A3	W58	200 Kt	1964-1981
SNARK	W39	Mt range	1958-1965
SPARTAN	W71	5 Mt	(1974)
SPRINT	W66	low Kt	(1974)
TITAN I	W35, W38, W49	1 Mt range	1960-1965

1 Yield for aircraft is estimate of yield of each bomb the aircraft was certified to carry and not total yield.
2 With nuclear weapons.
3 Aircraft information is derived from *Jane's All The World's Aircraft 1953-54, 1966-67*; Robert Jackson, *World Military Aircraft Since 1945* (NY: Scribners, 1974); Lloyd S. Jones *U.S. Bombers* (CA: Aero Publishers, 1974).
4 British version was called "CANBERRA."
5 The F-84F "THUNDERSTREAK" was not nuclear capable.

Fission warhead development from 1945-1955 was oriented toward replacements for the Mk-III and Mk-IV bombs (deployed in small numbers in 1947 and 1949, respectively). In the early 1950s, small, light weight "implosion" design bombs for tactical use (Mk-5, Mk-7, and Mk-12), strategic bombs of reduced weight, higher efficiency and yield (Mk-6, Mk-13, and Mk-18), and "penetrator" naval depth bombs (Mk-8 and Mk-11) were all produced and deployed. The first very high yield fusion bomb was the Mk-18, deployed in 1953 for interim use pending development of "deliverable" thermonuclear weapons (such as the Mk-15 and Mk-17 bombs which entered the stockpile in 1954-1955, and the Mk-14 and Mk-16, which were cancelled during the same period) (see Table 1.4).

As new nuclear bombs were being developed and deployed, the bomber force was also being upgraded and expanded. In March 1946, when the Strategic Air Command (SAC) was first established, it had 148 B-29 bombers. Two years later, two new bomber types were added to SAC, the B-36 and B-50, and the size of the bomber force grew to over 500. In 1951, the medium range B-47 was introduced, and SAC established forward bases closer to the Soviet Union, in Europe, North Africa, and Asia. The Korean War then further spurred weapons developments, and, in 1955, the new B-52 bomber was introduced, bringing the bomber force total to over 1500.

The first warhead types were all aircraft delivered bombs; it was not until 1952 that the first non-aircraft delivered "tactical" nuclear warhead was deployed (the Mk-9 atomic artillery projectile for a 280mm Army cannon). In 1953-1955, the Mk-7 bomb warhead was deployed on three different tactical weapons—as the warhead on the Army's CORPORAL and HONEST JOHN rockets, and as the first Atomic Demolition Munition (nuclear land mine).

Peak Production Years (1955-1967)

By 1955, both nuclear bombs and tactical weapons were firmly established as parts of the U.S. nuclear arsenal. The majority of the warheads in the stockpile were bombs, carried by a huge intercontinental and medium-range bomber force. Three factors in the 1950s contributed to a change in the size and composition of the stockpile. First, with Savannah River and Hanford reactors producing at full capacity, the supply of fissile material became sufficient for the production of a large number of warheads. Second, the breakthrough in the development of the thermonuclear weapon in 1953 created the ability to obtain very high yields with relatively

Nuclear Weapons Databook, Volume I   11

1

Stockpile Stabilization

high yield-to-weight ratios. Third, development of long range ballistic missiles received high priority and eventually displaced bombers as the central element of strategic nuclear forces.

Around 1955, a phenomenal jump in warhead production occurred as a result of a wide variety of nuclear delivery systems entering the military and the large number of nuclear bombs built to serve the strategy of massive retaliation. The number of warheads produced was massive; some 30,000 new warheads entered the stockpile from 1955-1967. By 1957, there were some 5000 warheads in the stockpile. From 1958-1960, approximately 11,000 warheads were added to the nuclear arsenal. In 1967, the number reached its all time peak at just over 32,000. As proven designs were mass produced and the features of the current U.S. military force structure began to take shape in the 1960s, rapid turnover of old designs ceased and attention was instead directed towards large scale production of new, smaller, safer, and more capable warheads.

During the 1955-1967 period, 54 warhead types and modifications entered the stockpile. They consisted of twelve new nuclear bomb designs (including three depth bombs), but by far the majority were warheads for tactical weapons and new strategic missile warheads. The new tactical warheads included four nuclear artillery warheads (W19, W23, W33, and W48), three warheads for air-launched missiles (W25 for GENIE, W45 for BULLPUP, and W54 for FALCON), five atomic demolition munitions (T4, W30, W31, W45, and W54), eight warheads for Army short-range missiles (W29 and W39 for REDSTONE, W31 for HONEST JOHN and NIKE-HERCULES, W40 for LA CROSSE, W45 for LITTLE JOHN, W50 for PERSHING 1, W52 for SERGEANT, and W54 for DAVY CROCKETT), and five warheads for naval anti-air or anti-submarine missiles (W30 for TALOS, W34 for ASTOR, W44 for ASROC, W45 for TERRIER, and W55 for SUBROC).

In September 1955, President Eisenhower assigned highest national priority to the development of ballistic missiles. Over the next ten years, eleven warheads would be deployed for strategic missiles: W27 for the REGULUS, W28 for the HOUND DOG and MACE, W38 for the ATLAS and TITAN, W39 for SNARK, W40 for BOMARC, W47 and W58 for POLARIS, W49 for THOR, JUPITER, ATLAS and TITAN, W53 for TITAN II, and W56 and W59 for MINUTEMAN. The liquid fuel ATLAS D was the first intercontinental ballistic missile (ICBM), deployed in April 1958, with a thermonuclear warhead in the megaton range. ATLAS D was first joined by two intermediate range missiles (IRBMs), the THOR and JUPITER in 1958-1959, and then in 1960 by the TITAN I ICBM, and the Navy's POLARIS. Strategic cruise missiles (REGULUS, MACE, SNARK) had received some attention prior to the accelerated development of the ballistic missile, but were eventually replaced by the longer range, higher payload, more accurate, and more reliable ballistic missiles.

Stockpile Stabilization and Refinement (1967-1980)

In 1967, after the number of nuclear warheads in the stockpile had reached its peak, a decrease in the number of warheads began to take place as the strategic force structure was fixed in numbers and missile delivery technology stabilized. Research and production efforts were oriented towards still smaller and more accurate warheads to supply the largely unchanged missiles or bomber delivery vehicles.[14] In 1968, the oldest weapon in the stockpile was 11 years and the average age was 7 years.[15]

The reduction of warheads in the stockpile came as strategic missile forces with fixed numbers of warheads began to partially replace the larger number of old strategic bombers with their duplicative bomb loads. During the 1960s, changes in nuclear strategy, particularly the incorporation of constant ground and air alert operations, forced the creation of more flexible arming systems and stricter reliability requirements for warheads. In 1968, after the two serious nuclear weapons accidents involving bombers loaded with nuclear weapons in Thule, Greenland, and Palomares, Spain (see Figure 1.2), air alert operations ceased and safety considerations in warhead design became a primary consideration.

Although there was a slight net increase in the stockpile in 1970-1973 with the introduction of multiple warheads in the POSEIDON (W68) and a portion of the MINUTEMAN (W62) strategic missile force, the stockpile again began to decrease. This was due to retirements of large numbers of bombs, U.S. based nuclear air defense warheads (NIKE-HERCULES (W31), GENIE (W25), and FALCON (W54)), and the retirement and reductions of some nuclear armed tactical air and sea launched missiles as new conventional weapons were deployed. In addition, the efficiency of new nuclear weapons designs contributed to further reductions in the stockpile through the 1970s. Short-range land-based Army missiles—SERGEANT (W52) and HONEST JOHN (W31)—and older aircraft bombs were replaced on a less

---

14 SASC, FY 1983 DOE, p. 54; SASC, FY 1982 DOE, p. 142.
15 HAC, FY 1982 EWDA, Part 7, pp. 106-107.

1
New Generation

than one-for-one basis by new, more capable, and versatile warheads using multiple yields rather than the previously required larger number of duplicative fixed yield warheads.[16] The reduction in numbers, therefore, was not a reduction in capability. In fact, the new variable yield warheads were directed against a larger number of potential targets than older single yield warheads.

Throughout the remainder of the 1970s, the number of warheads built was less than the number retired. In 1980, thirteen years after the stockpile peaked, the warheads were much older—the oldest was 23 years and the average age was 12.[17] It was not until after 1980 that a dramatic increase in warhead production again occurred.

New Generation of Nuclear Warheads (1980-Present)

During 1976-1978, the Department of Energy was "in a very low build mode."[18] The rate of production increased in FY 1980 and 1981 and "accelerated" in FY 1982.[19] The Nuclear Weapons Stockpile Memorandum signed by President Carter in October 1980 (for the period 1981-1983) called for a further "dramatic increase in warhead production"[20]—a "very sharp increase."[21] It was originally stated in 1981 that with this increase "the stockpile will remain well below the historic highs established in the late sixties"[22] and that the "total magnitude" of the stockpile would not change "in any great consequence."[23] More recently, in March 1982, a Defense Department official stated, "... over the next 5 years there will be an increase in the total number of nuclear warheads deployed, both strategic and tactical, on the order of several thousand."[24]

The magnitude of the increase in nuclear weapons production is reflected in the growth of the Department of Energy budget for warhead research, development, testing and production. The current FY 1984 nuclear weapons budget request is $6.8 billion, contrasting sharply with the FY 1980 level of $2.8 billion. The increase from FY 1981 ($3.7 billion) to FY 1982 ($5 billion) represents the largest single year increase in the history of the weapons program.[25]

The first Stockpile Memorandum of the Reagan Administration, signed in March 1982, approved changes in the mix of warheads, but authorized only a slight increase over the Carter plans. It is estimated that 16,000 new warheads will be produced through 1990 (see Table 1.7) and an additional 12,000 are identified in current research and development programs through the 1990s. An increasing gap between the production and retirement rate is expected, due primarily to two factors: many older weapons will be upgraded as part of a "stockpile improvement program" in which warhead safety and security will be increased, and many old warheads will remain in the stockpile while a determination is made on the deployment of their replacements.[26]

As in the 1950s, the availability of nuclear fissile materials is claimed to be a constraining factor in the current plans for large growth in the stockpile. Actually, warhead production through at least the mid-1980s is possible with the present supply of materials from retired weapons and new materials production. It is only in the early 1990s that potential material shortages have even been *projected*. This is mainly due to four factors: one, many new warhead designs require a different mix in

Figure 1.2 B28RI nuclear bomb recovered from 2500 feet of water off the coast of Palomares, Spain. A B-52 bomber carrying four B28s collided with its aerial refueler on 17 January 1966 and dropped its four bombs, scattering nuclear materials in the Spanish countryside. Three bombs were recovered on land, and the last was recovered in the sea.

---

16 Warhead design in the older weapons could only produce one yield per warhead. This meant that each yield desired required a different warhead. Newer designs could produce variable yields in one warhead (the so-called "dial-a-yield") and thus newer warheads with "selectable" yields could replace a larger number of less versatile older, single yield warheads.
17 HAC, FY 1982 EWDA, Part 7, pp. 106-107.
18 HAC, FY 1982, EWDA, Part 7, p. 100.
19 Ibid., p. 105.
20 HASC, FY 1982 DOE, p. 55.
21 HAC, FY 1982 EWDA, Part 7, p. 100.
22 HASC, FY 1982 DOE, p. 55.
23 HAC, FY 1982 EWDA, Part 7, p. 160.
24 SASC, FY 1983 DOD, Part 7, p. 4235.
25 HASC, FY 1982 DOE, pp. 32-35; HAC, FY 1982 EWDA, Part 5, p. 3.
26 This is particularly the case with large numbers of W33 8-inch artillery warheads deployed in Europe. They are being kept in the stockpile pending the resolution of a European deployment decision on the new enhanced radiation yield W79. According to the Senate Appropriations Committee (Report No. 97-673, 6 December 1982), "the DOE and DOD have grave concerns about the safety and the military usefulness of this atomic projectile."

1
Future Developments

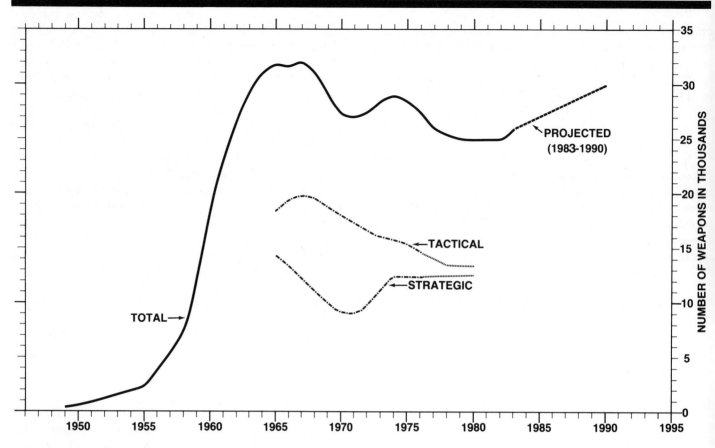

Figure 1.3 U.S. nuclear stockpile, 1949 to present (with future projections).

the materials utilized (more plutonium and tritium; see Chapter Two); two, the contingency plans for production exceed the maximum historic high of the stockpile in 1967; three, many warheads which may be built (and represent additional planning requirements) are still only being considered (a new anti-ballistic missile system, for instance); and four, present plans are to build up a reserve of fissile material should large scale production of warheads be necessary in the future.

Future Nuclear Weapons Developments

A separate agency of the government, the Department of Energy (DOE),[27] is responsible for nuclear weapons development. DOE's relationship to the Department of Defense is very intimate, and it has shown a direct interest in lobbying and supporting continuing nuclear weapons development and production. As such, its independent position may not accomplish what it was originally intended to accomplish, namely, to keep the critical resource of nuclear weapons under civilian control.

The typical life of a nuclear warhead extends through seven "phases," covering some 30 years (see Table 1.8).[28] The research and engineering phases (phases 1-4) typically take as much as nine years. The formal research phases draw upon a continuing advanced concepts and basic scientific research program within the laboratories. Production and stockpiling can take place over as much as an 8-25 year period. Underground testing initially occurs during the first three phases. The initial outlay of large amounts of research, production, and construction money occurs during phase 3. Once a warhead has been approved for production, it enters phase 4 and then advances to phase 5 when full scale production actually begins.

The current programs of nuclear weapons research continues the trend towards greater miniaturization, accuracy, and concurrently lower yields. Development of new warheads incorporating upgraded safety, control, and security features is also a high priority. In recent years, two new innovations have been applied to the stockpile. The first innovation is the widespread adoption of the enhanced radiation (ER) capability (see

---

[27] The DOE is the successor agency to the Energy Research and Development Administration (ERDA) and the Atomic Energy Commission (AEC).

[28] HASC, FY 1981 DOE, p. 185; HASC, FY 1980 DOE, p. 56-57.

1
Future Developments

Table 1.6
U.S. Nuclear Weapons Stockpile

Year	Number of Warheads
1945	2
1946	9
1947	13
1948	50
1949	250
1950	450
1951	650
1952	1000
1953	1350
1954	1750
1955	2250
1956	3550
1957	5450
1958	7100
1959	12,000
1960	18,500
1961	23,000
1962	26,500
1963	29,000
1964	31,000
1965	31,500
1966	31,500
1967	32,000
1968	31,000
1969	29,000
1970	27,000
1971	27,000
1972	27,500
1973	28,500
1974	29,000
1975	28,500
1976	27,500
1977	26,000
1978	25,500
1979	25,000
1980	25,000
1981	25,000
1982	25,000
1983	26,000

Stockpile numbers for 1945-1948 are taken from an undated State Department memorandum circulated in 1982, and David Alan Rosenberg, "U.S. Nuclear Stockpile, 1945-1950," *The Bulletin of the Atomic Scientists*, May 1982; Authors' estimates of the current size of the stockpile and historical trends are derived from SAC, FY 1981 EWDA, Part 2, pp. 798-799, 808; JCAE, *Development, Use and Control of Nuclear Energy for the Common Defense and Security and For Peaceful Purposes*, Second Annual Report, 30 June 1976, pp. 135-136; HASC, FY 1982 DOE, p. 142; DOD, FY 1984 Annual Report, p. 55. For the years 1949-1958 the estimate is rounded to the nearest multiple of 50 warheads, and for the years following 1958 to the nearest 500 warheads.

Chapter Two), which has been built into the W70-3 LANCE warhead (production completed in 1982) and the new W79 8-inch artillery warhead. An ER yield is also planned for the W82 155mm artillery warhead (scheduled to begin production in 1984) and under consideration as the warhead for the SENTRY anti-ballistic missile. The second development is in the W87 warhead planned for the PEACEKEEPER/MX (and possibly TRIDENT II) missile, which allows a quick conversion to significantly higher yield (an increase from 300 to 475 kilotons). The W87 also utilizes much smaller amounts of fissile material for equivalent yield compared to reentry vehicle warheads developed as late as in the 1970s.

Nine warheads will be in production in 1983-1984: B61-3 and -4 bombs, the W76 warhead for the TRIDENT I, the W78 warhead for the Mk-12A reentry vehicle on the MINUTEMAN III, the W79 enhanced radiation warhead for 8-inch artillery, the W80-0 warhead for the Sea-Launched Cruise Missile, the B83 bomb, the W84 warhead for the Ground-Launched Cruise Missile (GLCM), and the W85 warhead for the PERSHING II missile. One of these weapons, the W78, will complete its production run in 1983. Four other warheads will be in the engineering development phase in 1983: the W81 warhead for the STANDARD-2 missile, the W82 enhanced radiation warhead for 155mm artillery, the W87 warhead for the PEACEKEEPER/MX missile, and the SENTRY/Low Altitude Air Defense System anti-ballistic missile warhead (see Table 1.9).

According to the Department of Energy, 10-20 percent of a weapon system's cost is for the nuclear warhead,[29] but this estimate can vary greatly. The cost of each new nuclear artillery warhead in 1973 was $400,000 (8-inch) and $462,000 (155mm). This is far less than the total system cost of artillery, including thousands of guns with high levels of manning and support, all capable of firing the same nuclear projectiles.[30] The cost of the new W84 warhead for the Ground-Launched Cruise Missile (GLCM) was recently estimated at $1.1 million each, or approximately 17 percent of the $3.678 billion GLCM program.[31] In contrast, each new W82 enhanced radiation 155mm artillery warhead under development is estimated to cost over $3 million.

Many warheads, both in the research and production phase, have problems associated with need, scheduling, cost, and effectiveness. These problems rarely receive public attention, but on occasion some have been revealed in Congressional reports. For example, both the House and Senate Appropriations Committees stated in FY 1983 that "some build levels appear excessive in relation to military capabilities and requirements, as well as realistic assessments of deployment

---

29 HAC, FY 1982 EWDA, Part 7, p. 163.
30 *Military Applications of Nuclear Technology*, Part 2, p. 101.

31 The W84 program was quoted in Congressional hearings at $630 million for 560 warheads; HAC, FY 1982 DOD, Part 7, p. 749; program cost in Defense Department estimate as of 30 September 1982.

1
Projected Warhead Production

Table 1.7
Projected Nuclear Warhead Production, 1983 to mid-1990s

In Production (1983)	Number Planned
B61 Bomb	1000
W76 TRIDENT I	1600
W79 8-inch artillery shell (ER warhead)	800
W80 Air-Launched Cruise Missile	4000[1]
W80 Sea-Launched Cruise Missile	1000
B83 Bomb	2500
W84 Ground-Launched Cruise Missile	560
W85 PERSHING II	300
SUBTOTAL	11,760[2]
Planned (1983-1988)	
W81 STANDARD-2	500
W82 155mm artillery shell (ER Warhead)	1000
W87 MX Warhead	1055[3]
Surface and Air Delivered ASW Weapon	1250
Subsurface Delivered ASW Standoff Weapon	400
SUBTOTAL	4205[4]
Future Systems (Late 1980s-1990s)	
TRIDENT II	5000*
SENTRY (ABM)	500
New Strategic Air-Launched Missile	1200
Corps Support Weapon System	500
Advanced Tactical Air Delivered Weapon	2500
Advanced Cruise Missile Technology	(3000)[5]
Advanced Mobile ICBM	3000
Bomber Defense Missile	?
SUBTOTAL	12,700
Alternate Systems	
Tactical Air-to-Surface Munition	(1500)
MaRV for TRIDENT II	(7500)*
TOTAL WARHEAD PRODUCTION	28,665[6]

1 Number includes Advanced Cruise Missile Technology Warhead, which will replace ALCM warheads on a one-for-one basis.
2 Not all of these warheads will be produced in the 1980s.
3 Does not include W87 production for TRIDENT II.
4 Not all of these warheads will be produced in the 1980s.
5 Number includes Advanced Cruise Missile Technology Warhead, which will replace ALCM warheads on a one-for-one basis.
6 This number does not include Alternate Systems.
* Competing warhead programs for TRIDENT SLBM upgrade and TRIDENT II (instead of W87).

requirements in the current world political climate."[32] The committees also questioned the rate of the retirement program "particularly for those systems that would alleviate materials production requirements and those systems that are considered to be near or at a state of obsolescence," and the "unrealistic scheduling requiring mid course corrections" of PERSHING II (W85), cruise missiles, and the W82 155mm artillery warhead.

Funding for individual warheads has also been held up by Congress. Only minimal caretaker funds have been appropriated for the SENTRY anti-ballistic missile warhead, thus stopping production plans. Funding for the W82 155mm artillery warhead has been reduced by the Armed Services and Appropriations Committees in the FY 1983 budget. The Senate Appropriations Committee cited "uncertainty of deployment" and "extremely high costs" and noted that "it is premature to proceed to spend billions of dollars on these nuclear artillery warheads at this time."[33] Funding for the W87 MX warhead has been reduced by Congress "for reasons related to test status and capability to produce."[34] The FY 1983 Appropriations Conference Report provided no funds for proposed W31 NIKE-HERCULES warhead modifications. Finally, the House Appropriations Committee deleted funds for the W81 STANDARD-2 missile warhead request in FY 1983 "pending resolution of differences with respect to the adequacy of the design."[35]

---

32 House Report No. 97-859, 21 September 1982, p. 58; Senate Report No. 97-673, 6 December 1982, p. 88.
33 SAC, Report No. 97-673, 6 December 1982, p. 92.
34 HAC, Report No. 97-345, 19 November 1981, p. 29.
35 Senate Report No. 97-850, 21 September 1982, p. 61.

1
Strategic Developments

Table 1.8
Nuclear Weapons Development Phases

Phase		Activity
Phase 1	Weapons Conception	Studies by DOE/DOD/interested services generating interest in new weapon idea or concept warranting formal program review.
Phase 2	Program Study or Feasibility Study	DOE impact report to DOD, Draft Military Characteristics (MC) and Stockpile-to-Target (STS) sequences prepared by DOD; Phase 2A: Design definition and cost study; form DOE/DOD project offices, select laboratory design team.
Phase 3	Development Engineering or Full-Scale Development	Approved DOD development request, with approved MCs and STS; nomenclature assigned; quantitative requirements set with development and production milestones.
Phase 4	Production Engineering	Tooling and processing; prototyping, construction of production facilities.
Phase 5	First Production	Evaluation and testing for weapon acceptance.
Phase 6	Quantity Production and Stockpile	Weapons produced and deployed and stored by DOD.
Phase 7	Retirement	Disposal of weapons and related material and recovery of nuclear materials.

1 These phases are defined in a joint AEC-DOD agreement dated 21 March 1953; see also HAC, FY 1980 DOD, Part 4, p. 658.

Strategic Weapons Developments

Six hundred MINUTEMAN III missiles, each armed with W62 or W78 warheads of 170 and 335 kilotons, will remain deployed through the 1990s. Fifty MINUTEMAN II missiles will be replaced with MINUTEMAN III missiles, but the remaining 400 will retain their W56 one megaton warheads until the PEACEKEEPER/MX missile is deployed. Plans are to deploy 100 PEACEKEEPER missiles, each with 10 W87 warheads, initially with a yield of about 300 kilotons. A new small, single warhead strategic missile is being developed as a successor to the MX (see Chapter Five).

Plans to deploy a large scale ballistic missile defense system have been accelerated, although the initial nuclear armed system, called SENTRY (formerly the Baseline Terminal Defense System (BTDS) or Low Altitude Air Defense System (LoADS)), has been terminated.

Deployment of 31 POSEIDON submarines, 19 with POSEIDON missiles carrying 10 W68 warheads and 12 with TRIDENT I missiles carrying eight W76 warheads, will continue through the end of the 1980s, when some of the 30 year old submarines will be retired. Deployment of at least 20 TRIDENT submarines will continue through the 1980s and 1990s. The first eight TRIDENT submarines will be initially deployed with 24 TRIDENT I missiles. In 1988-1989, the remaining submarines will be deployed with the TRIDENT II D5 missile.

The bomber force will continue through the 1980s with about 250 B-52G/H and about 50 FB-111 aircraft armed with nuclear bombs and Short-Range Attack Missiles (SRAM) (W69). The bomber force will increasingly be supplied with Air-Launched Cruise Missiles (ALCM) (W80-1) until a new "Advanced Cruise Missile" is deployed in the late 1980s. The first of 100 B-1B bombers capable of carrying gravity weapons, SRAMs, and ALCMs will be deployed in FY 1985. In the early 1990s, 130-150 nuclear armed "Advanced Technology" (STEALTH) bombers (ATB) are planned for deployment. B-52G bombers will begin phasing out in 1990. B-52Hs will remain through the 1990s and the FB-111 will be transferred to the tactical inventory as the ATB is deployed.[36] A new nuclear armed missile, a versatile

Figure 1.4 Advanced Strategic Air-Launched Missile (**ASALM**) propulsion technology prototype.

36 DOD, FY 1984 Annual Report, pp. 222-224.

1

Theater Developments

Table 1.9
Nuclear Weapons Research and Development Programs (1983)

Warhead Program	Status[1]	First Deployment Planned	Number Planned	Weapon Application
W80 Sea-Launched Cruise Missile	Phase 3/4	1984	1000	New weapon
W81 STANDARD-2 Missile	Phase 3/4	1984-5	500	Replacing W45 and for AEGIS shipboard air defense systems
W82 155mm Artillery Projectile	Phase 3	1986	1000	Replacing W48 (ER warhead)
B83 Modern Strategic Bomb	Phase 3/4	1984	2500	Replacing B28, B43 and B53
W84 Ground-Launched Cruise Missile	Phase 3/4	1983	560	New weapon
W85 PERSHING II Missile	Phase 3/4	1983	300	Replacing W50
W87 MX Warhead	Phase 3	1986	1055	Warhead for MX/Mk-21 Advanced Ballistic Reentry Vehicle (ABRV)
Surface and Air Delivered Anti-Submarine Warfare Weapon*[2]	Phase 2	late 1980s	1250	Replacing B57 and W44 in new ASW standoff weapons
Maneuvering Reentry Vehicle (MaRV)	Phase 2	late 1980s	(5000)[3]	Alternate for Navy MaRVs, option to replace W68, W76, or W87
SENTRY ABM Warhead[4]	Phase 2/3	1988	500	New ABM weapon
New Strategic Air-Launched Missile[5]	Phase 2	late 1980s	1200	Air-to-Air/Ground Missile Warhead, replacing W69
Corps Support Weapon System	Phase 2	1988	500	Replacing W70
Advanced Tactical Air Delivered Weapon	Phase 1	1990	2500	New Multi-Purpose Guided Tactical Bomb
TRIDENT II Submarine-Launched Warhead	Phase 2/3	1989	(5000)[6]	Alternate for replacing W76 and W87; for TRIDENT II/Mk-5 Advanced Ballistic Reentry Vehicle (ABRV)
Tactical Air-to-Surface Munition Warhead (TASM)	Phase 1	early 1990s	1500	New weapon
Advanced Cruise Missile Technology Warhead (ACMT)	Phase 1/2	1986-1987	3000	Augmenting and replacing W80
Subsurface Delivered ASW Standoff Weapon*	Phase 2	late 1980s	400	Replacing W55
Advanced Mobile ICBM*	Phase 2	late 1980s	3000	New weapon for small ICBM
Bomber Defense Missile*	Phase 1	1990s	?	New weapon

1 Status in FY 1983-84. Phases refer to stage of development (see Table 1.8).
2 Warhead program was formerly called Anti-Submarine Warfare Weapon for Common ASROC and SUBROC replacement. Surface and Subsurface Warhead development has now been split (see Subsurface Delivered ASW Standoff Weapon).
3 Alternative warhead program competing for TRIDENT II SLBM programs.
4 Formerly Low Altitude Air Defense System (LoADS); also referred to as the Baseline Terminal Defense System (BTDS).
5 Warhead program was formerly called Lethal Neutralization System and Advanced Strategic Air-Launched Missile (ASALM).
6 Alternative warhead program competing for TRIDENT II SLBM programs.
* New warhead development program in FY 1984.

long-range air-to-air and air-to-ground bomber defense weapon called the Advanced Strategic Air-Launched Missile (ASALM), is under development to replace SRAM, with possible deployment in the early 1990s. In FY 1984, another new nuclear warhead program for a "Bomber Defense Missile" was started.

The most significant development within strategic forces is the planned addition of a second sea based system, the cruise missile. As many as 1000 Sea-Launched Cruise Missiles (SLCMs), armed with the W80-0 nuclear warhead, will be deployed on submarine and surface ships as part of a "strategic reserve force"[37] starting in the summer of 1984.

Strategic defensive forces will be upgraded during the 1980s with the replacement of older F-106 and F-4 interceptors with newer model F-4, F-15, and F-16

37 Ibid., p. 54.

Theater Developments

are also being deployed, including the F-16, F/A-18, and TORNADO fighters. These aircraft are all capable of being armed with the new versions of the 300-500 kiloton high speed delivery nuclear bomb, the B61, which is in production. An interim replacement for the nuclear strike F-4 and F-111 tactical fighters is under development. An enhanced version of the F-16 or F-15 will be chosen as the Derivative Fighter Aircraft, pending development of an Advanced Technology Fighter in the 1990s (see Chapter Seven).

The development of more capable precision battlefield conventional weapons has had little influence on the reduction or retirement of the bulk of the short-range tactical nuclear weapons. About 5000 low yield nuclear artillery warheads—the W33 8-inch and W48 155mm projectiles—are currently a part of the stockpile. These two warheads will be replaced in the 1980s with the W79 and W82 warheads. The new 155mm artillery warhead (W82) is under development with an enhanced radiation yield. The short-range LANCE missile, which has both fission and enhanced radiation versions of the W70 warhead, was introduced into U.S. and NATO forces in the mid-1970s, replacing the HONEST JOHN and SERGEANT missiles. The LANCE will continue in active forces until it is itself replaced with a "Corps Support Weapon System" being jointly developed in the Joint Tactical Missile System program by the Army and the Air Force. The NIKE-HERCULES air defense missile with the W31 nuclear warhead will be withdrawn from Europe during the 1980s and replaced with conventional "improved HAWK" and PATRIOT missiles.

Figure 1.5 MX missile undergoing stress test.

interceptors. The nuclear armed GENIE anti-bomber missile with the low yield W25 warhead will continue in use and a potential replacement in the form of a nuclear warhead for the Navy's PHOENIX air-to-air missile is possible.

Theater and Tactical Weapons Developments

Theater nuclear modernization trends include increased long-range capability, mobility and dispersal, and more precise guidance and targeting capabilities. These new theater weapons "permit the use of lower yield nuclear weapons while attaining military effects commensurate with the earlier generation of less accurate higher yield weapons."[38]

Beginning in late 1983, unless political or arms control developments intervene, 572 new long-range, accurate, low yield theater missiles will be deployed in Europe— 108 operational PERSHING II missiles with W85 warheads and 464 Ground-Launched Cruise Missiles with W84 warheads. New tactical nuclear capable aircraft

A new generation of tactical nuclear weapons for the Navy is under development and will be deployed during the 1980s. The Sea-Launched Cruise Missile (SLCM), armed with both nuclear (W80-0) and conventional warheads, will be widely deployed on Navy ships and submarines starting in 1984. In its nuclear role, the SLCM will be a long-range strike weapon for attacking land targets, and will thus introduce a new capability to the surface and subsurface Navy. Two anti-submarine warfare (ASW) nuclear warheads are under development as late 1980s replacements for ASROC (W44) deployed aboard surface ships, SUBROC (W55) aboard attack submarines, and B57 nuclear depth bombs. The new ASW weapons will be longer range and more accurate than either ASW missile currently deployed and will provide a standoff capability to replace the B57 nuclear depth bomb currently used in maritime patrol and anti-submarine operations. Another alternative for B57 replacement is to arm the versatile HARPOON cruise missile with a nuclear warhead.

---

38 ACDA, FY 1981 ACIS, p. 243.

1
Theater Developments

An earlier trend within the Navy toward reducing reliance on nuclear weapons for fleet air defense has been reversed. A significant cutback in fleet levels in 1970, a decision to reduce nuclear loading factors because of the marginal utility of the older systems, and phase-out of the nuclear armed TALOS in 1979, led to a steady decline from the late 1960s onward in the number of ships equipped with nuclear air defense systems. Since 1975, the number of nuclear warheads for surface-to-air missiles have been significantly reduced to only a small percentage of the ships' storage capacities.[39] Now, ships intermittently carry a nuclear warhead (W45) for the TERRIER missile, but beginning in the mid-1980s, large numbers of the new nuclear armed STANDARD-2 missile, with the low yield W81 warhead, will begin deployment.

---

39 ACDA, FY 1980 ACIS, p. 189.

2
Fission

Chapter Two
Nuclear Weapons Primer

Nuclear Fission and Fusion

A nuclear weapon is a device in which most or all of the explosive energy is derived from either fission, fusion, or a combination of the two nuclear processes:[2]

(1) *NUCLEAR FISSION* is the splitting of the nucleus of an atom into two (or more) parts. Certain isotopes of uranium and plutonium (and some other heavier elements), when bombarded by neutrons, will split into atoms of lighter elements and in the process will emit, on average, two or more neutrons from each nucleus and considerable energy—about ten million times as much, atom for atom, as is obtained from ordinary chemical combustion.[3]

(2) *NUCLEAR FUSION* is the joining (or fusing) of the nuclei of two atoms to form a single heavier atom. At extremely high temperatures—in the range of tens of millions of degrees—the nuclei of isotopes of hydrogen (and some other light elements) can readily combine to form heavier elements and in the process release considerable energy.[4] While a number of thermonuclear reactions are identified in the literature,[5] the most relevant to nuclear weapons is the reaction between deuterium (H-2 or D) and tritium (H-3 or T), two hydrogen isotopes, because the deuterium-tritium (D-T) reaction proceeds more rapidly at realizable temperatures than the other fusion reactions.[6]

Atom for atom, the energy released in fusion is less than that released in fission,[7] but the atoms involved in fusion are much lighter, so in theory, the maximum energy obtainable from fusion is about three or four times as great per unit weight as the maximum energy obtainable from nuclear fission.[8] Pure fission weapons of low yield are the simplest practical nuclear weapons to design and construct. Some low yield weapons, e.g., older nuclear artillery shells, are the pure fission type. Weapons of very high yield and weapons that have the highest yield-to-weight ratio use a combination of fission and fusion reactions.[9] Most weapons in the U.S. stockpile are probably of this type.

Fission Weapons

Chain Reaction

In a fission device (a weapon or a reactor), it is necessary to achieve a *chain reaction*, whereby neutrons emitted by fissioning nuclei induce fission in other fissionable nuclei. The neutrons from the fissions, in turn, induce fission in still other fissionable nuclei, and so on. When uranium-235 (U-235) fissions, an average of about 2.56 neutrons are released; an average of about 2.9 to 3.0 neutrons are released when a nucleus of plutonium-239 (Pu-239) fissions.[10] A portion of these neutrons is captured by nuclei that do not fission, and others

---

1 This section is derived from Ted Greenwood, George W. Rathjens and Jack Ruina, "Nuclear Power and Weapons Proliferation" (London: International Institute for Strategic Studies, *Adelphi Paper* No. 130, Winter, 1976), pp. 2-6. Other useful primers, although not an exhaustive list, include: Samuel Glasstone and Philip J. Dolan, *The Effects of Nuclear Weapons*, 3rd ed. (Washington: U.S. DOD and U.S. DOE, 1977), Chapter 1; Philip Morrison, "The Physics of the Bomb," *Atomic Energy* (Harmondsworth, Middlesex, U.K.: Penguin Books, 1950), pp. 2-29, 101-125, 194-201; Edward Teller, *et al.*, *The Constructive Use of Nuclear Explosives* (N.Y.: McGraw Hill, 1968), pp. 1-91; R. Serber, *et al.*, "The Los Alamos Primer," LA-1, April 1943; Robert W. Selden, "An Introduction to Fission Explosives," LLL, UCID 15554, July 1969; J. Carson Mark, *et al.*, "Nuclear Weapons," *Nuclear Proliferation and Safeguards* (N.Y.: Praeger Publishers, 1977), Chapter VI, pp. 139-145; M.B. Neyman and K.M. Sadilenko, "Thermonuclear Weapons," translation by Technical Information Center (Wright Patterson AFB, OH, October 1960); Torsten Magnusson, "Design and Effects of Atomic Weapons" (U.S. Joint Publication Research Service, 8295, 22 May 1961); Howard Morland, "The H-bomb Secret," *The Progressive*, November 1979, pp. 14-23; (read with "Errata," *The Progressive*, December 1979, p. 36) and *The Secret that Exploded* (New York: Random House, 1981); numerous encyclopedia articles by authors including Edward Teller, Hans Bethe, John S. Foster.
2 While all nuclear weapons deployed contain chemical high explosive material to initiate the nuclear reaction, it is theoretically possible to use electromagnetic energy (i.e., light from lasers) instead of chemical explosives.
3 Approximately 200 MeV (3.2×10^{-11} joules) is released in each fission, but only about 180 MeV is immediately available as explosive energy (from gamma rays and the kinetic energy of fission products and neutrons, and from only a small fraction of the decay energy of fission products). Based on the latter value, an explosion equivalent to 1 kiloton (Kt) of TNT (defined as the release of 10^{12} calories) is obtained by the fission of 1.45×10^{23} nuclei. Thus, the complete fission of one kg of fissionable material produces a 17.5 Kt explosion, or 1 Kt is released from the complete fission of 0.057 kg (57 grams or 2 ounces); see, Glasstone and Dolan, *op. cit.*, pp. 12-13.

4 The principal thermonuclear reactions involving isotopes of hydrogen include:
(i) $D + T \rightarrow He^4 (3.52 \text{ MeV}) + n (14.07 \text{ MeV})$
(ii) $D + D \rightarrow \begin{cases} T (1.01 \text{ MeV}) + p (3.02 \text{ MeV}) \\ He^3 (0.82 \text{ MeV}) + n (2.45 \text{ MeV}) \end{cases}$
(iii) $T + T \rightarrow He^4 + 2n + 11.4 \text{ MeV}$
The (D + D) reaction proceeds with approximately equal probability via the two channels.
5 See for example, Neyman and Sadilenko, *op. cit.*, p. 8.
6 The (D-T) reaction is reaction (i) in footnote 4. It is about 100 times more probable than the (D + D) reaction (ii) in the temperature range 10-100 KeV (1 KeV = 11.6 million °K). Alternately, a given reaction rate can be achieved at a lower temperature for the (D-T) reaction than for other fusion fuels; Booth, *et al.*, "Prospects of Generating Power with Laser-Driven Fusion," *Proceedings IEEE*, 64, October 1976, p. 1461.
7 17.6 MeV for the D-T reaction compared to 200 MeV for fission of a uranium or plutonium atom.
8 Compared to 200 MeV for fission, approximately four times for fusion of deuterium with tritium nuclei and approximately three times for fusion of deuterium nuclei or deuterium and tritium nuclei from lithium-6 deuteride; see Glasstone and Dolan, *op. cit.*, p. 21, for derivation.
9 Although possible in principle, pure fusion weapons—that is, where the high temperature necessary for fusion is not obtained from a fission explosion—have not as yet been developed in practical form and may never be. One of the objectives of the U.S. inertial confinement fusion research program probably is to determine the feasibility of using lasers to produce a practical pure fusion weapon.
10 These values are for fission induced by 1 MeV neutrons. The average number of neutrons per fission decreases slightly as the energy of the neutron inducing the fission drops; see USAEC, *Reactor Physics Constants*, ANL-6800, July 1963, pp. 20-23, and A.M. Weinberg and E.P. Wigner, *The Physical Theory of Neutron Chain Reactors* (Chicago: University of Chicago Press, 1958), p. 129.

2
Fissionable Materials

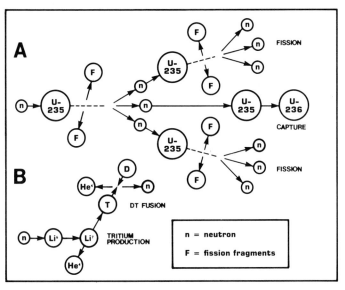

Figure 2.1 (a) Fission chain reaction in uranium-235. (b) Fusion of deuterium and tritium into helium (upper right). Here, the tritium is produced by absorption of a neutron in lithium-6 (lower left).

escape the material without being captured. What is left can cause further fissions. If more than one neutron per fission remains for the chain reaction, more fissions are achieved in the next "generation" than in the previous one. To achieve a high efficiency in a nuclear explosion, a very rapid growth in the number of fissions is sought—that is, a rapidly *multiplying chain reaction*. This means, among other things, that an effort must be made to keep down the leakage of neutrons out of the fissile material and to avoid neutron absorbing impurities in the fissionable material.

Fissionable Materials

Many heavy atomic nuclei are capable of being fissioned; but only a fraction of these are *fissile*, which means fissionable by slow (or zero energy) neutrons, as well as fast (highly energetic) neutrons. Since the neutrons resulting from nuclear fission are emitted with a wide range of energies, nuclei which fission only from the capture of fast neutrons would generally not be able to sustain a chain reaction.[11] From a practical point of view, fission weapons must be made using fissile materials,[12] principally U-235, Pu-239, U-233, or some combination of these.

U-238 and Thorium-232 (Th-232), both abundant in nature, are also fissionable but only by fast neutrons, so they cannot sustain a chain reaction by themselves. Nevertheless, these two materials can contribute to the yield in both fission and thermonuclear explosions where the many excess high-energy neutrons generated by other fission and by fusion reactions can cause them to fission (see below: Fusion Weapon Design, Thermonuclear Weapon Design).

Uranium as found in nature consists primarily of two isotopes, U-235 and U-238, with U-235 (the fissile isotope) occuring only 0.711 percent in abundance, while U-238 constitutes 99.3 percent. Plutonium does not occur naturally except in minute concentrations. Therefore the fissile isotope Pu-239 is made artificially in nuclear reactors from U-238.[13]

To date only U-235 and Pu-239 seem to be used as the fissile material in stockpiled nuclear weapons. Other fissile isotopes of uranium and plutonium, e.g., U-233 and Pu-241, occur only in trace amounts in nuclear weapons. Because U-233 and Pu-241 are more difficult and costly to produce in quantity than U-235 and Pu-239, they are not used as primary fissile material. U-233 and Pu-241 are also more radioactive, can complicate weapons fabrication, and can degrade the reliability of other weapon components.[14] U-238 is used to contribute to the yield of some fission and thermonuclear weapons; Th-232 does not appear to be used.

Fission weapons are made using grades of enriched uranium (from an enrichment plant) or plutonium (produced in a reactor) which contain the fissile isotopes U-235 and Pu-239, respectively, in levels which provide efficient reactions and use a minimum of materials. However, fission weapons do not require uranium or plutonium pure in the isotopes U-235 and Pu-239 to make an explosion, nor do they require uranium or plutonium in the form of a metal. Theoretically, uranium weapons cannot be made using uranium enriched to less than about 5-6 percent U-235. In the range 6-10 percent U-235, very large quantities—thousands of kilograms—of uranium would be required. The fissile uranium material of current nuclear weapons is 93.5 percent enriched U-235.[15]

In theory plutonium compounds containing 6-10 percent (or even less) Pu-239 are usable for weapons. In practice, plutonium is produced from U-238 in a reactor

---

[11] Generally these would include isotopes of heavy elements with an even isotope number, e.g., U-236; U-238. Pu-240 is an exception in that it is not fissile and yet can sustain a chain reaction with fast neutrons.
[12] Generally these would include isotopes of heavy elements with an odd isotope number, e.g., U-235; Pu-241.
[13] In the reaction: U-238 $\xrightarrow{(n, gamma)}$ U-239 $\xrightarrow[23.5m]{beta^-}$ Np-239 $\xrightarrow[2.35d]{beta^-}$ Pu-239
[14] U-233 is produced in nuclear reactors from Thorium-232 in the reaction:

Th-232 $\xrightarrow{(n, gamma)}$ Th-233 $\xrightarrow[22.2m]{beta^-}$ Pa-233 $\xrightarrow[27d]{beta^-}$ U-233

The U-233 is used as the primary fuel in some types of nuclear reactors. Interest in U-233 as a nuclear weapons usable material stems from concern that, like highly enriched uranium (U-235) and plutonium, U-233 may be diverted from its use in civil activities and used in weapons. (The U.S. has tested U-233 weapons.)
[15] SASC, FY 1983 DOD, Part 7, p. 4979.

2
Critical Mass

to a purity of about 93.5 percent Pu-239 for weapons use.[16] The element Pu-240, a byproduct of plutonium production, is an undesirable element for weapons design because of its high spontaneous fission rate.[17] Consequently, in reactors used for the production of plutonium for weapons, the period of time that the U-238 is left in the reactor is restricted to limit buildup of Pu-240 (to about 6 percent) while creating the fissile Pu-239.

Although one basic difference between U-235 and Pu-239 for weapons design is that U-235 occurs in nature, a larger amount of U-235 is required to make an explosion of equal yield to a plutonium weapon. Plutonium-239 (Pu-239) is more expensive to produce and must be made artificially, but it can be used to obtain a higher yield-to-weight ratio, smaller weapons size, and decreased weight.

From 1945-1947, the U-235/Pu-239 production ratio in the U.S. was approximately 8 to 1. It was therefore highly desirable to utilize U-235 and achieve the maximum efficiency in the use of both U-235 and Pu-239. Consequently, composite fission cores containing both U-235 and Pu-239 were developed; these fission cores were actually stockpiled at the end of 1947 for use in the Mark III implosion type bomb, although the percentage of plutonium needed to achieve maximum effect was then unknown. There is still some six to seven times more U-235 in the U.S. weapons stockpile than Pu-239 (see *Nuclear Weapons Databook, Volume II*). Most U.S. nuclear weapons contain both Pu-239 and U-235.

Critical Mass

Small amounts of fissile material will not sustain a chain reaction because a large fraction of neutrons leak out, making them unavailable to cause fission in other nuclei. The minimum mass of material necessary to sustain a chain reaction is called the *critical mass* and is dependent on the type of fissile material, its density, and its geometry. A mass that is less than the critical amount is said to be *subcritical*, while a mass greater than the critical amount to achieve a multiplying chain reaction is referred to as *supercritical*.

Because a sphere has the highest volume-to-surface ratio of any solid shape and, therefore, the least number of escaping neutrons per unit of material, it is the shape for which the critical mass is smallest. The critical mass of a bare sphere of U-235 at normal density is approximately 52 kilograms (kg), that of U-233 about 16 kg, and that of certain dense metallurgical phases of Pu-239 as low as 10 kg.[18]

The critical mass can be lowered in several ways. The fissile material may be surrounded by a shell of other material to reflect some of the neutrons which would otherwise escape. Practical *reflectors* can reduce the critical mass by a factor of two or three so that about 5-10 kg of either Pu-239 or U-233 and about 13-25 kg of U-235 at normal density can be made critical.[19]

The critical mass is also lowered if the material is compressed to increase its density.[20] Consequently, an efficient practical fission bomb, which depends on extremely high compression of the nuclear core, could use significantly smaller amounts of fissile materials

---

16 Weapon-grade plutonium is defined as plutonium containing less than 7 percent Pu-240; fuel-grade plutonium is 7 percent to less than 19 percent Pu-240; and reactor-grade is 19 percent or greater in Pu-240 content.

17 Pu-240 is the most troublesome of plutonium isotopes for bomb design, but not because of its inability to sustain a chain reaction. Pu-240, in fact, has a bare sphere fast neutron critical mass of 40 kg, less than that of U-235. Although when Pu-240 is mixed with Pu-239 it raises the critical mass of the mixture, the undesirable aspects of Pu-240 arise primarily from the fact that it fissions spontaneously with a much shorter fission half-life than Pu-239.

For one kilogram of U-235, spontaneous fission produces approximately one neutron per second. The spontaneous fission rates of weapons-grade plutonium are 60,000 and 300,000 times higher. Another [smaller] source of neutrons is the alpha-n reaction. In this case, radioactive decay of the fissile isotope yields alpha particles, some of which then collide with impurities such as boron, carbon, or oxygen to yield neutrons.

The classic problem presented by background neutrons is that of *preinitiation* of the nuclear-fission chain reaction. In order to assemble fissionable material to produce a nuclear explosion, a subcritical mass (or masses) of material must be rapidly moved into a configuration which has a level of supercriticality sufficient to produce a significant nuclear yield before it blows itself apart. Preinitiation in a nuclear explosive is defined as the initiation of the neutron chain reaction before the desired degree of supercriticality has been achieved. Because the nuclear yield depends upon the degree of supercriticality at the time the chain reaction is initiated, preinitiation will result in a lower yield. However, initiation is a statistical process and can be understood using statistical techniques.

Preinitiation, by itself, does *not* necessarily make an explosive unreliable. Preinitiation *does* result in a statistical uncertainty in the yield. Another way to state this is that the probable nuclear yield is statistically distributed between predictable upper and lower limits, which are likely to be more than a factor of 10 apart. For a well-understood design properly constructed, however, the most probable yield range could be predicted within much closer limits. [Mark, *op. cit.*; p. 141.]

Because of their particular sensitivity to preinitiation, gun devices (see Fission Weapon Design) are never designed with plutonium of any quality.

For low-technology [implosion] devices using high neutron background materials the probable yields could be lower by a factor of 3 to 10 or more (depending on the design) than using low-neutron background materials (i.e., U-233 , U-235 and weapons-grade plutonium). *Military useful weapons with reliable yields in the kiloton range can be constructed using low technology.* (emphasis added) [Mark, *op. cit.*, p. 142.]

The U.S. tested a weapon constructed with reactor-grade plutonium in 1957. Had the first U.S. nuclear device (*Trinity* test, 16 July 1945) been constructed with reactor-grade plutonium its yield would have exceeded 1 kiloton. Using high-technology, or sophisticated design techniques, the problems presented by preignition can be largely overcome. Fuel-grade and reactor-grade plutonium are not used in U.S. weapons, but not primarily because of preinitiation. Pu-240 (and Pu-241) are more radioactive than Pu-239 and therefore generate more heat that must be dissipated if the integrity of the device is to be maintained for extended periods of time. Pu-240 is also more hazardous to handle than Pu-239, thus complicating further the manufacture of weapons using reactor-grade and fuel-grade rather than weapon-grade plutonium.

18 Pu in alpha phase (M_c = 17 kg for Pu in delta phase). The critical mass of Pu-239 is lower than that of U-235 because it has a higher fission cross-section—that is, each Pu-239 nucleus is more likely than a U-235 nucleus to capture a neutron and fission—and it produces on the average more neutrons per fission than U-235.

19 M_c = 26.6 kg for sphere of 94 percent U-235 surrounded by 1.74 inches natural uranium; M_c = 8.4 kg for sphere of alpha phase Pu-239 (4.5 percent Pu-240) surrounded by 1.6 inches natural uranium; and M_c = 7.6 kg for sphere of 98 percent U-233 surrounded by 2.1 inches natural uranium. Lower values of M_c can be achieved with other reflecting materials and/or thicker reflectors.

20 For a spherical mass of fissile material of radius, R, and uniform density, ρ:

$$(\rho \cdot R)_{critical} = \text{constant}.$$

If a fixed core mass, M, is uniformly compressed, the density is given by

$$\rho = 3M/4\pi R^3$$

consequently, the critical mass is approximately proportional to the reciprocal of the square of the density, i.e.,

$$M_{critical} = K/\rho^2.$$

2
Fission Weapon Design

Figure 2.2 Interior components of *Fat Man* type implosion bomb. The spherical shell of twelve pentagonal sections contains explosive "lenses" surrounding a uranium tamper and plutonium core.

than mentioned above. On the other hand, to obtain an appreciable fission yield more than one critical mass may be necessary. Thus, different types of nuclear weapons use different amounts of nuclear materials, and the reflected critical mass values discussed above—about 15 kg of U-235 and 5 kg of Pu-239—indicate only the order of magnitude of the actual amount of fissile material that may be required for a nuclear weapon. Most weapons in the U.S. arsenal are believed to use only a fraction of a critical mass (at normal density)—a "fractional crit"—as the fissile component.[21] Other things being equal, fission weapons of higher yield require larger quantities of fissile material; therefore, the actual amount of fissile material in a weapon depends on the desired yield and the sophistication of the design.

Fission Weapon Design

An explosion is the release of a large quantity of energy in a small volume in a short period of time. There are numerous ways to assemble nuclear fissile materials to make them explode. For an efficient bomb or weapon the goal is to achieve a rapidly multiplying chain reaction that within a very short time—a few microseconds—involves a very large number of atomic nuclei.

The fission chain reaction can be viewed as a sequence of stages or "generations," each marked by the fissioning of nuclei by neutrons produced in the preceding generation. The "generation time" is the average time between the emission of a fast neutron and its absorption by another fissionable nucleus, taking into account that neutrons are also lost by leakage and capture in other materials. The value of the generation time is roughly 0.01 microsecond (a "shake")[22] and varies depending on the kind of fissionable material used, the design of the weapon, and the densities achieved during the explosion.[23]

The energy release from a fission device takes place over a number of generations, depending on how many neutrons from fission in one generation remain to produce fissions in the next generation. In the fission of a single nucleus, between 2.5 and 3 fast neutrons are emitted. If, for example, 2 of these survive to produce other fissions, then the energy release from a device with a yield in the range 1 to 100 Kt would occur in about 53 to 58 generations.[24] Moreover, 99.9 percent of the energy release occurs in about the last 7 generations, which is roughly the last .07 microsecond of the explosion.

The energy release, or yield, is proportional to the number of nuclei fissioned, which is equivalent to the number involved in the chain reaction. To achieve a significant yield, the mass of the assembled fissile material must be several critical masses in order to obtain and maintain a rapidly multiplying chain reaction and avoid disassembly (i.e., becoming subcritical) before much material has been fissioned. This can be achieved by assembling two (or more) subcritical mass elements, or by changing the density and geometry of a fissile mass, initially subcritical, to reduce its critical mass. Higher yields (larger explosions) are obtained by design techniques which increase the mass above critical, and increase the time the fissile material is held together before the energy released by the nuclear explosion blows the weapon materials apart, stopping the chain reaction.

To keep the weapon from exploding and becoming subcritical before a reasonable fraction of its fissionable

---

21 The idea of using a fraction of a critical mass ("fractional crit") for an atomic explosion was originated by Hans A. Bethe from implosion calculations during the Manhattan project. After fission bombs had been thoroughly developed by postwar Los Alamos Laboratory the fractional crit became a practical possibility. It was strongly advocated by the Laboratory and the AEC in 1948-1949; see Hans A. Bethe, "Comments on the History of the H-Bomb," 1954, reprinted in *Los Alamos Science*, Fall 1982, p. 45.
22 Glasstone and Dolan, *op. cit.*, pp. 16, 17.
23 For a bare sphere critical assembly of U-235 the generation time is .66 shake. It is 2.0 shakes for a critical assembly with a U-235 spherical core and a thick natural uranium reflector, and .35 shakes for a bare sphere critical assembly of plutonium; John D. Orndoff, *Nuclear Science and Engineering*: 2, 450-460 (1957).

24 A yield of 1 Kt requires the fissioning of 1.45×10^{23} nuclei. If $1 + x$ neutrons per fission from one generation produce fissions in the next generation, the population of fission producing neutrons after n generations is given by $\exp(xn)$ for a chain initiated by a single neutron (Glasstone and Dolan, *op. cit.*). In the example in the text above, $x = 1$. For *Little Boy* and *Fat Man*, the fission devices detonated over Japan, x is estimated to have been less than 0.7 neutrons per fission. For neutron leakage data, see George D. Kerr in V.P. Bond and J.W. Thiessen, eds., *Reevaluations of Dosimetric Factors, Hiroshima and Nagasaki*, *op. cit.*, pp. 64, 65.

2

Fusion Weapons

nuclei undergoes a nuclear reaction, the fissile material is surrounded by a heavy material called a *tamper*. The same material may be used for both the tamper and the reflector; consequently, these terms are often used interchangeably. If constructed of fissionable material, e.g., U-238, the tamper will contribute to the fission yield as a result of the fissioning of its nuclei by the fast neutrons coming from the interior (see also Thermonuclear Weapon Design).

Two basic nuclear weapon design approaches that are used to achieve a supercritical mass of fissile material are the implosion technique and the gun assembly technique. In the *implosion technique*, a peripheral charge of chemical high explosive (HE) is uniformly detonated in a manner designed to compress (implode) a subcritical mass into a supercritical configuration. For example, an implosion device might consist of a spherical core of fissile material (a fraction of a critical mass) surrounded by a material that acts as both a tamper and a reflector; the tamper is then surrounded by high explosives. When detonated, the explosive sets up an implosion, or ingoing shock wave, that can create overpressures of millions of pounds per square inch in the core, increasing the density by a factor of two or more and thereby making the previously subcritical mass supercritical. The material occupying the volume within the surrounding high explosive charge, including the fissile core and the tamper/reflector, is commonly referred to as the "pit."

The implosion technique is commonly used in nuclear weapons where the fissionable material is Pu-239, U-235, or a composite of the two. It was used in the first U.S. nuclear test (*Trinity*, 16 July 1945) and also in *Fat Man*, the second nuclear weapon dropped on Nagasaki, Japan; the spherical assembly consisted of a thick charge of high explosive to compress a subcritical mass of Pu-239 surrounded by a thick layer of U-238 that served as both the tamper and reflector.

The *gun device* involves the assembly of two (or more) masses of fissionable material, each less than a critical mass. A conventional explosive is used to propel the subcritical pieces of fissionable material together, thereby assembling a supercritical mass. For example, a subcritical mass of fissionable material can be propelled down a tube, i.e., shot down a "gun barrel," into a second subcritical mass. This approach was used in *Little Boy*, the U-235 weapon dropped on Hiroshima, Japan. For that weapon, the second mass was held in a tamper consisting of a thin layer of tungsten carbide on the inside and steel on the outside. The W33 artillery fired atomic projectile now in the U.S. weapons stockpile

Figure 2.3 The *Mike* device, the first successful thermonuclear test, had a reported weight of 62 tons, due in part to the cryogenic equipment needed to maintain its thermonuclear fuel, deuterium, at liquid temperatures. Tested 31 October 1952 in Operation Ivy, at Elugelab, Enewetak Atoll, *Mike* had a yield of 10.4 megatons.

utilizes the gun assembly technique. Gun devices are conceptually relatively simple and can be designed with high confidence. Their design almost of necessity requires the use of U-235, rather than plutonium. Because an implosion device can generally be made with higher efficiency (higher yield for same amount of material) than a gun device, most fission devices in the U.S. stockpile, even those utilizing U-235 alone, use the implosion rather than the gun assembly technique.

Fusion Weapons

Fusion Reactions

"Thermonuclear" weapons, also referred to as "fusion" or "hydrogen" weapons, are usually defined as atomic weapons in which at least a portion of the release of energy occurs through nuclear fusion. The fusion reaction rates are extremely sensitive to temperature and are extremely small at normal temperatures. Only at 10-100 million degrees Kelvin—the interior of the sun is 14 million degrees Kelvin—are the rates sufficiently high to make fusion weapons (or reactors) possible; hence the term "thermonuclear." In thermonuclear weapons, the required temperatures and the required density of the fusion materials are achieved with a fission explosion.

2
Thermonuclear Weapon Design

Figure 2.4 The Mk-17, the first U.S. droppable thermonuclear bomb to be tested, weighed 21 tons and had a yield in the megaton range.

Thermonuclear Fuels

As noted previously, the D-T reaction is the principal source of fusion energy in thermonuclear weapons.[25] It is not necessary, however, to use elemental deuterium and tritium, which are gases at ordinary temperatures, directly in a thermonuclear weapon. The principal thermonuclear material is likely to be lithium-6 deuteride, which is a solid chemical compound at normal temperatures. In this case the tritium is produced in the weapon itself by neutron bombardment of the lithium-6 isotope during the course of the fusion reaction.[26] Since tritium decays radioactively (5.5 percent is lost each year),[27] lithium-6 deuteride has the added advantage of a longer storage life compared to tritium. Once fusion burn has been initiated, the action of fast neutrons on the isotope lithium-7, in the material lithium-7 deuteride, could be the source of additional tritium.[28]

Thermonuclear Boosted Fission Weapons

By incorporating thermonuclear fuel, typically deuterium and tritium gas (or lithium hydrides) directly into (or proximate to) the core of fissile material, the efficiency of the fission bomb can be improved; that is, one can obtain a much higher yield from a given quantity of fissile material, or alternatively the same yield with a much smaller quantity. This process is called "boosting."[29] The fusion process itself may add only slightly to the yield of the device. Far more important to the yield is the extra quantity of free neutrons produced as a result of the fusion reaction.[30] These in turn produce additional fissions in the plutonium or uranium in the weapon, resulting in the increased efficiency. Thus, in boosted weapons, the thermonuclear fuel is used primarily as a source of neutrons to help the fission reactions, rather than as a direct source of yield. Boosted weapons are therefore basically fission weapons.

Because tritium decays radioactively, the effectiveness of the boosting process can degrade with time. Consequently, in stockpiled weapons which use tritium gas, the tritium is periodically replaced to ensure that a sufficient amount will be available.

Thermonuclear Weapon Design

In thermonuclear weapons, the fusion material can be incorporated directly into (or proximate to) the fissile core—for example in the boosted fission device—or external to the fissile core, or both. In the latter cases, radiation from a fission explosive is contained and used to transfer energy to compress and ignite a physically separate component containing the fusion material (and in some cases fissile material). The fissile core is referred to as the primary, and the component with the fusion material external to the primary is called the secondary. The weapon in this case would be said to have two stages.

The radiation from the fusion secondary can be contained and used to transfer energy to compress and ignite a third, or tertiary, stage, and the tertiary could similarly ignite a fourth, and so on. There is no theoreti-

---

25 A notable exception is the first full-scale American thermonuclear explosion (Mike Shot, Operation Ivy, Enewetak Atoll, 31 October 1952), which used liquid deuterium and had a yield of 10.4 megatons (Mt), but this was a nuclear "device" designed for experimental purposes, not a prototype for an operational bomb.

26 Tritium is bred from lithium-6 in the reaction:

$$Li^6 + n \rightarrow He^4 + T + 4.8 \text{ MeV.}$$

When this reaction is combined with reaction (i) in footnote 4 of this chapter, the net thermonuclear reaction is:

$$Li^6 + D \rightarrow 2 He^4 + 22.4 \text{ MeV.}$$

27 Tritium has a half-life of 12.33 years.

28 $Li^7 + n \rightarrow T + He^4 + n.$

This tritium bonus was verified in 1954; see Lee Bowen, *A History of the Air Force Atomic Energy Program 1943-1953* (Washington: USAF Historical Division), Vol. IV, p. 40.

29 The boosting principle was recognized at least as early as November 1945 when possibilities of this general type were included in a patent application filed at Los Alamos. The designation "booster" only became general after its use by Edward Teller in September 1947; J. Carson Mark, "A Short Account of Los Alamos Theoretical Work on Thermonuclear Weapons, 1946-1950," LA-5647-MS, (Los Alamos: LASL, July 1974), p. 9. Teller in 1947 invented a "booster" design using liquid deuterium and tritium as the thermonuclear fuel. The design of this device was frozen in October 1950, and it was tested on 24 May 1951 in Shot *Item* of the Greenhouse series; J. Carson Mark, *The Bulletin of the Atomic Scientists*, March 1983, p. 47. The detonation of this 45.5 Kt device was a major contribution to the development of thermonuclear weapons. Shot *George*, an earlier detonation of this series (on 8 May 1951) produced the first significant U.S. thermonuclear reaction. *George* was an experiment using a fission bomb to ignite a small quantity of deuterium and tritium that contributed only a small amount to the 225 Kt yield.

30 Complete fusion of 1 kg of D-T releases about 25 times as many free neutrons as the complete fission of 1 kg of uranium or plutonium. Alternatively, fusion produces up to 6 times more free neutrons than fission for the same energy.

2

Enhanced Radiation Weapons

cal limit to the number of stages that might be used and, consequently, no theoretical limit to the size and yield of a thermonuclear weapon. A thermonuclear weapon with a separate primary and secondary may, but does not necessarily, take advantage of boosting the primary.

While uranium-238 cannot maintain a self-sustaining fission explosion, it can be made to fission by an externally maintained supply of fast (highly energetic) neutrons from the fission or fusion reactions. Thus the yield of a nuclear weapon can be increased by surrounding the device with U-238, in the form of either natural or depleted uranium.[31] This approach is particularly advantageous in a thermonuclear weapon where there is an abundance of fast neutrons from the fusion reaction. In a thermonuclear device, this U-238 blanket is sometimes referred to as the third stage of what would otherwise be a two stage weapon.

In general, the energy released in the explosion of a large thermonuclear weapon stems from three sources—a fission chain reaction, the first stage; "burning" of thermonuclear fuel, the second stage; and the fission of the U-238 blanket (if one exists), the third stage—with, very roughly, half the total energy stemming from fission and the other half from fusion. However, to obtain tailored weapons effects or to meet certain weight or space constraints, different ratios of fission-yield-to-fusion-yield may be employed, ranging from nearly pure fission yield weapons to a weapon where a very high proportion of the yield is from fusion.

Enhanced Radiation Weapons (Neutron Bombs)[32]

The "neutron bomb" is a thermonuclear device designed to maximize the lethal effects of high energy neutrons produced by the fusion of deuterium and tritium and to reduce the blast (the kinetic energy of charged particles) from the explosion. In this weapon the burst of prompt nuclear radiation (neutrons and gamma rays) is enhanced by minimizing the fission yield relative to the fusion yield. This is accomplished, in part, by the elimination of (or substitution for) U-238 components, particularly the U-238 tamper and blanket. Thus, the neutron bomb is referred to as an "enhanced radiation" (ER) weapon.

The neutron bomb uses deuterium with tritium, rather than lithium-6 deuteride, as the fusion material to maximize the release of fast neutrons. Each reaction of one deuterium nucleus with a tritium nucleus gives rise to a neutron with an energy of approximately 14 million electron-volts (or a "14 MeV neutron"). Such neutrons from fusion are up to six times more numerous per kiloton of energy release than are neutrons escaping from the fission chain reaction in a fission bomb.[33]

Consequently, for a given explosive energy release, it is possible for a lethal dose of nuclear radiation to be delivered on the battlefield for a somewhat greater distance in the case of a neutron bomb than for a fission weapon. The lethal radius of a neutron bomb's radiation dose for a low altitude burst is about 700 meters. This is about twice the lethal radius of a fission weapon with an equal yield and about the same lethal radius as a fission weapon with ten times the yield.[34]

For intermediate and large yields, the destructive radius of the blast generally far exceeds that of nuclear radiation. The blast damage radius drops off more rapidly with decreasing yield than the lethal radius of the radiation. Only in the energy range of 1 Kt and below would the radiation kill radius of an ER weapon (due to high-energy neutrons) considerably exceed the destructive radius of the same weapon due to blast.[35] Although a 1 Kt ER weapon has a lethal radius for nuclear radiation equal to that of a 10 Kt fission weapon, the fission weapon has a considerably greater radius for blast damage to urban structures.[36]

---

31 While other fissionable materials such as U-234, U-236, and Th-232 could also be used, U-238 is used because it is readily available as tails from the enrichment plants (in quantities far more plentiful than U-234 or U-236), and it has a higher fission cross section than Th-232.

32 See Richard L. Garwin, "Trends in the Technological Development of Nuclear Weapons Systems," Third Draft of Chapter 2 for 1980 UN Report on Nuclear Weapons, 27 March 1980, pp. 10-11; Herbert Scoville, Jr., "The Neutron Bomb," *The Arms Race and Arms Control*, SIPRI, 1982, p. 115; and S.T. Cohen, *The Neutron Bomb* (Cambridge, Mass.: Institute for Foreign Policy Analysis, Inc., November 1978), p. 66 ff.

33 The energy released from a typical (low-yield) fission bomb consists of 50 percent blast, 35 percent thermal radiation (heat), 5 percent prompt radiation (gamma-rays, x-rays, and neutrons), and 10 percent residual radiation (fallout); whereas an ER weapon consisting of half fission and half fusion releases energy as 40 percent blast, 25 percent thermal, 30 percent prompt, and 5 percent residual radiation; Fred Kaplan, "The Neutron Bomb, *The Bulletin of the Atomic Scientists*, October 1981, p. 6.

34 Estimates used by the U.S. Army (and the Joint Chiefs of Staff) are that a dose of 8000 rads is necessary to inflict "immediate permanent incapacitation" on a human being. The "military radius" (or radiation-kill radius) is the distance from ground zero inside of which the dose is greater or equal to this value. Following are the dose radii for tank crews for doses of 8000 and 650 rads. At 650 rads personnel will become functionally impaired within 2 hours of exposure. Radii are in meters for a burst height of 150 meters. [Cohen, *op. cit.*; Scoville, *op. cit.*]

Weapon	Radius (m)	
	8000 rad	650 rad
1 Kt fission	360	690
1 Kt ER	690	1100
10 Kt fission	690	1100

35 Theoretically, the pressure at a given distance from an explosion is proportional to the cube root of the energy yield; see Glasstone and Dolan, *op. cit.*, p. 100. Thus, a doubling of the blast kill radius results from an 8-fold increase in yield; the same 8-fold increase of yield adds only about 400 m to the prompt lethal radius of nuclear radiation effects for a neutron bomb. The difference in scaling occurs because the neutrons and gamma rays are strongly absorbed in the atmosphere while the blast wave is only slightly attenuated.

36 A pressure of 4 psi (pounds per square inch) produces a moderate level of blast damage to urban structures. Lethal radiation and blast damage radii in meters for low altitude burst (150 meters) are as follows [Cohen, *op. cit.*, and S.T. Cohen, *Strategic Review*, Winter 1978, p. 9]:

Weapon	Radius (m)	
	8000 rad	4 psi
1 Kt ER	690	550
1 Kt fission	360	610
10 Kt fission	690	1220

A 1 Kt neutron weapon is more costly to manufacture than a 10 Kt fission weapon. A neutron weapon also has more constraints on its delivery, in part because of the need to maintain the tritium supply.

Third Generation Weapons

A new "third generation"[37] of advanced nuclear weapon concepts is now emerging.[38] These weapons are described as being highly selective in their effects and suited to "purely defensive"[39] use for destroying an adversary's offensive systems. Development is being led[40] by the Lawrence Livermore National Laboratory[41] with participation of other weapons laboratories. The new designs are to be ready for introduction in the 1990s.[42]

Among third-generation concepts are the following:

- An x-ray laser pumped by x-rays from a nuclear explosion for use as a defense against incoming enemy ballistic missiles above the earth's atmosphere, or as an anti-satellite weapon.[43]
- Enhanced radiation (ER) weapons of very low yield (50-100 tons) guided by radars into the path of incoming ballistic missiles high in the atmosphere. Compared to these weapons, the ER weapon for battlefield use is characterized by officials as a "crude forerunner"[44] of a third-generation ER weapon; the ER warhead of the SENTRY anti-ballistic missile system (now under intensive development at Livermore) is described as a device "at the beginning of the defensive use of nuclear weapons."[45]
- EMP weapons specially designed to create a large electromagnetic pulse (EMP) to burn out enemy communications, utilizing directed or non-directed EMP created by a nuclear explosion above the atmosphere.

Figure 2.5 B-52 sits atop the TRESTLE electromagnetic pulse simulator at Kirtland Air Force Base, New Mexico. Ten million volts of EMP energy are created to simulate effects from nuclear explosions on aircraft and electrical equipment.

The realm of advanced technologies now includes a *directed* form of EMP using a "high power microwave coherent beam of immense peak power." The Department of Defense Directed Energy Program (under DARPA) includes "radio frequency weapons" as one of its three major programs.[46]

The feasibility of reliable "defensive" nuclear weapons has been questioned by some familiar with weapons design. Skepticism about the use of the x-ray laser as a new weapon has been particularly keen.[47] One critic has noted that "in the nuclear area, the offense will continue to have the advantage and can negate any defensive weapon with relatively little effort."[48] It has been suggested that should a defensive breakthrough occur, "it will probably not involve either nuclear fission or fusion in an essential way."[49]

For both ER and fission weapons, prompt radiation is the predominant battlefield effect. The radiation-kill radius for tank crews is significantly larger for both the 1 Kt ER and 10 Kt fission weapons than the radius at which the tanks themselves will be damaged by blast. (For a discussion of battlefield strategies see Scoville, *op. cit.*)

37 The first generation was in the 1940s and 1950s when the first nuclear bombs and warheads were developed. The second generation came in the 1960s with the development of compact high-yield warheads.
38 Judith Miller, *New York Times*, 29 October 1982; *FAS Newsletter*, October 1982, p. 6.
39 Edward Teller, *Science*, 24 December 1982, p. 1270.
40 *Ibid.*
41 As stated in the Directors' Comments of the FY 1982-1987 Lawrence Livermore Institutional Plan (1982, p. 2., later recalled), "Increased demand for modern nuclear weapon systems, coupled with significant developments in the design of a new nuclear weapon for ballistic missile defense applications and a concept for directing the energy of a nuclear weapon, strengthens our recommendation for additional attention in the areas of nuclear design and testing."
42 Dr. Richard Wagner, HASC 97-41, April 1982.

43 The nuclear pumped x-ray laser was reported to have been successfully tested underground at the Nevada Test Site by Lawrence Livermore scientists; see AW&ST, 23 February 1981.
44 Quote from Judith Miller, *New York Times, op. cit.*
45 Edward Teller as quoted in *FAS Newsletter, op. cit.*, p. 9.
46 Letter of Charles E. Kinney, *Science 83*, March 1983, p. 21.
47 For example, by Richard Garwin, as reported by David Perlman, *San Francisco Chronicle*, 25 September 1982. Hugh E. DeWitt states that a variety of countermeasures could make the x-ray laser ineffective. For example, the enemy ballistic missile could release large pieces of metal chaff having the same radar image as the missile itself; see Hugh E. DeWitt, Lawrence Livermore National Laboratory, in letter to David Saxon, President, University of California, 3 October 1982.
48 Hans A. Bethe, *Science*, 24 December 1982, p. 1270. Bethe is reported to have later said, after reviewing theoretical designs and studies for the x-ray laser at LLNL, that "this is the one and only one proposal that makes scientific sense," but "to translate this into an operational device is a fantastic business"; R. Jeffrey Smith, *Science*, 1 July 1983, p. 30.
49 Herbert F. York, *Science*, 21 January 1983, p. 236.

2

Warhead Features

Warhead Features

A nuclear weapon consists of a nuclear warhead package of fissile and/or fusion materials. Typically, it also contains a sequencing microprocessor; chemical high explosives; a neutron actuator; an arming system, the component which serves to ready (prearm), safe, or resafe (disarm) the warhead; a firing system; a fuzing mechanism (radar, pressure sensitive, time) which regulates the detonation of the warhead; and control and safety devices; all of which may or may not be an integral part of a warhead package, depending on its design and age.

Warhead Safety and Control

Depending upon its age and deployment location, each warhead contains safety and control devices which are to prevent accidental, unauthorized use (prearming, arming, launching, firing) or inadvertent use of nuclear weapons. In some cases, the safeguard and arming features are no more than a wire seal, a switch, and a lock; however, the modern weapons (and all weapons deployed outside the United States) have a series of complementary features, including locked containers and code controlled arming and fuzing systems. In addition, the "two-man rule"—which requires a minimum of two authorized personnel (each capable of detecting incorrect or unauthorized procedures) present during any functions where people come in contact with nuclear weapons or code materials related to their release—enforces adherence to safety and security procedures.[50]

The most significant control feature of the warhead, the Permissive Action Link (PAL), is the incorporation of a coded switch or lock device in the arming line.[51] A PAL is a coded "lock" that requires the insertion of a proper number (manual combination or electronic digits) in order to "open" (unlock) circuits to arm the weapon.[52] With the newer PALs, after the repeated unauthorized entry of false numbers, the weapon locks, certain key components are made unusable, and the warhead would have to be returned to the assembly plant for repair.[53] Each typical PAL device (Category D) costs about $50,000.

Each successive generation of PAL devices since their introduction—the latest is the "Category F" PAL—has represented more than just an increase in safety. The evolution of PAL from the original single code combination lock device for warheads (still deployed on some older weapons) to the present 12 digit electronic "multiple-code coded switch" made it possible to control the release of individual warheads and specified yields, each with a uniquely generated code. Thus the "safety" features also serve to support the evolving limited nuclear war strategies.

The other safety and control systems incorporated into the warheads or containers besides the PAL—command disable feature, electrical safety, weak link/strong link, unique signal generator, etc.—are described in Table 2.1, Warhead Safeguard Features. Many newer weapons also contain "insensitive high explosives" (IHEs), a more stable chemical compound than previously used high explosives to "guard against detonation from fire, small arms, aircraft crash, or inadvertent release in flight."[54] IHEs, which are more expensive and

Table 2.1
Warhead Safeguard Features

Command Disable[1]	inertial nonviolent code activated disabling device which destroys critical warhead components, rendering the warhead useless, integrated into the storage container
Electical Safety	surge capacitors, exclusion region, for weak link/strong link system
Environmental Sensing Device (ESD)	barometric pressure sensor for arming bombs and artillery projectiles
Permissive Action Link[2] Category A	mechanical combination lock
Category B	ground and airplane operated 4 digit coded switch, later version with limited try
Category C	single 6 digit coded switch, with limited try
Category D	multiple code 6 digit coded switch, with limited try
Category F	multiple code 12 digit coded switch, with limited try
Unique Signal Generator	arming-safing signal encoder for firing system requiring unique electrical signal
Weak Link/ Strong Link	exclusion region for warhead electrical system preventing inadvertent or accidental electrical surges and firing

1 ACDA, FY 1978 ACIS, p. 96; ACDA, FY 1979 ACIS, p. 121; SASC, FY 1978 ERDA, p. 109.
2 ACDA, FY 1979 ACIS, pp. 92, 130; SASC, FY 1981 DOE, pp. 74-76.

---

50 U.S. Army, *Operations for Nuclear Capable Units*, FM 100-50 (March 1980), p. 11.
51 SASC, FY 1982 DOE, pp. 266-267; background is also provided in *Military Applications of Nuclear Technology*, Part 1, pp. 44ff.
52 ACDA, FY 1979 ACIS, p. 92; ACDA, FY 1978 ACIS, p. 96.
53 ACDA, FY 1979 ACIS, p. 136.
54 ACDA, FY 1979 ACIS, p. 92.

2
Weights and Yields

heavier than previously used high explosives, will only be integrated into weapons which are transported frequently. According to the Sandia Corporation, "insensitive high explosives minimize the risk of detonation ... thereby reducing the risk of scattering of plutonium."[55] The IHEs, first incorporated into the B61 Mod 3 and 4 bombs, are now in the W80 cruise missile warhead and programmed for most future warheads.[56]

Selectable Yields

There is little in the open literature describing how the yields of current nuclear weapons are varied. The actual designs probably incorporate one (or more) of the following procedures.

In a pure fission device, the yield can be varied by varying the timing of the initiation of the chain reaction or by interchanging pits. In modern weapons, the chain reaction is initiated by a neutron gun.

In a boosted fission weapon, or a thermonuclear weapon with a boosted primary, the yield can be varied readily by carefully selecting the amount of tritium gas bled into the fissile core from an external reservoir.[57]

In thermonuclear weapons with one (or more) fusion stages, the yield is varied by tritium control or by interchanging the pits. In addition, mechanical measures that dictate whether or not additional fusion stages ignite could be used, although there is no evidence in the open literature that this is done. The yield of some older bombs in the U.S. stockpile, e.g., the B28, appears to be varied by interchanging the pits. This procedure, which is performed on the ground, is less flexible than selecting the yield by tritium control, the approach used in more modern weapons.

Weights and Yields of Nuclear Weapons

The first nuclear explosive device, called the *Gadget*, had a plutonium core weighing about 13.5 lb (6.1 kg) which was imploded by some 5000 lb of high explosive.[58] The *Gadget* was tested at the Trinity Site's 100 foot tower in Alamogordo, New Mexico (16 July 1945,

Figure 2.6 The 100 foot tower at ground zero of the Trinity Site at Alamogordo, New Mexico. It was built to test the "Gadget" contained (without its plutonium core) in the crate at the base of the tower.

Figure 2.7 The "Gadget"—prototype for the *Fat Man* weapon design—at the top of the tower at the Trinity Site shortly before its test on 16 July 1945.

---

55 SASC, FY 1981 DOE, p. 74.
56 HASC, FY 1980 DOE, p. 140.
57 Both tritium gas and deuterium gas in a selected volume ratio could be bled into the weapon's "pit" during the arming sequence.

58 Major Gen. Leslie R. Groves, Memorandum for the Secretary of War, 18 July 1945, TOP SECRET (DECLASSIFIED) reprinted as Appendix P to Martin J. Sherwin, *A World Destroyed*, (New York: Alfred A. Knopf, 1975); see also James W. Kunetka, *City of Fire*, (Englewood Cliffs, N.J.: Prentice-Hall, Inc., 1978), p. 164.

2
Weights and Yields

Figure 2.8 Plutonium fission weapon of the *Fat Man* type. *Fat Man* was the atomic bomb that was dropped on Nagasaki, Japan on 9 August 1945. Its yield was 22 kilotons.

Figure 2.9 Nuclear weapon of the *Little Boy* type. *Little Boy* was dropped over Hiroshima, Japan on 6 August 1945, the first nuclear weapon ever delivered. It was 120 inches long, 28 inches in diameter and weighed 9000 pounds.

5:29 am), with a yield of 22±2 Kt.[59] The following description is from Kunetka.[60] "While the actual size of the plutonium core was about that of a grapefruit, the uranium tamper and explosive charges added considerably to the bomb's size. The core, tamper, and high explosives were held in place by a metal sphere made of twelve pentagonal sections. These were bolted together to form a sphere . . ."

Fat Man, the nuclear weapon which was dropped on Nagasaki, Japan (estimated 503 meters or 1650 ft,[61] 9 August 1945, 11:02 am), was based on the same design as the *Gadget*; the weapons' names are often interchanged. *Fat Man* included stabilizing fins and a protective egg-shaped outer shell, or bomb casing, 60 inches in diameter. The bomb was 12 feet long and weighed 10,800 lb with fins.[62] The yield of the *Fat Man* has been well-established at 22±2 Kt.[63]

Little Boy, the nuclear weapon dropped on Hiroshima, Japan (estimated 580 meters or 1903 ft,[64] 6 August 1945, 8:15 am), contained 60 kg of highly enriched uranium and utilized the gun assembly technique.[65] The gun barrel had a diameter of over 6 inches, was 6 feet in length, and weighed about one-half ton (or less than one-fifth the weight of standard guns of that calibre). The bomb with its casing was 10 feet long, 28 inches in diameter, and weighed 8900 lb.[66] The yield of the Hiroshima *Little Boy* is uncertain, with estimates ranging from 12 Kt to 15 Kt[67].

The explosive yields of the first two implosion devices, the *Gadget* and *Fat Man*, both about 22±2 Kt, corresponded to an efficiency[68] of about 17 percent in utilization of the 6 kg of plutonium,[69] whereas the 12 to 15 Kt *Little Boy* gun device had an efficiency of only about 1.3 percent. In order to achieve these efficiencies in the first U.S. designs, considerable weight was allocated to the chemical explosive and tamper. The yield-to-weight ratios were 0.0045 Kt/kg for the *Fat Man*, and 0.003 Kt/kg for the *Little Boy*, both very low compared to modern designs (see Figure 2.15).

The Mark III (based on *Fat Man*) was the only stockpiled implosion bomb from 1945 to 1948. In late 1948 the Mark IV assembly began entering the stockpile. Unlike

---

59 Reevaluation by D. Eilers; see John Malik in V.P. Bond and J.W. Thiessen, eds., *Reevaluation of Dosimetric Factors, Hiroshima and Nagasaki*, CONF-810928 (DE 81026 279) (Washington: U.S. DOE, Technical Information Center, 1982), p. 100. Nevada Operations Office gives 19 Kt from an earlier evaluation. "Announced United States Nuclear Tests July 1945-December 1981," NVO-209 Rev-2 (U.S. DOE, Office of Public Affairs, Nevada Operations Office, January 1982).
60 Kunetka, *op. cit.*, p. 123.
61 Malik, *op. cit.*, p. 105.
62 Robert T. Duff, Director, Office of Classification, DOE, letter to David A. Rosenberg, 4 December 1980.
63 Malik, *op. cit.*, pp. 98, 100, 107. Nevada Operations office gives 23 Kt based upon an earlier evaluation; see U.S. DOE, Nevada Operations Office, *op. cit.*, p. 22.
64 Malik, *op. cit.*
65 John McPhee, *The Curve of Binding Energy* (New York: Farrar, Strauss and Giroux, 1974), p. 14.

66 Duff, *op. cit.*
67 Malik, *op. cit.*, pp. 98, 107. Malik gives values of 12±4 Kt deduced from blast effects, 15±3 Kt from thermal effects and a "suggested best value" of 15±3 Kt; Kerr (*op. cit.*, p. 88) prefers an earlier value of 12.5±1 Kt; U.S. DOE, Nevada Operations Office, *op. cit.*, quotes a value of 13 Kt. "[Robert] Oppenheimer gave the *Little Boy* a good chance of 'optimal performance'; only a 12 percent chance of less than this; a 6 percent chance of an explosion under 5000 tons; and a 2 percent chance of one under 1000 tons of TNT"; Memorandum from Robert Oppenheimer to General Leslie Groves and William Parson, 23 July 1945; quoted by Kunetka, *op. cit.*, p. 178.
68 Since complete fission of 1 kg of Pu or U corresponds to 17.5 Kt,
$$\epsilon = Y/17.5M$$
where ϵ = efficiency, Y = yield in Kt and M = mass in kg.
69 For *Fat Man* it is assumed that 80 percent of the yield was from fission occurring in the Pu-239 core and 20 percent in the U-238 tamper. See Kerr, *op. cit.*, p. 81.

2
Weights and Yields

Figure 2.10 The *Davy Crockett*, the smallest yield nuclear weapon ever deployed by the U.S., on display at the National Atomic Museum. It is 31 inches in length (25.5 inches excluding fins) and 11 inches in diameter.

Figure 2.11 The Mk-5 nuclear weapon on display at the National Atomic Museum. The Mk-5 was the first light-weight (3000 pounds) tactical gravity bomb.

Figure 2.12 The Mk-23, a 16-inch projectile for Navy battleship guns, was the first and only Navy nuclear artillery shell.

the Mark III, the fissile core of the Mark IV could be inserted in flight. Also, from 1945 to 1951 the only gun-type weapons assembly in the stockpile was the *Little Boy*.[70]

Most of the early explosions of other nuclear weapons nations have been in the same 10 to 20 Kt range as *Fat Man* and *Little Boy*. The first designs were undoubtedly conservative, with a sufficient amount of fissile material to assure a good yield without placing undue demands on the implosion or gun device used.[71]

The smallest nuclear weapon (in terms of size, weight, and yield) in the U.S. stockpile is the W54, a fission implosion weapon which is currently deployed as the warhead in the Special Atomic Demolition Munition (SADM). It was also utilized as the warhead for the FALCON missile (retired 1972), and the DAVY CROCKETT rocket (retired 1971). The reported yield of the DAVY CROCKETT was 0.25 Kt. The SADM may have yield options even smaller, i.e., in the 0.01 Kt range. The DAVY CROCKETT weighed 51 lb (23.2 kg), only 0.5 percent of the weight of the *Fat Man* bomb, and its dimensions were 30.97 inches (25.5 inches excluding fins) in length and 11 inches in diameter; in comparison, the *Fat Man* was 120 inches in length and 62 inches in diameter.[72]

---

70 David Allen Rosenberg, "U.S. Nuclear Stockpile 1945 to 1950," *The Bulletin of the Atomic Scientists*, May 1982, p. 26.

71 If substantially less than a critical mass of material at normal pressure is used, a much greater burden is placed on the performance of the chemical implosion device to achieve the compression needed for reasonable efficiency. In other words, the risk of a fizzle, or at least of a very inefficient explosion, is increased if there is an attempt to conserve fissile material in a first test.

72 U.S. Army, "Operation and Employment of the DAVY CROCKETT Battlefield Missile," XM-28/29, FM 9-11, June 1963. The weight of the SADM may be somewhat heavier. The DNA, "Motion Picture Catalog" January 1981, for example, lists the weight of the W54 as 26.6 kg; the W54 warhead was based on a design by Theodore Taylor at LASL and has also been referred to as weighing "less than 50 lbs," in McPhee, *op. cit.*, p. 8.

Nuclear Weapons Databook, Volume I **33**

2
Weights and Yields

Figure 2.13 The Army's **280mm atomic cannon**, which was designed to fire the Mark-19, the first nuclear artillery shell.

The largest *fission* device ever detonated, the Super Oralloy Bomb, had a yield of 500 Kt.[73] This is larger than most *thermonuclear* weapons currently in the U.S. stockpile.[74] At yields over approximately 50 Kt, thermonuclear weapons can be produced at much lower cost and much less weight than pure fission weapons. Consequently, one does not expect to see high yield fission weapons stockpiled.

The first successful test of a thermonuclear device was on 31 October 1952 with the 10.4 Mt *Mike* shot at Enewetak Atoll.[75] The largest thermonuclear device tested was reportedly 58 Mt, a Soviet atmospheric test at approximately 12,000 feet at Novaya Zemlya, on 30 October 1961.[76]

As noted previously, only the practical considerations of weight and size limit the yield of deliverable thermonuclear weapons. The first deliverable thermonuclear warhead in the U.S. stockpile was the B17 bomb (retired in 1957) which weighed almost 21 tons and had a yield of several megatons.[77] The largest warhead presently in the U.S. stockpile is the B53 bomb (and W53 warhead on the TITAN II missile) which has a yield of 9 Mt and a weight of 8850 lb, or a yield-to-weight ratio of about 2.2 Kt/kg, some 400 times that of *Fat Man*.

---

[73] Defense Nuclear Agency, "The Radiological Cleanup of Enewetak Atoll" (Washington, D.C.: 1981), p. 49. Event King, during Operation Ivy, 15 November 1952 (air drop detonation at 1500 feet). The Super Oralloy Bomb was designed by Theodore Taylor at LANL; McPhee, *op. cit.*, p. 8.

[74] For example, the thermonuclear W78 warhead—used on 300 MINUTEMAN III ICBMs—has a yield of 335 Kt, the highest yield of any currently deployed U.S. strategic missile, except for the 9 Mt TITAN II (W53) warheads which are soon to be retired.

[75] U.S. DOE, Nevada Operations Office, *op. cit.* The weight of the *Mike* device has been reported as 62 tons; Neyman and Sadilenko, *op. cit.*, p. 20.

[76] Herbert F. York, *The Advisors* (San Francisco: W.H. Freeman and Company, 1976), p. 93; see also Glasstone, 1964 Ed., Appendix B, p. 681A.

[77] National Atomic Museum, Albuquerque, NM. See Table 1.4 for list of U.S. warheads; the B17 may have been tested as the *Cherokee* Shot in Operation Redwing (airdrop 4320 feet, at Bikini Atoll, 20 May 1956). The *Cherokee* Shot was the first air drop by the U.S. of a thermonuclear weapon. See U.S. DOE, *op. cit.*

2
Weights and Yields

Figure 2.14 W19, a projectile for the Army's 280mm howitzer, was the first nuclear artillery shell. It was an oralloy weapon utilizing the gun assembly technique.

Modern thermonuclear weapons with yields above 100 Kt can be expected to have yield-to-weight ratios in the 1-3 Kt/kg range. This is still far below the theoretical limit of 80 Kt/kg represented by the complete fusion of deuterium-tritium material.[78] Figure 2.15 shows a plot of yield-to-weight ratio versus both yield and weight for nuclear warheads and bombs currently in the stockpile. The strategic missiles and bombs with yields greater than about 100 Kt (concentrated in the upper right portion of the figure) show yield-to-weight ratios in the range 0.3-2.5 Kt/kg. The low-yield tactical nuclear weapons (in the lower left portion of the figure) all have substantially smaller yield-to-weight ratios in the 0.004 to 0.1 Kt/kg range.

---

[78] See discussion of fusion energy release earlier in this chapter.

2
Yield-to-Weight Ratios

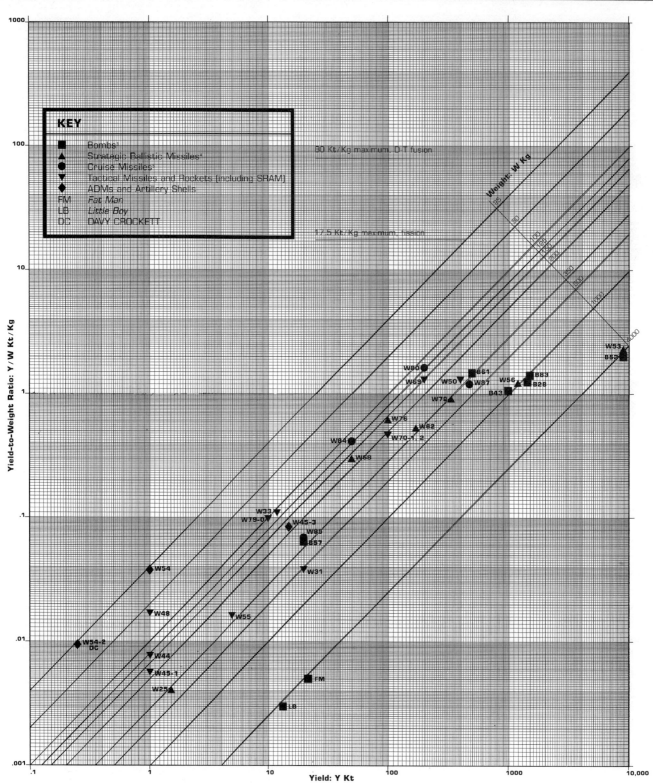

1 Warheads shown were in the nuclear stockpile as of 1982, with the following exceptions. Retired: W54-2 DAVY CROCKETT (DC), *Little Boy* (LB), *Fat Man* (FM). Planned additions: W84 GLCM, W85 PERSHING II, W87 MX, and B84 Modern Strategic Bomb.

2 Maximum yields are used where warheads have variable yields. Data for yields and weights are subject to some degree of uncertainty.

3 Bomb weight includes the bomb casing.

4 Ballistic missile weight consists of the warhead package with the reentry vehicle and a fraction of the bus.

5 For cruise missiles, weight is for warhead package only.

Figure 2.15 Yield-to-Weight Ratios vs. Yield and Weight for Current Stockpile Nuclear Weapons and Bombs.[1,2]

3
Stockpile

Chapter Three
U.S. Nuclear Stockpile

This section describes the 24 types of warheads currently in the U.S. nuclear stockpile. As of 1983, the total number of warheads was an estimated 26,000. They are made in a wide variety of configurations with over 50 different modifications and yields. The smallest warhead is the man-portable nuclear land mine, known as the "Special Atomic Demolition Munition" (SADM). The SADM weighs only 58.5 pounds and has an explosive yield (W54) equivalent to as little as 10 tons of TNT. The largest yield is found in the 165 ton TITAN II missile, which carries a four ton nuclear warhead (W53) equal in explosive capability to 9 million tons of TNT.

The nuclear weapons stockpile officially includes only those nuclear missile reentry vehicles, bombs, artillery projectiles, and atomic demolition munitions that are in "active service."[1] Active service means those which are in the custody of the Department of Defense and considered "war reserve weapons." Excluded are nuclear devices under development, test models, those in production, in inactive storage,[2] and devices which have been withdrawn from active service but have not yet been dismantled. The nuclear device contained in the weapon is commonly refered to as the "warhead."

A total of 85 warhead types have been designed and tested since 1945 (see Chapter One). Of that number, 60 have reached the operational stockpile and the remainder have been cancelled before deployment.[3] Five operational warhead types were in production during FY 1983,[4] while some ten warhead types were being either partly or fully retired. Production of five new warheads (W80-0 Sea-Launched Cruise Missile warhead, B83 bomb, W84 Ground-Launched Cruise Missile warhead, W85 PERSHING II missile warhead, W87 PEACEKEEPER/MX warhead) begins in 1983, with deployments starting in December 1983.

The variations in weapons yield, weight, materials, and delivery systems are evident from this chapter's description of each operational warhead. The oldest in the arsenal is the W33, a gun assembly low yield fission warhead for an 8-inch diameter artillery shell, first deployed by the U.S. Army in 1956. The W33 uses highly enriched uranium (oralloy) as its nuclear fissile material and is considered volatile and unsafe. As a result, its nuclear materials and fuzes are kept separately from the artillery projectile. The W33 can be used in two different yield configurations and requires the assembly and insertion of distinct "pits" (nuclear materials cores) with the amount of materials determining a "low" or "high" yield.

In contrast, the newest of the nuclear warheads is the W80,[5] a thermonuclear warhead built for the long-range Air-Launched Cruise Missile (ALCM) and first deployed in late 1981. The W80 warhead has a yield equivalent to 200 kilotons of TNT (more than 20 times greater than the W33), weighs about the same as the W33, utilizes the same material (oralloy), and, through improvements in electronics such as fuzing and miniaturization, represents close to the limits of technology in building a high yield, safe, small warhead. Unlike the W33, the ALCM is fully assembled with its warhead and fuzing system and can be remotely armed from the cockpit of its carrying airplane. Although it is not clear whether the ALCM has one or more warhead yields options, the technology exists with most of the newest warheads to select a variable yield by merely turning a dial.

The stockpile contains weapons of different categories for different "delivery systems": strategic missiles, defensive missiles, tactical/theater/intermediate range missiles, artillery projectiles, atomic demolition munitions, and nuclear bombs. Warheads are allocated generally to Unified and Specified Commands (see Chapter Four) for strategic or theater warfare, or to the Joint Chiefs of Staff "weapons reserve" for future allocation in contingencies not yet anticipated by established war plans. Warheads are utilized by all three military services and the Marine Corps. They are deployed in the United States and abroad and are held for use ("custodial warheads") by allied countries.

The total number of nuclear warheads in the stockpile peaked at over 32,000 in 1967 and has since then generally decreased, with some small interim increases (see Table 1.6, Figure 1.3, and Chapter One).[6] The cur-

---

1 HASC, FY 1981 DOE, p. 325.
2 Two warheads from the old SAFEGUARD ABM System—the W66 and the W71—are in inactive storage and are being retired beginning in 1982..
3 HASC, FY 1982 DOE, p. 5.

4 These warheads are the B61-3/4 bomb, the W76 TRIDENT I warhead, the W78 Mk-12A MINUTEMAN III warhead, the W79 enhanced radiation artillery warhead, and the W80-1 Air-Launched Cruise Missile warhead.
5 The warheads are consecutively numbered and prefixed with a "Mk," "W," or "B."
6 HASC, FY 1982 DOE, p. 142.

3
Stockpile

Table 3.1
U.S. Nuclear Weapons Stockpile (1983)

Warhead/Weapon	Year First Deployed	Yield (Kt)	User	Number in Stockpile†	Status
W25/GENIE	1957	1-5	AF	200	To be possibly replaced by new air-to-air missile
B28/bomb*	1958	70-1450	AF, NATO	1200	To be replaced by B83
W31/NIKE HERCULES*	1958	1-20	A,NATO	500	To be replaced by PATRIOT; being partially withdrawn
/HONEST JOHN*	1958	1-20	NATO	200	Retired from US use, only left in Greek and Turkish armies
W33/8-inch artillery*	1956	Sub 1-12	A,MC,NATO	1800	To be replaced by W79
B43/bomb	1961	1000	AF,MC,N,NATO	2000	To be replaced by B83
W44/ASROC	1961	1	N	850	To be replaced by Surface Delivered ASW Weapon
W45/TERRIER*	1956	1	N	310	To be replaced by W81
/Medium ADM	1964	1-15	A,MC,NATO	300	
W48/155mm artillery	1963	0.1	A,MC,NATO	3000	To be replaced by W82
W50/PERSHING 1a*	1962	60-400	A,NATO	410	To be replaced by W85
B53/strategic bomb*	1962	9000	AF	150	To be replaced by B83
W53/TITAN II*	1963	9000	AF	49	Being withdrawn starting late 1982
W54/Special ADM	1964	Sub 1	A,MC,NATO	300	No planned replacement
W55/SUBROC*	1965	1-5	N	400	To be replaced by Subsurface Delivered ASW Standoff Weapon
W56/MM II	1966	1200	AF	540	To be partially replaced with MINUTEMAN III/MX
B57/depth bomb	1964	Sub-20	AF,MC,N,NATO	1000	To be replaced by Air Delivered ASW Weapon
B61/bomb	1968	100-500	AF,MC,N,NATO	3000	In production
W62/Mk-12 MM III	1970	170	AF	900	Being partially replaced by W78
W68/POSEIDON	1971	40-50	N	3480	Being partially replaced by W76
W69/SRAM	1972	170	AF	1140	To be replaced by Advanced Strategic Air-Launched Missile or other missile
W70/LANCE	1972	1-100	A,NATO	945	In production 1981-1982, circa 380 neutron types produced
W76/TRIDENT I	1979	100	N	2028	In production, circa 3600 planned
W78/Mk-12A MM III	1979	335	AF	1083	
W79/8-inch artillery	1981	1+	A,MC,NATO	120	In production, circa 800 planned
W80/Air-Launched Cruise Missile	1981	200	AF	350	In production, 1499 planned before transfer to advanced missiles

* Weapons scheduled in present plans for complete or partial retirement in 1982-1987. Other weapons may begin retirement or partial removal in the mid-1980s, but are not included.

† Authors' estimates of stockpile breakdown within range of 26,000 warheads as of 1983.

3
Stockpile

rent rate of new warhead production is estimated to average approximately five a day, and the rate of old warhead retirement averages about three warheads per day.[7] While the size of the stockpile has reduced since its peak in the mid 1960s, the explosive yield of the weaponry has decreased by two-thirds as new, smaller warheads were introduced.[8]

As new weapons are produced and enter the stockpile, older weapons are retired. Table 3.1, The U.S. Nuclear Stockpile, provides a summary of the current warheads in the stockpile as of 1983 and the future status of each warhead. Of the 24 warheads in the stockpile, at least 10 are slated for replacement with newer generation nuclear warheads over the next decade. Only one, the W31 NIKE-HERCULES, is slated for replacement by a conventional weapon (PATRIOT). Only two warhead types will be completely retired with no replacement planned. According to the Department of Defense, "over the next five years ... there will be an increase in the total number of nuclear warheads deployed, both strategic and tactical, on the order of several thousand."[9] An estimated 19 new warhead types and approximately 30,000 warheads will enter an expanding arsenal through the early 1990s, creating almost a new generation of warhead types. Descriptions of the warheads known to be in the development stage are contained in subsequent chapters. Table 1.9, Nuclear Weapons Research and Development Programs (Chapter One) provides a summary of those programs.

---

[7] The average age of warheads in the stockpile in 1981 was 12 years. This suggests that in recent years approximately 4% of the existing inventory is retired annually, which is equivalent to approximately 1000 warheads per year or about 3 per day.

[8] HASC, FY 1982 DOE, p. 142; Senate Report No. 97-173, 30 July 1981; According to one DOD report, "the total number of megatons was four times as high in 1960 than in 1980"; DOD, FY 1984 Annual Report, p. 55.

[9] SASC, FY 1983 DOE, Part 7, p. 4235.

3
W25

Nuclear Warheads in the Stockpile
W25

Figure 3.1 GENIE (AIR-2A) rocket.

FUNCTION: Warhead for GENIE (AIR-2A) unguided strategic air defense air-to-air rocket

WARHEAD MODIFICATIONS: none known

SPECIFICATIONS:
Yield: 1.5 Kt; often referenced in the 1.5 Kt range; "a few kilotons,"[1] "less than 2 Kt"[2]

Weight: 833 lb (missile weight)[3]

Dimensions:
Length: 9 ft 6 in
Diameter: 17.4 in[4] (warhead section only)

Materials: probably oralloy fission weapon; probably Cyclotol (75 percent RDX) as primary HE[5]

SAFEGUARDS AND ARMING FEATURES: probably armed by inflight insertion of nuclear materials

DEVELOPMENT:
Laboratory: LANL

History:
1954   Lab assignment (Phase 3)
1957   initial deployment (Phase 5)

Production Period: 1956-1962

DEPLOYMENT:
Number Deployed: approximately 200 (1983)

Delivery Systems: F-106, F-101, F-4, and possibly F-15;[6] under wing or in missile bay of F-101 and F-106[7]

Service: Air Force

Allied User: Canada

Location:[8] United States and Canada with strategic interceptor units of the Tactical Air Command and Canadian Air Command under the control of NORAD (see Table 4.6)[9]

Retirement Plans: A nuclear warhead is being designed for the Navy's tactical PHOENIX air-to-air missile, which could be used as a replacement for the GENIE.

COMMENTS: Thousands of dual capable GENIE missiles were produced before production ended in 1962. Versions of GENIE missile include training rocket (ATR-2N), simulator (ATR-2A), and conventional trainer (ATR-2L).

---

1 *Military Applications of Nuclear Technology*, Part 1, p. 24.
2 Test results of March and May 1956 tests.
3 Fact sheet prepared by National Atomic Museum, Albuquerque, NM.
4 Measurements of missile on display at National Atomic Museum, Albuquerque, NM.
5 Letter from P.R. Wagner, PANTEX, to J.F. Burke, 23 May 1979.

6 The F-15 is nuclear certified, but it is not known whether those aircraft assigned air defense missions would carry the GENIE.
7 Aircraft carry one or two missiles; F-101 and F-4 have capability of carrying 2 GENIEs.
8 For detailed information on deployment, see Chapter Four.
9 ACDA, FY 1982 ACIS, p. 443.

3
B28

Figure 3.2 B28IN bomb.

FUNCTION:	Strategic and tactical thermonuclear bomb built in numerous modifications and carried by a wide variety of aircraft.
WARHEAD MODIFICATIONS:	B28 Mod 0 originally deployed 1958; B28 Mod 1 originally deployed 1960; B28 Mod 2 originally deployed 1961; Mod 0-3 versions inactive; Current B28 versions are the B28 Mod 4 RE/EX and the B28 Mod 5 FI.[1]
SPECIFICATIONS:	
Yield:	each individual bomb has only one yield;[2] five different yields exist, four are known: 70 Kt, 350 Kt, 1.1 Mt, 1.45 Mt[3]
Weight:	2540 lb;[4] B28RE: 2170 lb; B28EX: 2027-2040 lb; B28FI: 2340 lb[5]
Dimensions:	
Length:	warhead section is approximately 3 ft in length; B28EX: 170 in; B28RE: 166 in[6]
Diameter:	20 in[7]
Materials:	thermonuclear bomb, contains plutonium;[8] lithium-6 deuteride, tritium for fusion; probably PBX-9505[9] or cyclotol as primary HE[10]
SAFEGUARDS AND ARMING FEATURES:	Does not provide the same levels of security and safety as the B61.[11] B28FI will be modified under DOE's Stockpile Improvement Program to incorporate new electrical equipment, safety features, and high explosive material more resistant to detonation in fire or crash.[12]
FUZING AND DELIVERY MODE:	fuzing option must be selected on the ground by maintenance personnel;[13] air or surface burst. Only one Mod of the B28 (FI) has a laydown option;[14] minimum altitude of delivery is 300-600 feet, can be delivered over the shoulder and at low or medium angle loft.[15]
DEVELOPMENT:	
Laboratory:	LANL
History:	
1955	Lab assignment (Phase 3)
1958	initial deployment (Phase 5)
Production Period:	1957-? (not in production: 1970s on)
DEPLOYMENT:	
Number Deployed:	approximately 1200 (1983)
Delivery Systems:	internal and external carriage; B-52, B-1B;[16] wide variety of dual capable tactical aircraft: A-4, A-6, A-7, F-4, F-100, F-104[17]

3
B28

Table 3.2
B28 Modifications

B28EX	(Externally carried, free fall, supersonic capable, radar fuzing for airburst/ground burst) Category B PAL
B28IN	(Internally carried, free fall, radar fuzing for airburst/ground burst) Category B PAL
B28FI	(Internally carried, parachute-retarded full fuzing including laydown)
B28RE	(Externally carried, free fall or parachute-retarded radar fuzing for airburst/ground burst) Category B PAL
B28RI	(Internally carried, free fall or parachute-retarded laydown fuzing)

Figure 3.3 Four **B28FI** bombs loaded into bomb rack which will fit into B-52 bombers.

Service: Air Force

Allied User: NATO Air Forces

Location: United States; Europe[18]

Retirement Plans: Warhead Improvement Program (see above) is an interim fix until the B83 replacements become available starting in FY 1984.[19] B28s are earmarked for retirement (with B43 and B53) starting in 1984; bombs in B-52 units will be retired in order of 1.1 Mt, 350 Kt, 1.45 Mt and 70 Kt versions.[20]

COMMENTS: W28 was also used in the HOUND DOG and MACE missiles, both of which have been retired.

---

1 Letter, National Atomic Museum, to Authors, 13 October 1981; SASC, FY 1981 DOE, p. 76.
2 AW&ST, 10 May 1976, p. 137.
3 SASC, FY 1983 DOD, Part 7, p. 4172; five different yields are known, Y1 = 1.1 Mt, Y2 = 350 Kt, Y3 = 70 Kt, Y5 = 1.45 Mt. Y4 is unknown.
4 GAO, Draft Study for B-1.
5 Information provided by National Atomic Museum, Albuquerque, NM; F-4C *Flight Manual* (1 September 1963).
6 *Ibid.*
7 *Ibid.*
8 SASC, FY 1982 DOE, p. 282.
9 Film at the National Atomic Museum, Albuquerque, NM.
10 Letter from P.R. Wagner, PANTEX, to J.F. Burke, 23 May 1979; See also "The Thule Affair," *USAF Nuclear Safety*, January-March 1970.
11 ACDA, FY 1979 ACIS, p. 93.
12 HAC, FY 1982, DOE, Part 7, pp. 179, 279; SASC, FY 1981 DOE, p. 76.
13 *Military Applications of Nuclear Technology*, Part 1, p. 7
14 ACDA, FY 1980 ACIS, p. 169.
15 ACDA, FY 1979 ACIS, p. 92.
16 GAO, Draft Study for B-1.
17 The B28 also was carried on a number of retired aircraft: B-47, B-66, A-3, A-5, F-101, F-105.
18 B28RE bombs are deployed in Europe, AFM 50-5, Volume II, p. 3-87.
19 HAC, FY 1982 DOE, Part 7, p. 279.
20 SASC, FY 1983 DOD, Part 7, p. 4172.

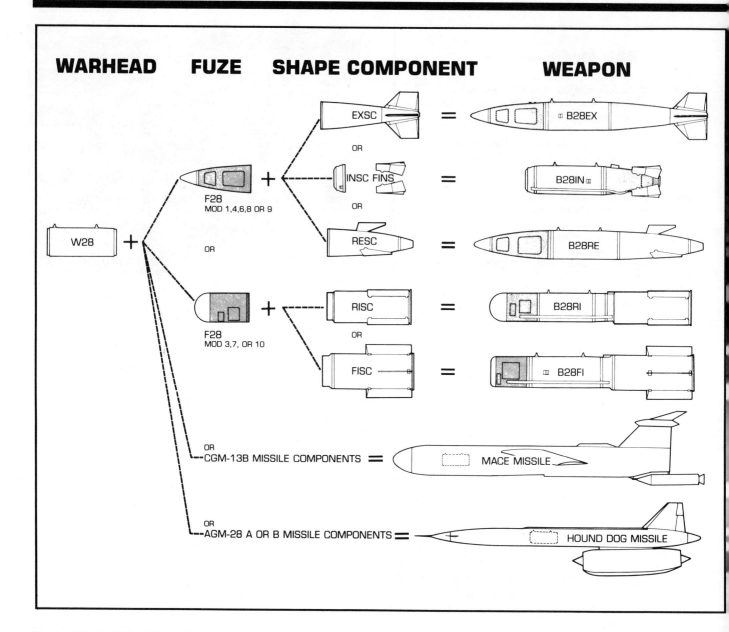

Figure 3.4 Building Block Concept. A graphic portrayal of a single warhead (W28) having several delivery applications.

W31

Figure 3.5 **HONEST JOHN (MGR-1B)**, left, and **NIKE-HERCULES (MiM-14)**, right.

FUNCTION: Warhead for the HONEST JOHN (MGR-1B) battlefield support surface-to-surface missile (SSM) and the NIKE-HERCULES (MiM-14) surface-to-air missile (SAM)

WARHEAD MODIFICATIONS: W31 Mod 2 is used on NIKE-HERCULES[1]

SPECIFICATIONS:
Yield: three yield options;[2] 1-20 Kt range, possibly 2-20-40 Kt, most often referred to as 20 Kt[3]
  HONEST JOHN: three separate warhead sections, the M27, M47, and M48; each with a different yield[4]
  NIKE-HERCULES: two separate warhead sections deployed, the M22 and M97; "warheads are interchangeable."[5]

Weight:
  HONEST JOHN: 1238 lb[6]
  NIKE-HERCULES: 1123 lb[7]

Dimensions:
  HONEST JOHN: dimensions of the M480E1 steel shipping container for the warhead section are: length, 134.5 in; width, 44 in; height, 52 in[8]
  NIKE-HERCULES: dimensions of the M409 steel shipping container for the warhead section are: length, 99.25 in; width, 54.25 in; height, 62 in.[9]

Materials: oralloy as fissile material;[10] tritium contained in M47 and M48 HONEST JOHN and M97 NIKE-HERCULES warhead sections;[11] probably cyclotol (75 percent RDX) as primary HE[12]

SAFEGUARDS AND ARMING FEATURES: mechanical combination lock PAL;[13] improvement to NIKE HERCULES proposed for FY 1982 for electrical safety and command and control;[14] M7 timer fuze, burst option air burst-ground, height of burst on HONEST JOHN[15]

DEVELOPMENT:
Laboratory: LANL

History:
  1954 Lab assignment (Phase 3)
  1958 initial deployment (Phase 5)

Production Period: 1953-? (not in production: 1970s to the present)

DEPLOYMENT:
Number Deployed: approximately 1000 retired NIKE-HERCULES and HONEST JOHN W31s were returned from Europe during 1980.[16]
  HONEST JOHN: approximately 200 (1983)
  NIKE-HERCULES: approximately 500 (1983)

3
W31

Delivery Systems:
- HONEST JOHN: dual capable mobile transporter erector vehicle
- NIKE-HERCULES: dual capable launchers at fixed battery firing sites

Service:
- HONEST JOHN: Army National Guard[17]
- NIKE-HERCULES: Army

Allied User:
- HONEST JOHN: Greece and Turkey; South Korea(?)
- NIKE-HERCULES: NATO Air Forces of Belgium, Netherlands, Greece, Italy, and West Germany; South Korea (?). Non-nuclear missiles in Denmark, Japan, Norway, Taiwan and Turkey.

Location:
- HONEST JOHN: United States, Greece, Turkey, South Korea (?)
- NIKE-HERCULES: Greece, Italy, West Germany, South Korea (?)

Retirement Plans: No plans currently exist to replace NIKE-HERCULES with new nuclear weapons.[18] PATRIOT conventionally armed SAM will begin deployment in the mid-1980s. W31s will remain in the inventory until past FY 1992.[19]

COMMENTS: NIKE-HERCULES is also capable of being used in surface-to-surface mode. The HONEST JOHN was originally deployed with the Mk-7 warhead. The M72 is an inert training warhead for the HONEST JOHN.[20]

---

1 U.S. Army, *Nuclear Weapons Maintenance Specialist, Soldier's Manual*, FM 9-55G4 (June 1980), pp. 3-118, 3-139.
2 *Military Applications of Nuclear Technology*, p. 9.
3 See for instance, Fred M. Kaplan, "Enhanced Radiation Weapons," *Scientific American*, 238, May 1978, p. 48; *Tactical Nuclear Weapons: European Perspectives* (London: SIPRI, Taylor & Francis, 1978), p. 111.
4 *Military Applications of Nuclear Technology*, Part 1, p. 9. Three separate warhead sections with three separate yields are deployed. U.S. Army, *Field Artillery Honest John Rocket Gunnery*, FM 6-40-1 (June 1972), p. 3-1.
5 *Military Applications of Nuclear Technology*, Part 1, p. 9. NIKE-HERCULES is often referred to as having a yield of one kiloton, but it probably has yields identical to the HONEST JOHN.
6 U.S. Army, *Special Ammunition Unit Operations*, FM 9-47 (October 1970), p. C-2; see also U.S. Army, *Field Artillery Honest John Rocket Gunnery, op. cit.*, (June 1972), p. 3-3.
7 U.S. Army, *Special Ammunition Unit Operations, op. cit.*, p. C-4. The M22/M23/M97 nuclear warhead section which contains the W31 warhead weighs 1123 pounds.
8 *Ibid.*, p. C-2.
9 *Ibid.*
10 HASC, FY 1982 DOE, p. 103.
11 U.S. Army, *Nuclear Weapons Maintenance Specialist, Soldier's Manual, op. cit.*, pp. 3-118, 3-139.
12 Letter from P.R. Wagner, PANTEX to J.F. Burke, 23 May 1979.
13 U.S. Army, *Nuclear Weapons Maintenance Specialist, Soldier's Manual, op. cit.*, pp. 3-135.
14 HAC, FY 1982 DOE, Part 7, p. 279.
15 U.S. Army, *Nuclear Weapons Maintenance Specialist, Soldier's Manual, op. cit.*, p. 3-145.
16 HASC, FY 1982 DOE, p. 104; DOD, FY 1982 Annual Report, p. 125.
17 Last active U.S. Army HONEST JOHN battalion deactivated in 1979. Only remaining U.S. HONEST JOHN units are in the reserves; HASC, FY 1981 DOE, Part 1, p. 931.
18 DOD, FY 1983 RDA, p. VII-14.
19 HAC, FY 1982 DOE, Part 7, p. 279.
20 FM 6-61, p. 47.

W33

Figure 3.6 8-inch conventional artillery projectiles. The W33 nuclear projectile is similar in size and appearance when assembled.

FUNCTION: Warhead for the M422 8-inch (203mm) artillery-fired atomic projectile (AFAP)

WARHEAD MODIFICATIONS: two warhead modifications deployed; one with low yield and one with high yield.[1]

SPECIFICATIONS:
Yield: two yield options contained in two separate versions;[2] sub Kt - 12 Kt range; often referenced as 5-10 Kt[3]

Weight: 243 lb, 215 lb (114 kg),[4] 264 lb[5]

Dimensions: dimensions of the M500 container for M422 projectile are: length, 49.5 in; diameter, 11.5 in; dimensions of the M102 "birdcage" for the nuclear materials are: length, 16 in; width, 16 in; height, 25 in[6]
Length: 37 in
Diameter: 8 in

Materials: gun assembly oralloy fission weapon with insertable materials capsule

SAFEGUARDS AND ARMING FEATURES: mechanical combination lock PAL, no command disable feature;[7] mechanical fuze[8]

DEVELOPMENT:
Laboratory: LANL
History:
1954 Lab assignment (Phase 3)
1956 initial deployment (Phase 5)

Production Period: 1955-? (not in production during 1970s to the present)

DEPLOYMENT:
Number Deployed: circa 1000 in Europe;[9] approximately 1800 estimated overall (1983)

Delivery Systems: dual capable M110 self propelled 8-inch howitzers, older 8-inch howitzers in Allied forces (M55 and M115)

Service: Army and Marine Corps

Allied User:[10] Belgium, Italy, Greece, Netherlands, Turkey, United Kingdom, West Germany

Location: United States, Greece, Italy, Netherlands, South Korea, Turkey, West Germany

3
W33

Retirement Plans: The W33 will eventually be replaced by the enhanced radiation W79, which began production in 1981.

COMMENTS: Three nuclear materials capsule cores (992 T-Z nuclear package, 992 P-Z nuclear package, and 994 P-W nuclear package) with two separate yields exist (992 and 994); 992 T-Z is a modernized core which contains a limited life component, probably tritium.[11] The warhead must be assembled in the field. Maximum range of M422 is 18,200 meters.[12] It is not ballistically similar to the conventional 8-inch round and requires a special "spotting" round to line the sights of the gun. This reduces accuracy and slows response time.[13] The M423 is the nuclear training round and the M424 is the high explosive spotter.[14]

---

1 SASC, FY 1982 DOD, Part 7, p. 3886.
2 DOD, FY 1981 RDA, p. VII-5; SASC, FY 1982 DOD, Part 7, p. 3886; SASC, FY 1983 DOD, Part 7, p. 4387.
3 Kaplan, *Scientific American*, op. cit., p. 48; Fred Kaplan, "The Neutron Bomb," *The Bulletin of the Atomic Scientists*, October 1981, p. 7.
4 *New Encyclopedia*, 15th Ed., 19, p. 694.
5 The weight of the projectile is 243 lb; "armed" weight is higher, possibly 264 lb; National Atomic Museum, Albuquerque, NM.
6 U.S. Army, *Special Ammunition Unit Operations*, FM 9-47 (October 1970), pp. C-5 - C-6.
7 ACDA, FY 1981 ACIS, p. 271.
8 ACDA, FY 1980 ACIS, p. 151.
9 Walter Pincus, *Washington Post*, 12 August 1981, p. A12.
10 Most 8-inch units in NATO are currently certified for nuclear rounds; ACDA, FY 1980 ACIS, p. 154.
11 U.S. Army, *Special Ammunition Unit Operations*, op. cit.; see also U.S. Army, *Nuclear Weapons Maintenance Specialist, Soldier's Manual*, FM 9-55G4 (June 1980), pp. 3-64, 3-71 - 3-73.
12 Information provided by U.S. Army Armament Research and Development Command, Dover, NJ.
13 ACDA, FY 1979 ACIS, p. 130.
14 LTC Robert B. Rosenkranz, "The 'nuclear' ARTEP in USAREUR— an idea whose time has come," *Field Artillery Journal*, July-August 1979, pp. 16-19.

B43

Figure 3.7 **B43** bomb on display in the National Atomic Museum, Albuquerque, New Mexico.

FUNCTION:	High-yield nuclear bomb capable of being delivered by most of the nuclear capable strategic and tactical aircraft in the U.S. inventory.
WARHEAD MODIFICATIONS:	B43 Mod 0: without PAL; B43 Mod 1: fuzing radar set, with PAL
SPECIFICATIONS:	
Yield:	1 Mt[1]
Weight:	2060-2330 lb;[2] B43-0: 2060 lb; B43-1: 2120 lb
Dimensions:	
Length:	144 in;[3] 150/165 in[4]
Diameter:	18 in[5]
Materials:	thermonuclear bomb; oralloy as fissile material;[6] lithium-6 deuteride and tritium for fusion
SAFEGUARDS AND ARMING FEATURES:	does not provide the same levels of security and safety as the B61;[7] does not contain IHE, weak link/strong link, CAT D PAL, or command disable[8]
FUZING AND DELIVERY MODE:	laydown mode, retarded or freefall ground burst, retarded or freefall air burst;[9] fuzing option must be selected on the ground by maintenance personnel;[10] minimum altitude of delivery is 300-600 feet, delivery can be over the shoulder and at low or medium angle loft.[11]
DEVELOPMENT:	
Laboratory:	LANL
History:	
1956	Lab assignment (Phase 3)
1961	initial deployment (Phase 5)
Production Period:	1959-? (not in production during the 1970s to the present)
DEPLOYMENT:	
Number Deployed:	approximately 2000 (1983)
Delivery Systems:	internal or external carriage; B-52, FB-111, F-4, F-16, F-111, A-4, A-6, A-7[12]
Service:	Air Force, Marine Corps, and Navy[13]
Allied User:	NATO Air Forces

3
B43

Location: United States, Europe;[14] South Korea (?), Philippines (?)

Retirement Plans: to be replaced in strategic forces by B83; being replaced in tactical forces by B61.

COMMENTS: Numerous designs and modifications. Practice bombs are designated BDU-6, BDU-8, BDU-18 and BDU-24. B43 was originally designed to destroy "high-value urban-industrial targets and moderately hard military targets."[15]

---

1 SASC, FY 1983 DOD, Part 7, p. 4172; yield is Y1 of B43, other yields may also exist.
2 U.S. Navy, *Loading and Underway Replenishment of Nuclear Weapons*, NWP 14-1, Rev. A (November 1979). The loaded weight of B43 on the H695A bomb truck is given as 2330 lbs. Flight manual for F-4C gives B43-0 weight as 2060; B43-1 as 2120; NATOPS Manual, September 1963, gives B43 weight as 2140.
3 B-58A *Flight Manual* (USAF TO 1B-58A-1).
4 U.S. Navy, *Loading and Underway Replenishment of Nuclear Weapons, op. cit.* The dimensions of the B43 on the H695A bomb truck are given as: length, 150/165 in; width, 31 in; height, 31 in.
5 National Atomic Museum, Albuquerque, NM.
6 SASC, FY 1979 DOE, p. 43.
7 ACDA, FY 1979 ACIS, p. 93.
8 SASC, FY 1979 DOE, pp. 42, 47.
9 *Ibid.*
10 *Military Applications of Nuclear Technology*, Part 1, p. 7.
11 ACDA, FY 1979 ACIS, p. 92.
12 The B43 was also carried by the following retired aircraft: B-66, A-3, A-5, B-57, B-58, F-100, F-101, F-105.
13 U.S. Navy, *Loading and Underway Replenishment of Nuclear Weapons, op. cit.*, p. 1-5.
14 AFM 50-5, Volume II, p. 3-87.
15 ACDA, FY 1979 ACIS, p. 89.

W44

Figure 3.8 ASROC (RUR-5A) being fired from a destroyer.

FUNCTION: Warhead in nuclear depth bomb (Mk-17) fitted to the ASROC (RUR-5A) Anti-Submarine Warfare (ASW) rocket system aboard surface ships.

WARHEAD MODIFICATIONS: none

SPECIFICATIONS:

Yield: 1 Kt[1]

Weight: less than 280 lb[2]

Dimensions: dimensions of the H-651 shipping container are: length, 35 in; width, 20 in; height, 22 in.;[3] external diameter of ASROC warhead section is 13.8 in[4]

Materials: probably fission weapon

SAFEGUARDS AND ARMING FEATURES: unknown

DEVELOPMENT:

Laboratory: LANL

History:
 1956 Lab assignment (Phase 3)
 1961 initial deployment (Phase 5)

Production Period: 1960-? (not in production during 1970s to the present)

DEPLOYMENT:

Number Deployed: approximately 850 on 170 ships (1983); over 20,000 ASROC missiles were produced; many more than the number of nuclear warheads.[5]

Delivery Systems: numerous dual capable ASROC launching systems[6]

Service: Navy

Allied User: none

Location: surface combatant ships (cruisers, destroyers, frigates)[7]

Retirement Plans: to be replaced by the Surface Delivered Anti-Submarine Warfare Weapon, under development, probably starting in the late 1980s.

---

1 United Nations, Report of the Secretary General, "Comprehensive Study on Nuclear Weapons," A/35/392 (12 September 1980), p. 22.
2 The loaded weight of the H-651 shipping container for the W44 ASROC warhead is 280 lbs. U.S. Navy, *Loading and Underway Replenishment of Nuclear Weapons*, NWP 14-1, Rev. A (November 1979), p. 1-5.
3 U.S. Navy, *op. cit.*
4 Measurements of ASROC rocket system on display at National Atomic Museum, Albuquerque, NM.
5 Norman Polmar, *The Ships and Aircraft of the U.S. Fleet*, 12th Ed. (Annapolis, MD: United States Naval Institute, 1981), p. 332.
6 See Chapter Eight for a complete listing of missile launching systems on naval ships.
7 Not deployed on newest class of frigates, the OLIVER HAZARD PERRY (FFG-7) class.

3

W45

Figure 3.9 TERRIER (RIM-2D) launch.

Figure 3.10 Medium Atomic Demolition Munition (MADM) with (from left) packing container, warhead, coder-decoder unit, and firing unit.

FUNCTION:	Warhead used in two configurations, in the TERRIER (RIM-2D) naval surface-to-air missile (SAM) and in the Medium Atomic Demolition Munition (MADM)
WARHEAD MODIFICATIONS:	Mod 1: TERRIER; Mod 3: MADM; M167, M172, and M175 "atomic demolition charges" are known configurations

SPECIFICATIONS:
Yield:
   TERRIER: reported as 1 Kt[1]
   MADM: 1-15 Kt range, probably three yields

Weight:
   TERRIER: less than 365 lb[2]
   MADM: less than 391 lb[3]

Dimensions: dimensions of the MADM shipping container (H815) are: length, 42.5 in; width, 24.5 in; height, 28 in[4]

Materials: probably fission weapon

SAFEGUARDS AND ARMING FEATURES:	locking pin, mechanical combination lock PAL; M3 and M4 coder-transmitters, M5 decoder-receiver, M96 firing device on MADM[5]

DEVELOPMENT:
Laboratory: LLNL

History:
   Nov 1956   Lab assignment (Phase 3)
   1965   initial deployment of MADM (Phase 5)[6]

Production Period: unknown

DEPLOYMENT:
Number Deployed:
   TERRIER: approximately 310 (1983)
   MADM: approximately 300 (1983)

Delivery Systems:
   TERRIER: numerous dual capable TERRIER launching systems[7]
   MADM: vehicle and air/helicopter portable

3
W45

Service:
- TERRIER: Navy
- MADM: Army and Marine Corps, Probably Navy

Allied User: Some NATO nations are trained to use MADM.

Location:
- TERRIER: guided missile cruisers, 3 aircraft carriers[8]
- MADM: United States, Italy, West Germany, South Korea(?)

Retirement Plans: nuclear TERRIER is planned for retirement with deployment of the new STANDARD-2/W81.

COMMENTS:

TERRIER "BTN" (Beam-riding, Terrier, Nuclear) is obsolescent and the only nuclear SAM left in naval service.[9] MADM is emplaced by an engineer team below the ground surface, or near bridges, tunnels, or other important structural targets. W45 was also used as the warhead for the Army's LITTLE JOHN missile, and the Air Force BULLPUP B (AGM-12D) missile, both of which have been retired from active service.

---

1 *Tactical Nuclear Weapons* (SIPRI). See also W50.
2 The Mk-22 warhead section weighs 365 lb; U.S. Navy, *Loading and Underway Replenishment of Nuclear Weapons*, NWP 14-1, Rev. A (November 1979), p. 1-6.
3 The loaded weight of the H815 shipping container for the MADM is 386-391 lbs. U.S. Navy, op. cit., p. 1-7; and U.S. Army, *Special Ammunition Unit Operations*, FM 9-47 (October 1970), p. C-1.
4 U.S. Navy, op. cit., p. 1-7; U.S. Army, op. cit., p, C-1.
5 U.S. Army, *Nuclear Weapons Maintenance Specialist, Soldier's Manual*, FM 9-55G4 (June 1980), pp. 3-108 - 3-117.
6 *Tactical Nuclear Weapons*, op. cit., p. 131.
7 See Chapter Eight for a complete listing of missile launching systems on naval ships.
8 The nuclear TERRIER does not appear to be deployed on destroyers; ACDA, FY 1980 ACIS, p. 272.
9 JCS, FY 1981, p. 48.

3
W48

W48

Figure 3.11 155mm nuclear artillery projectile with inert warhead.

FUNCTION: Warhead for the M454 155mm Artillery Fired Atomic Projectile (AFAP).

WARHEAD MODIFICATIONS: W48 Mod 1 deployed[1]

SPECIFICATIONS:
Yield: one sub kiloton yield option[2] "very small"; probably 0.1 Kt; often incorrectly referred to as 1 Kt[3]

Weight: 119.5 lb[4]

Dimensions: M467 container dimensions are: length, 57 in; width, 22 in; height, 21 in[7]
Length: 34 in[5]
Diameter: 6 in[6]

Materials: probably plutonium fission weapon

SAFEGUARDS AND ARMING FEATURES: mechanical combination lock PAL, no command disable feature;[8] variable time (VT)/mechanical fuze[9]

DEVELOPMENT:
Laboratory: LLNL

History:
Aug 1957 — Lab assignment (Phase 3)
1963 — initial deployment (Phase 5)[10]

Production Period: 1962-late 1960s

DEPLOYMENT:
Number Deployed: approximately 3000 (1983)

Delivery Systems: dual capable M198 and M109 155mm howitzers; older 155mm howitzers in NATO and U.S. use

Service: Army and Marine Corps

Allied User: Greece, Italy, Netherlands, Turkey, United Kingdom, West Germany, Belgium(?)

Location: United States, Greece, Italy, Turkey, West Germany, South Korea

3
W48

Retirement Plans: to be replaced by the W82 enhanced radiation projectile starting in 1986.[11]

COMMENTS: The yield of the W48 is lower than that of the W33.[12] The projectile has a more limited range (1.6-14.0 km),[13] is not considered accurate, and is not ballistically exact to 155mm conventional rounds.[14] M455 is training atomic projectile for M454.

---

1 U.S. Army, *Nuclear Weapons Maintenance Specialist, Soldier's Manual*, FM 9-55G4 (June 1980), p. 3-43.
2 *Nuclear Weapons and Foreign Policy*, p. 201; SASC, FY 1982 DOD, Part 7, p. 3886.
3 *Military Applications of Nuclear Technology*, Part 1, p. 10.
4 Weight stenciled on "inert" projectile shown in Figure 3.11; the M-454 projectile which contains the W48 warhead weighs 128 pounds, according to U.S. Army, *Special Ammunition Unit Operations*, FM 9-47 (October 1979), p. C-4.
5 Information provided by U.S. Army Armament Research and Development Command, Dover, NJ.
6 *Ibid.*
7 U.S. Navy, *Loading and Underway Replenishment of Nuclear Weapons*, NWP 14-1, Rev. A (November 1979), p. 1-7; and U.S. Army, *Special Ammunition Unit Operations, op. cit.*, p. C-4.
8 United Nations, *op. cit.*, p. 271.
9 ACDA, FY 1980 ACIS, p. 151.
10 *Tactical Nuclear Weapons* (SIPRI), p. 131.
11 JCS, FY 1981, p. 47.
12 ACDA, FY 1981 ACIS, p. 280.
13 SASC, FY 1982 DOD, Part 7, p. 3886.
14 HASC, FY 1981 DOD, Part 4, Book 2, p. 2305.

3
W50

Figure 3.12 **PERSHING 1a** (MGM-31A/B) missile.

FUNCTION:	Warhead for the PERSHING 1a (MGM-31) tactical ballistic missile
WARHEAD MODIFICATIONS:	Mod 1 deployed in 1963 and retired in 1978; present warhead is Mod 2; three warhead sections (M28, M141, and M142) with warheads of three different yields.[1]
SPECIFICATIONS:	
Yield:	three yield options[2] in three warhead sections, reportedly of 60, 200, and 400 Kt[3]
Weight:	less than 697 lb[4]
Dimensions:	dimensions of the M483 shipping container for the warhead section are: length, 168 in; width, 52.5 in; height, 53 in;[5] length of warhead section is 146.7 in[6]
Materials:	possibly D-T boosted fission weapon[7]
SAFEGUARDS AND ARMING FEATURES:	unknown
DEVELOPMENT:	
Laboratory:	LANL
History:	
1958	Lab assignment (Phase 3)
1963	initial deployment of Mod 1 (Phase 5)[8]
Production Period:	1960s
DEPLOYMENT:	
Number Deployed:	approximately 410 (1983)
Delivery Systems:	mobile missile transporter-erector-launcher (TEL) vehicle
Service:	Army
Allied User:	West Germany (Air Force)
Location:	United States, West Germany

3
W50

Retirement Plans: Mod 1 retired in 1978; current warhead planned to be replaced by W85 on PERSHING II starting in December 1983. PERSHING 1a in the U.S. Army will be replaced by the longer-range PERSHING II which will carry the W85 warhead. Future of missiles in West German Air Force is still uncertain, but development of PERSHING II Reduced Range (RR) missile (P1b) is proceeding for possible use by West Germany.

COMMENTS: M70 and M95 are training warheads for P1a.

Figure 3.13 **PERSHING 1a** missile in down position, with warhead section container to right.

---

1 U.S. Army, *List of Applicable Publications for PERSHING 1a Field Artillery Missile System*, TM 9-1425-380-L (February 1972), p. 2-5.
2 *Military Applications of Nuclear Technology*, Part 1, p. 9; ACDA, FY 1979 ACIS, p. 115.
3 *Tactical Nuclear Weapons: European Perspectives* (SIPRI, London: Taylor and Francis, Ltd., 1978), p. 111.
4 The M28/M141/M142 nuclear warhead sections which contain the W50 weigh 697 pounds. U.S. Army, *Special Ammunition Unit Operations*, op. cit., p. C-5.
5 Ibid.
6 Information provided by Pershing Program Office, Redstone Arsenal, AL.
7 U.S. Army, *Nuclear Weapons Maintenance Specialist, Soldier's Manual*, FM 9-55G4 (June 1980), pp. 3-30 - 3-31.
8 *Tactical Nuclear Weapons*, op. cit.

3

B53

Figure 3.14 B53 bomb.

FUNCTION: High yield thermonuclear, B-52 internally carried, heavy strategic bomb (see also W53)

WARHEAD MODIFICATIONS: reportedly in two configurations

SPECIFICATIONS:
Yield: 9 Mt[1]

Weight: 8850 lb[2]

Dimensions:
 Length: 12 ft 4 in
 Diameter: 50 in[3]

Materials: all oralloy (no plutonium) weapon;[4] lithium-6 deuteride as fusion material

SAFEGUARDS AND ARMING FEATURES: does not provide the same degree of security and safety as the newer B61 and B83 bombs.[5]

FUZING AND DELIVERY MODE: fuzing option must be selected on the ground by maintenance personnel.[6] Airburst, contact burst, laydown,[7] free fall, or retarded delivery.

DEVELOPMENT:
Laboratory: LANL

History:
 1958 Lab assignment (Phase 3)
 1962 initial deployment (Phase 5)

Production Period: 1961-late 1960s

DEPLOYMENT:
Number Deployed: approximately 150 (1983)

Delivery System: B-52[8]

Service: Air Force

Allied User: none

Location: United States

Retirement Plans: B53 will be phased out as B83 and ALCMs equip the strategic bomber force.

COMMENTS: B53 warhead is similar to W53, the warhead in the Mk-6 reentry vehicle carried by the TITAN II missile. Bomb has 5 parachutes—three each 48 ft, one 16 ft, and one 5 ft "pilot" chute. Free fall delivery could be accomplished by blowing out parachute can and jettisoning all chutes.

---

1 SASC, FY 1983 DOD, Part 7, p. 4172.
2 Fact Sheet prepared by National Atomic Museum, Albuquerque, NM.
3 Ibid.
4 SASC, FY 1982 DOE, p. 282.
5 ACDA, FY 1980 ACIS, p. 169.
6 *Military Applications of Nuclear Technology*, Part 1, p. 7.
7 ACDA, FY 1980 ACIS, p. 169; Fact Sheet prepared by National Atomic Museum, Albuquerque, NM.
8 Also formerly carried by B-58, B-70, and B-47 bombers.

W53

Figure 3.15 W53 warhead.

FUNCTION: High yield thermonuclear warhead in the Mk-6 reentry vehicle deployed on TITAN II ICBMs (see also B53)

WARHEAD MODIFICATIONS: none known

SPECIFICATIONS:
Yield: 9 Mt

Weight: 8275-8800 lb; 8140 lb[1]

Dimensions: slightly smaller configuration than the B53 bomb
Length: 102 in overall[2]
Diameter: 36.52 in at warhead mid-section[3]

Materials: all oralloy system (no plutonium);[4] lithium-6 deuteride for fusion; standard HE (probably cyclotol) (75 percent RDX) as primary HE[5]

SAFEGUARDS AND ARMING FEATURES: unknown

DEVELOPMENT:
Laboratory: LANL
History:
1960 Lab assignment (Phase 3)
1962 initial deployment (Phase 5)
Oct 1982 retirement of TITAN II missiles (one per month) begins

Production Period: 1961-mid 1960s

DEPLOYMENT:
Number Deployed: 65 total (1982);[6] 52 active missiles deployed in silos prior to beginning of retirement program in October 1982; 49 deployed (January 1983)

Delivery System: TITAN IIs in underground silos

Service: Air Force

Allied User: none

Location: Kansas, Arizona, Arkansas

Retirement Plans: TITAN II missile is being retired under the Reagan Administration Strategic Program with the first missile withdrawal started in October 1982, with one missile withdrawn per month thereafter.[7]

COMMENTS: W53 warhead is similar to basic warhead used in the B53 bomb.[8] Mk-6 has three-target selection capability and elaborate penetration aids.[9]

---

1 Mk-6 RV weighs 3700 kg; *Jane's Weapon Systems.*
2 Measurements given in inventory card at National Atomic Museum, Albuquerque, NM. Author's measurement at the museum is 104 inches for overall length.
3 *Ibid.*
4 SASC, FY 1982 DOE, p. 282.
5 Letter from P.R. Wagner, PANTEX, to J.F. Burke, 23 May 1979.
6 USAF, *Missile Procurement Justification, FY 1981* (January 1980), p. 184.
7 HASC, FY 1983 DOD, Part 2, p. 163.
8 Display at National Atomic Museum, Albuquerque, NM.
9 *Projected Strategic Offensive Weapons Inventories of the U.S. and U.S.S.R. An Unclassified Estimate,* CRS 77-59F (24 March 1977), p. 150.

3

W54

Figure 3.16 Container for W54 **Special Atomic Demolition Munition (SADM).**

FUNCTION: Low yield warhead once used in several configurations, currently used in the Special Atomic Demolition Munition (SADM)

WARHEAD MODIFICATIONS: two separate configurations, M129 and M159 "atomic demolition charges" are known.

SPECIFICATIONS:

Yield: very low; variable yield options of SADM in .01-1 Kt range, probably two yields; DAVY CROCKETT (W54-2) had a yield on the order of 0.25 Kt;[1]

Weight: warhead weighs 58.6 lb;[2] ADM containing W54 weighs less than 163 lb[3]

Dimensions: dimensions of the SADM shipping container (H913) are: length, 35 in; width, 26.2 in; height, 26.6 in[4]

Materials: probably plutonium fission weapon

SAFEGUARDS AND ARMING FEATURES: mechanical combination lock PAL,[5] M96 firing device, internal timer, M3 and M4 coder-transmitters, M5 decoder-receiver.

DEVELOPMENT:

Laboratory: LANL

History:
1960 Lab assignment (Phase 3) (SADM)
1964 initial deployment (Phase 5) (SADM)

Production Period: 1960-1963 (DAVY CROCKETT); 1963-late 1960s (SADM)

DEPLOYMENT:

Number Deployed: approximately 300 (1983)

Delivery System: can be carried in a backpack by a single soldier

Service: Army and Marine Corps

Allied User: some NATO nations are trained to use SADM.

Location: United States, Italy, West Germany, South Korea (?)

Retirement Plans: no new ADMs are known to be in research and development at this time.

COMMENTS: W54 was used as the warhead for the FALCON missile (W54-0) and DAVY CROCKETT (W54-2) rocket, retired from active service in 1972 and 1971, respectively. W54 is also identified as the B54.

---

1 The "hypothetical" yield given by U.S. Army, *Operation and Employment of the Davy Crockett Battlefield Missile, XM-28/29*, FM 9-11 (June 1963).

2 As listed in DNA, "Motion Picture Catalog," January 1981; the DAVY CROCKETT weight, which also used the W54 (Mod 2) is 51 lb; the warhead was designed by LASL and has also been referred to as weighing "less than 50 lbs," in John McPhee, *The Curve of Binding Energy*, (New York: Farrar, Strauss and Giroux, 1974), p. 8.

3 The loaded weight of the H913 container for the SADM is 162-163 lb. U.S. Navy, *Loading and Underway Replenishment of Nuclear Weapons*, NWP 14-1, Rev. A (November 1979), p.1-7; and U.S. Army, *Special Ammunition Unit Operations*, FM 9-47 (October 1970), p. C-1.

4 *Ibid.*; Dimensions of DAVY CROCKETT were, maximum diameter: 11 in; length: 25.5 in (excluding fins); measurements of display at National Atomic Museum, Albuquerque, NM.

5 U.S. Army, *Nuclear Weapons Maintenance Specialist, Soldier's Manual*, FM 9-55G4 (June 1980), pp. 3-86.

W55

Figure 3.17 SUBROC (UUM-44A) launch.

FUNCTION: Low yield warhead contained in the Mk-57 warhead section for the SUBROC (UUM-44A) ASW depth charge rocket system, a nuclear-only weapon aboard attack submarines

WARHEAD MODIFICATIONS: none known

SPECIFICATIONS:

Yield: 1-5 Kt range, often referred to as 1Kt[1]

Weight: less than 675 lb[2]

Dimensions: dimensions of the SUBROC shipping container (H863) are: length, 49 in; width, 21 in; height, 23 in;[3] the diameter of the missile is 13 in[4]

Materials: probably fission weapon

SAFEGUARDS AND ARMING FEATURES: depth pressure fuze

DEVELOPMENT:

Laboratory: LLNL

History:
Mar 1959 — Lab assignment (Phase 3)
1965 — initial deployment (Phase 5)

Production Period: 1964-1968, 1972-1974[5]

DEPLOYMENT:

Number Deployed: approximately 400 (1983)

Delivery System: 63 attack submarines (SUBROC capable)[6]

Service: Navy

Allied User: none

Location: submarines homeported in Pearl Harbor, HI; Groton, CT; San Diego, CA; Charleston, SC; and Norfolk, VA

Retirement Plans: to be replaced by the Subsurface Delivered ASW Standoff Weapon in the late 1980s

COMMENTS: SUBROC is launched from a torpedo tube, emerges from underwater and reenters the water to attack submerged submarines.

---

1 United Nations, 1980, op. cit., p. 22.
2 The loaded weight of the H863 container for the SUBROC W55 warhead is 675 lbs. U.S. Navy, *Loading and Underway Replenishment of Nuclear Weapons*, NWP 14-1, Rev. A (November 1979), p. 1-7.
3 Ibid.
4 Measurements of external diameter of warhead section of missile on display at National Atomic Museum, Albuquerque, NM.
5 After producing SUBROCs from 1964-1968, the production line was reopened in 1972; SASC, FY 1982 DOD, Part 7, p. 3896.
6 SAC, FY 1983 DOD, Part 1, p. 100.

3
W56

Figure 3.18 **MINUTEMAN II** (LGM-30F) missile in silo.

FUNCTION:	High yield thermonuclear warhead in the Mk-11 reentry vehicle on the MINUTEMAN II ICBM
WARHEAD MODIFICATIONS:	Mod 1 initially deployed; it has since been modified, hardened and upgraded to Mod 4.[1]
SPECIFICATIONS:	
Yield:	1.2 Mt;[2] 1-2 Mt range often referenced[3]
Weight:	1600 lb; 2200 lb
Dimensions:	unknown
Materials:	plutonium as fissile material; lithium-6 deuteride as fusion material
SAFEGUARDS AND ARMING FEATURES:	unknown
DEVELOPMENT:	
Laboratory:	LLNL
History:	
Dec 1960	Lab assignment (Phase 3)
1965	initial deployment (Phase 5)
Production Period:	unknown
DEPLOYMENT:	
Number Deployed:	approximately 540 (1983), 450 MM IIs actively deployed with one W56 warhead per missile
Delivery System:	MINUTEMAN IIs in underground silos
Service:	Air Force
Allied User:	none
Location:	Montana, Missouri, South Dakota
Retirement Plans:	Fifty MM II missiles are being replaced by MM III under the Reagan Strategic program. The remainder will probably be withdrawn with planned deployment of the MX missile, starting in 1986.
COMMENTS:	Mk-11C reentry vehicle is latest modification with eight target selection capability, penetration aids (Mk-1 or Mk-1A canister), and hardened against nuclear weapons effects.[4]

---

1 SASC, FY 1978 ERDA, p. 111.
2 AW&ST, 16 June 1980, p. 178.
3 *Military Balance 1980-1981*, p. 88; Heritage Foundation, *SALT Handbook*, p. 75 lists 1-2 Mt; Collins, op. cit., p. 446, and *Projected Strategic Offensive Inventories of the U.S. and U.S.S.R. An Unclassified Estimate*, CRS, p. 151, list 1 Mt. 2 Megatons is also referenced in *The World's Missile Systems*, 6th Ed., p. 286 and *Jane's Weapons Systems*; appears high based upon yield-to-weight ratio.

4 *Projected Strategic Offensive Weapons Inventories of the U.S. and U.S.S.R.*, op. cit., p. 151.

B57

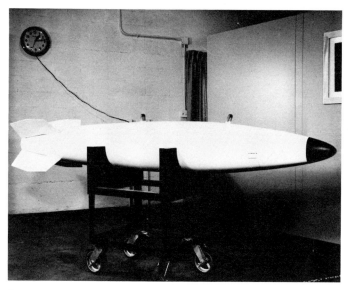

Figure 3.19 B57 bomb.

FUNCTION: Lightweight, multi-purpose nuclear depth charge and nuclear bomb used for anti-submarine warfare (ASW) and land warfare.

WARHEAD MODIFICATIONS: Mod 0 retired in 1967; Mod 1 presently deployed

SPECIFICATIONS:
Yield: sub Kt-20 Kt range; variable yield options; often referred to as 5-10 Kt[1]

Weight: under 710 lb;[2] 500-510 lb[3]

Dimensions:
Length: less than 119 in[4]
Diameter: 14.75 in[5]

Materials: unknown, probably fission weapon

SAFEGUARDS AND ARMING FEATURES: does not provide the same degree of security and safety as the newer B61 and B83.[6]

FUZING AND DELIVERY MODE: air or surface burst;[7] laydown;[8] over the shoulder and low or medium angle loft; minimum altitude for laydown is 300-600 ft;[9] depth pressure fuze for underwater detonations.

DEVELOPMENT:
Laboratory: LANL

History:
1960   Lab assignment (Phase 3)
1964   initial deployment (Phase 5)

Production Period: 1960s

DEPLOYMENT:
Number Deployed: approximately 1000 (1983); B57s for tactical (non ASW) use are being reduced in overall numbers as the B61 enters the stockpile.[10]

Delivery Systems: S-3, P-3, SH-3 maritime patrol aircraft and helicopters;[11] allied maritime patrol aircraft; wide variety of tactical aircraft.

Service: Marine Corps and Navy

Allied User: some NATO naval aviation components (Netherlands, United Kingdom, West Germany, et al.) are nuclear capable and trained to use nuclear depth bombs.[12]

3

B57

Location:[13]	aircraft carriers; United States, United Kingdom, Europe, Pacific region.	Retirement Plans:	to be replaced with the Air Delivered ASW Weapon in the late 1980s.
		COMMENTS:	practice bombs are designated BDU-12 and BDU-19.

---

1 United Nations, 1980, op. cit., p. 22.
2 The B57 with the H841 shipping frame weighs 710 lb; U.S. Navy, *Loading and Underway Replenishment of Nuclear Weapons*, NWP 14-1, Rev. A (November 1979), p. 1-5.
3 F-4C *Flight Manual* (1 September 1963) lists weight as 500 lb; *NATOPS Flight Manual* (A-7C/A-7E) cites 510 lb.
4 U.S. Navy, *Loading and Underway Replenishment of Nuclear Weapons*, op. cit. The dimensions of the B57 in the H841 shipping frame mounted on the H1012 dolly are: length, 119 in; width, 37 in; height, 33 in.
5 National Atomic Museum, Albuquerque, NM.
6 ACDA, FY 1980 ACIS, p. 169.
7 *Military Applications of Nuclear Technology*, Part 1, p. 15.
8 ACDA, FY 1980 ACIS, p. 169.
9 ACDA, FY 1979 ACIS, p. 92.
10 HASC, FY 1981 DOD, Part 4, Book 2, p. 2318.
11 JCS, FY 1981, p. 48.
12 SASC, FY 1980 DOD, Part 6, p. 3428.
13 B57 bombs are deployed to Europe and the Pacific; AFM 50-5, Volume II, p. 3-87.

3
B61

B61

Figure 3.20 B61 bomb with parachute drogue deployed.

FUNCTION: Lightweight, multipurpose thermonuclear "modern tactical bomb" in 6 modifications used on a wide variety of aircraft, including strategic bombers.

WARHEAD MODIFICATIONS: Mods 0-5 with differing safety features and delivery capabilities

SPECIFICATIONS:
Yield: four yield options,[1] 100-500 Kt range; possibly low (10 Kt) option; maximum of "a few hundred kilotons"[2]

Weight: less than 840 lb;[3] 718 lb;[4] 765 lb[5]

Dimensions:
Length: 142 in[6]
Diameter: 13.4 in[7]

Materials: oralloy as fissile material;[8] probably D-T boosted; primary HE of B61-0, 1, and 2 is probably PBX 9404. B61-3/4 utilize (PBX 9502) IHE[9]

SAFEGUARDS AND ARMING FEATURES:[10]

B61-0: CAT B PAL;[11] no command disable; no enhanced electrical safety

B61-1: No PAL;[12] no command disable; no enhanced electrical safety

B61-2: CAT D PAL; inertial command disable;[13] no enhanced electrical safety

B61-3: CAT F PAL;[14] command disable;[15] weak ling/strong link driven by unique signal generator[16]

B61-4: CAT F PAL;[17] command disable, weak link/strong link driven by unique signal generator[18]

B61-5: CAT D PAL; nonviolent command disable, no enhanced electrical safety;[19] weak link/strong link switches driven by unique signal generator[20]

All Mods of the B61 are one-point safe by the present criterion.[21]

Funds in FY 1982 DOE budget were to begin retrofit of B61-0/1 with new safety and command and control features.[22] All forward deployed B61 bombs were planned to be upgraded to B61-3/4, slated to begin in FY 1983.[23]

Nuclear Weapons Databook, Volume I **65**

3

B61

FUZING AND DELIVERY MODE:	In-flight fuzing selection and yield by merely turning a dial.[24] Can be delivered from as low as 50 feet;[25] B61 Mods 3/5: full fuzing options including time delay fuze;[26] laydown mode/retarded delivery; ground burst/air burst. B61 Mods 3/4: free fall, contact burst, or parachute retarded.[27] Can be delivered at supersonic speeds. "Penetration and environment sensing devices built into" B61 bombs.[28]	Production Period:[38]	
		B61-0:	production completed June 1969
		B61-1:	production completed April 1971
		B61-2:	production continued through FY 1978[39]
		B61-3:	full scale production, FY 1981-present
		B61-4:	full scale production, FY 1981-present
		B61-5:	production completed FY 1979; in production (Phase 6) in early 1979[40]
Accuracy:	highest accuracy, better than 600 ft CEP[29]	**DEPLOYMENT**: Number Deployed:	approximately 3000 (1983)
DEVELOPMENT:			
Laboratory:	LANL	Delivery Systems:	usable on any U.S. aircraft that can deliver a nuclear bomb,[41] both strategic and tactical aircraft: A-4, A-6, A-7, F-4, F-16, F-111, FB-111, B-52; planned for use on the B-1B.[42]
History:			
1963	Lab assignment (Phase 3)[30]		
Jan 1968	initial deployment of B61-0 (Phase 5)[31]		
Feb 1969	initial deployment of B61-1 (Phase 5)[32]	Service:	Air Force, Marine Corps, Navy
Jun 1975	initial deployment of B61-2 (Phase 5)[33]	Allied User:	Belgium, Greece, Italy, Netherlands, Turkey, West Germany
1976	Lab assignment of B61-3 (Phase 3)[34]	Location:[43]	United States, Aircraft carriers, Belgium, Greece, Italy, Netherlands, Turkey, United Kingdom, West Germany, South Korea
1976	Lab assignment of B61-4 (Phase 3)[35]		
1977	Lab assignment of B61-5 (Phase 3)[36]		
1979	initial deployment of B61-3 (Phase 5)[37]		

COMMENTS: B61-1 is reported to be a strategic version carried on B-52 and FB-111 aircraft. B61 Mods 3-5 are used for NATO delivery systems.[44] Practice mods of the B61 are BDU-36, BDU-38 and BDU-39E. B61 was designated TX-61 during development.

---

1. ACDA, FY 1979 ACIS, p. 92.
2. *Military Applications of Nuclear Technology*, Part 1, p. 7.
3. The weight of the B61 and the H1125 bomb cradle is 840 lb; U.S. Navy, *Loading and Underway Replenishment of Nuclear Weapons*, NWP 14-1, Rev. A (November 1979), p. 1-5.
4. GAO, Draft Study for B-1.
5. "The kevlar-29 parachute can slow the 765 lb [B61] vehicle from 1000 mph to 35 mph in two seconds"; caption of photograph at the National Atomic Museum, Albuquerque, NM.
6. U.S. Navy, *Loading and Underway Replenishment of Nuclear Weapons*, op. cit. The dimensions of the B61 and the H1125 bomb cradle are: length, 142 in; width, 33 in; height, 34 in.
7. Measurements taken at the National Atomic Museum, Albuquerque, NM.
8. *Military Applications of Nuclear Technology*, Part 1, pp. 52-53; later production B61s may use different materials than others; *Military Applications of Nuclear Technology*, Part 2, pp. 39-40.
9. HASC, FY 1982 DOE, p. 217; SASC, FY 1979 DOE, pp. 43-47.
10. HASC, FY 1980, DOE, p. 138; also HAC, FY 1982 EWDA, Part 7, p. 279; and also SASC, FY 1979, DOE, pp. 43, 46-47; ACDA, FY 1979 ACIS, p. 92; HASC, FY 1980 DOE, p. 140.
11. SASC, FY 1979 DOE, p. 46.
12. ACDA, FY 1979 ACIS, p. 92; ACDA, FY 1980 ACIS, p. 169.
13. ACDA, FY 1979 ACIS, lists with (p. 92); ACDA, FY 1980 ACIS, lists without (p.169); see also HASC, FY 1980 DOE, p. 140; SASC, FY 1978 ERDA, p. 109.
14. SASC, FY 1979 DOE, p. 58.
15. ACDA, FY 1979 ACIS, lists with (p. 92); ACDA, FY 1980 ACIS, lists without (p.169); see also HASC, FY 1980 DOE, p. 140; SASC, FY 1978 ERDA, p. 109.
16. SASC, FY 1979 DOE, p. 47.
17. SASC, FY 1979 DOE, p. 58.
18. SASC, FY 1979 DOE, p. 47.
19. ACDA, FY 1979 ACIS, lists with (p. 92); ACDA, FY 1980 ACIS, lists without (p.169); see also HASC, FY 1980 DOE, p. 140; SASC, FY 1978 ERDA, p. 109.
20. SASC, FY 1979 DOE, p. 47.
21. One-point safe by the present criterion means that "in the event of a detonation initiated at any one point in the high explosive system, the probability of achieving a nuclear yield greater than 4 pounds of TNT equivalent shall not exceed one in one million"; ACDA, FY 1979 ACIS, p. 92; ACDA, FY 1980 ACIS, p. 169.
22. HAC, FY 1982 EWDA, Part 7, p. 279; SASC, FY 1982 DOD, Part 7, p. 3880.
23. DOD, FY 1983 RDA, p. VII-14.
24. *Military Applications of Nuclear Technology*, Part 1, p. 7.
25. ACDA, FY 1980 ACIS, p. 169.
26. ACDA, FY 1979 ACIS, p. 92; SASC, FY 1980 DOE, p. 165.
27. HASC, FY 1982 DOE, p. 217.
28. ACDA, FY 1980 ACIS, p. 171.
29. *Aerospace Daily*, 28 December 1978, p. 263.
30. HASC, FY 1980 DOE, p. 137; SASC, FY 1980 DOE, p. 164.
31. ACDA, FY 1980 ACIS, p. 169.
32. *Ibid.*
33. *Ibid.*
34. HASC, FY 1980 DOE, p. 137; SASC, FY 1980 DOE, p. 164.
35. *Ibid.*
36. *Ibid.*
37. *Ibid.*
38. SAC, FY 1981 EWDA, Part 2, p. 825.
39. SASC, FY 1978 ERDA, p. 37.
40. HASC, FY 1980 DOE, p. 137; SASC, FY 1980 DOE, p. 164.
41. HAC, FY 1980 EWDA, Part 7, p. 2655.
42. GAO, Draft Study for B-1.
43. B61 bombs are deployed in both Europe and the Pacific region; AFM 50-5, Volume II, p. 3-87.
44. ACDA, FY 1979 ACIS, p. 93.

3
W62

W62

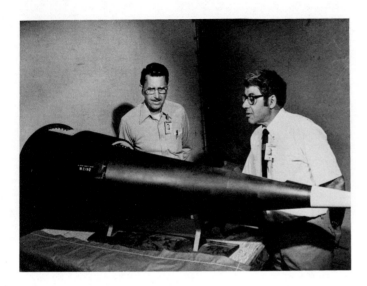

Figure 3.21 Mock-up of Mk-12 reentry vehicle, which contains W62 warhead.

FUNCTION: Warhead in the Mk-12 multiple independently targetable reentry vehicle (MIRV) on a portion of the MINUTEMAN III ICBM force.

WARHEAD MODIFICATIONS: none

SPECIFICATIONS:

Yield: 170 Kt (each missile carries 2 or 3 W62 at 170 Kt each), also reported as 200 Kt[1]

Weight: less than 800 lb; 733 lb

Dimensions: unknown

Materials: plutonium as fissile component,[2] probably D-T boosted[3]

SAFEGUARDS AND ARMING FEATURES: unknown

DEVELOPMENT:

Laboratory: LLNL

History:
  June 1964 — Lab assignment (Phase 3)
  June 1970 — initial deployment (Phase 5)

Production Period: circa 1969-FY 1978[4]

DEPLOYMENT:

Number Deployed: approximately 900 (1983); more than 750 (250 MM IIIs with Mk-12 MIRVs) deployed with active forces (after 1982), 150 additional Mk-12s will be added to the active force as 50 MM IIIs replace MM IIs during 1983-1985.[5]

Delivery System: MINUTEMAN III ICBMs in underground silos

Service: Air Force

Allied User: none

Location: Montana, North Dakota, Wyoming, Nebraska, Colorado

Retirement Plans: between 1980 and 1983 W62s on 300 of 550 MM IIIs will be replaced by W78s on Mk-12A reentry vehicles. W62 warheads are retired as the W78 is being built into the new Mk-12A RV.[6]

COMMENTS: U.S. has 123 extra MM IIIs in storage as of March 1982, for tests and spares.[7] These are probably supplied with W62 warheads.[8]

---

1 *Jane's Weapons Systems*.
2 *Military Applications of Nuclear Technology*, Part 1, p. 26.
3 AFM 50-5, Volume II, p. 3-87.
4 SASC, FY 1978 ERDA, p. 37.
5 SASC, FY 1983 DOD, Part 7, p. 4972.
6 SASC, FY 1981 DOE, p. 195.
7 Michael Getler, *Washington Post*, 5 May 1981, p. A12.
8 AF Public Affairs, information given to authors.

W68

Figure 3.22 POLARIS A3 (UGM-27C) missile, left, shown with its replacement, the **POSEIDON C3 (UGM-73A)** missile, right. POSEIDON is 3 feet longer and 1 foot 6 inches larger in diameter than POLARIS.

FUNCTION: Warhead in the Mk-3 multiple independently targetable reentry vehicle (MIRV) on the POSEIDON C3 SLBM

WARHEAD MODIFICATIONS: none

SPECIFICATIONS:
Yield: 40-50 Kt (each missile can carry up to 14 W68; 10 is average)[1]

Weight: 367 lb

Dimensions: unknown

Materials: plutonium as fissile material; probably D-T boosted;[2] LX-09 and LX-10 as primary HE[3]

SAFEGUARDS AND ARMING FEATURES: unknown

DEVELOPMENT:
Laboratory: LLNL

History:
Dec 1966 — Lab assignment (Phase 3)
Mar 1971 — initial deployment (Phase 5)

Production Period: 1970-late 1970s

DEPLOYMENT:
Number Deployed: approximately 3480 (1983); as high as 4256 possible remaining in stockpile (19 POSEIDON submarines with 304 C3 missiles, with as many as 4256 warheads)[4]

Delivery System: POSEIDON C3 SLBM on ballistic missile submarines, each of which carries 16 missiles

Service: Navy

Allied User: none

Location: submarines homeported and refitted in Holyloch, U.K.; Groton, CT; Charleston, SC; Kings Bay, GA

Retirement Plans: TRIDENT I C4 missiles carrying W76 warheads have been backfitted onto 12 of 31 POSEIDON SSBNs, replacing POSEIDON C3/W68s

3
W68

COMMENTS: Development costs for the W68 were $131 million.[5] (See TRIDENT I missile and POSEIDON submarine.) A POSEIDON missile was accidentally dropped during a winching operation at Holy Loch, Scotland on 2 November 1981.

---

1 The C3 missile has been tested with 14 Mk-3 MIRVs. The actual loading is less. The 1981 *SIPRI Yearbook* assumes 10 W68/Mk-3 MIRVs per C3 missile. Paul H. Nitze indicates 8-10 MIRVs per C3 missile and uses an average of nine.

2 *Military Applications of Nuclear Technology*, Part 1, p. 26; ACDA, FY 1983 ACIS, p. 41; ACDA, FY 1981 ACIS, p. 80; SAC, FY 1980 DOD, Part 4, p. 1037.

3 LX-09 was originally used as the primary HE in the W68. Production was completed in June 1975. Due to problems encountered with LX-09, DOE subsequently began replacing LX-09 with LX-10 when weapons were returned to PANTEX for routine maintenance; letter from David G. Jackson, DOE Albuquerque Operations Office, to Thomas B. Cochran, 26 October 1981. On 30 March 1977, 3 men were killed in an accident at the PANTEX Plant, while working in a bay containing two explosives, LX-09 and LX-14. An investigation by DOE indicated that LX-09 was probably the explosive that initially detonated, probable cause being an error in machining or handling of the explosive billet which was being worked on a lathe. It was reported (Dick Stanley, "Working With the Bomb," *Atlanta Constitution*, November 7, 1982, pp. 1, A12) that the explosion was 414 lb TNT equivalent from 126 pounds of LX-09. DOE claims that the decision to replace LX-09 was entirely unrelated to the question of sensitivity and safety, but was due to evidence from tests that the plastic binding components of the LX-09 released minute quantities of plasticizer which, over the lifetime of the warhead, could conceivably affect other warhead components, resulting in a degradation of the reliability of the warhead; DOE, "Report on the Sensitivity of the High Explosives in Poseidon Warheads," December 1981. Because of the high sensitivity of LX-09, this DOE claim has been questioned in an analysis by Norman Solomon; letter to Representative Ronald V. Dellums, 8 January 1982. "Several hundred" W68 warheads (in 1981) still contained LX-09; to be replaced by LX-10 during the next three to five years. Norman Soloman, Pacific News Service, *San Francisco Examiner and Chronicle*, 18 October 1981, p. A18. Quote attributed to Maj. Gen. William W. Hoover, Director OMA, DOE. Hoover is also quoted as saying some warheads containing LX-09 will remain in deployment for another "three to five years."

4 This figure assumes full loading of all POSEIDON missiles with the maximum of 14 warheads. Although "force loading" of deployed warheads on operational submarines is less, conversion of 12 POSEIDON submarines (192 missiles, circa 2000 warheads) to TRIDENT probably results in a large number of warheads in the stockpile, pending continued retirement.

5 HASC, FY 1979 DOD, Part 9, p. 6697.

W69

Figure 3.23 **Short-Range Attack Missiles (SRAM) (AGM-69)** mounted in bomb bay of B-52 bomber.

FUNCTION: Warhead for the Short Range Attack Missile (SRAM) (AGM-69) carried aboard B-52 and FB-111 strategic bombers

WARHEAD MODIFICATIONS: none

SPECIFICATIONS:

Yield: 170-200 Kt range[1]

Weight: reportedly greater than the W80[2]

Dimensions: unknown

Materials: plutonium as fissile material[3]

SAFEGUARDS AND ARMING FEATURES: unknown

DEVELOPMENT:

Laboratory: LANL

History:
1967 Lab assignment (Phase 3)
1970 initial deployment (Phase 5)

Production Period: 1970-1976

DEPLOYMENT:

Number Deployed: 1140 authorized missiles in 16 B-52G/H squadrons and 2 FB-111 squadrons[4] (1983); total number in stockpile is probably more

Delivery Systems: B-52G/H and FB-111

Service: Air Force

Allied User: none

Location: SAC bomber bases in United States (see Table 4.4)

Retirement Plans: to be replaced by Advanced Strategic Air-Launched Missile with a new warhead in the late 1980s to early 1990s.

COMMENTS: SRAM is a supersonic air-to-surface missile with a range of 200 km. Two SRAMs are carried on FB-111, and up to 20 are carried on B-52 bombers.

---

1 Yield estimated to be in the range of the W62; *Air Force Magazine*, May 1976, p. 124; see also *The World's Missile Systems*, 6th Ed., p. 116.
2 AW&ST, 22 November 1976, p. 15.
3 *Military Applications of Nuclear Technology*, Part 1, p. 26.
4 HAC, FY 1982 DOD, Part 2, p. 101.

3
W70

Figure 3.24 LANCE (MGM-52) missile at moment of ignition.

FUNCTION:	Warhead in the Army LANCE (MGM-52) short-range, highly mobile, guided surface-to-surface tactical ballistic missile.
WARHEAD MODIFICATIONS:	W70-1: presently deployed, selectable fission yield; W70-2: presently deployed, improved selectable fission yield;[1] W70-3: presently deployed, enhanced radiation version;[2] W70-4: dual capable fission/ER version, with insertable tritium reservoir to convert to enhanced radiation yield, deferred in favor of the W70-3

SPECIFICATIONS:

Yield:	W70-1/2: sub 1-100 Kt[3] in three yield options;[4] 10 Kt often referenced as typical fission yield;[5] 50 Kt as intermediate yield; W70-3: Two yield options, one slightly less than 1 Kt, the other slightly more than 1 Kt, both consisting of about 40 percent fission and 60 percent fusion.[6] A program was revealed in FY 1980 to give the LANCE (W70-1/2) warhead "a more useful spread of yields."[7]
Weight:	211 kg;[8] 450 lb;[9] 465 lb[10]
Dimensions:	
Length:	97 in[11]
Diameter:	22 in[12]
Materials:	plutonium as fissile material; Mod 3 (ER version) is tritium weapon. Non-ER versions (Mods 1/2) are probably D-T boosted fission warheads.
SAFEGUARDS AND ARMING FEATURES:	inertial nonviolent command disable system,[13] Cat D PAL built into the warhead section,[14] radiofrequency shielded against electronic countermeasures.[15] Nuclear explosion is initiated by the M1140 fuze.[16]

DEVELOPMENT:

Laboratory:	LLNL
History:	
Apr 1969	Lab assignment of W70-1 (Phase 3)
1973	initial deployment of W70-1 (Phase 5)
Apr 1976	Lab assignment of W70-3/4 (Phase 3)[17]
Oct 1978	production activities begin on W70-3 ER warhead[18]
1981	initial deployment of W70-3 (Phase 5)[19]

3
W70

Production Period:	1971-1977 (W70-1/2); 1981-1982 (W70-3)	Location:	United States, Netherlands, Italy, West Germany

DEPLOYMENT:

Number Deployed: approximately 945 (1983); at least 340 warheads for approximately 100 launchers in NATO plus missiles and 12 launchers in United States and test versions;[20] some 380 W70-3 ER warheads produced in 1981-1982.[21]

Delivery System: dual capable LANCE missile launcher on tracked vehicles

Service: Army

Allied User: Belgium, Italy, Netherlands, West Germany, United Kingdom

COMMENTS: nuclear warhead fitted into M234 nuclear warhead section; warhead section has external access cover for PAL connector, hazard indicator, command disable system, and sequential timer access cover. Warhead container is M511E2 with PAL connector.[22] Warhead section is of aluminum construction covered with an ablative skin that burns off in layers, preventing the warhead from overheating.

---

1 According to one source, the W70-3 was originally designed to "replace" the W70-2 warheads deployed in Europe; Col. William E. Serchak, "Artillery Fired Atomic Projectiles—A Field Artilleryman's Viewpoint," *Field Artillery Journal*, March-April 1980, pp. 7-12.
2 "Production of improved LANCE warheads, with ER/RB features, began earlier this year"; DOD, FY 1983 RDA, p. VII-12.
3 ACDA, FY 1982 ACIS, p. 253; *Nuclear Weapons and Foreign Policy*, p. 201; Kaplan, *Scientific American*, op. cit. p. 48; George B. Kistiakowsky, "The Folly of the Neutron Bomb," *Atlantic*, June 1978, p. 9; *Tactical Nuclear Weapons: European Perspectives* (SIPRI, London: Taylor and Francis, 1978), p. 111; Fred Kaplan, "The Neutron Bomb," *The Bulletin of the Atomic Scientists*, October 1981, p. 7.
4 *Military Applications of Nuclear Technology*, Part 1, p. 9; *Field Artillery Lance Missile Gunnery*, FM 6-40-4 (15 June 1979), p. 2-4. The three yields are contained in warheads designated M234A, M234B, and M234C.
5 "Lance," *Armies & Weapons*, 42, April 1978, pp. 55-62.
6 Fred Kaplan, "The Neutron Bomb," *The Bulletin of the Atomic Scientists*, October 1981, p. 7.
7 U.S. Army, "Equipping the United States Army, A Statement to the Congress on the FY 1980 Army RDTE and Procurement Appropriations," n.d., p. 28.
8 "Lance," *Armies & Weapons*, 42, op. cit., pp. 55-62.
9 U.S. Army, "Equipping the United States Army, A Statement to the Congress on the FY 1980 Army RDTE and Procurement Appropriations," n.d., p. 28.
10 *The World's Missile Systems*, 6th Ed., p. 284; *Field Artillery Battalion, Lance*, FM 6-42, p. 2-6; *System Description for Lance Guided Missile System*, TM 9-1425-485-10-1, p. 1-12.
11 *Field Artillery Battalion, Lance*, p. 2-6; TM 9-1425-484-10-1, p. 1-12.
12 *Ibid.*
13 SASC, FY 1978 ERDA, p. 109.
14 ACDA, FY 1982 ACIS, p. 244; *Field Artillery Battalion, Lance*, FM 6-42, p. 2-6; ACDA, FY 1980 ACIS, p. 151.
15 *Field Artillery Battalion, Lance*, p. 2-6; TM 9-1425-484-10-1, p. 1-12.
16 *Field Artillery Lance Missile Gunnery*, FM 6-40-4 (15 June 1979), p. 1-2.
17 SASC, FY 1979 DOE, p. 35; production activities on the W70-3 were halted in October 1977, and then began again in October 1978; HASC, FY 1980 DOE, p. 100.
18 HASC, FY 1980 DOE, p. 100.
19 SASC, FY 1981 DOE, p. 149.
20 DOD FY 1980 Annual Report, p. 137; ACDA, FY 1982 ACIS, p. 244.
21 Walter Pincus, *Washington Post*, 9 August 1981, p A8.
22 TM 9-1425-485-10-1, p. 1-35.

3
W76

W76

Figure 3.25 TRIDENT I C4 (UGM-93A) missile launch.

FUNCTION:	Warhead for the Mk-4 multiple independently targetable reentry vehicle (MIRV) on the TRIDENT I C4 (UGM-93A) SLBM
WARHEAD MODIFICATIONS:	none
SPECIFICATIONS:	
Yield:	100 Kt
Weight:	362.5 lb
Dimensions:	unknown
Materials:	probably plutonium as fissile material; possibly D-T boosted[1]
SAFEGUARDS AND ARMING FEATURES:	unknown
DEVELOPMENT:	
Laboratory:	LANL
History:	
1973[2]	Lab assignment (Phase 3)
1978	initial deployment (Phase 5)
Production Period:	1977-present (1983)
DEPLOYMENT:	
Number Deployed:	approximately 2028 (1983); 3600 planned; circa 5696 warheads planned before TRIDENT II procurement decision in October 1981;[3] 12 POSEIDONs backfitted with C4 plus 9 TRIDENTs planned as minimum TRIDENT I missile force[4]
Delivery Systems:	TRIDENT I C4 SLBM on POSEIDON and TRIDENT ballistic missile submarines[5]
Service:	Navy
Allied User:	none
Location:	TRIDENT missile submarines homeported in Charleston, SC; Kings Bay, GA; and Groton, CT
COMMENTS:	development costs for the W76 were $128 million.[6]

---

1 *Military Applications of Nuclear Technology*, Part 1, p. 26.
2 HASC, FY 1980 DOE, p. 137; SASC, FY 1980 DOE, p. 164.
3 712 missiles with 8 warheads each.
4 See also discussion under Chapter Five, TRIDENT I (C4) Missile, for further details on deployment.
5 JCS, FY 1981, p. 43. See also Chapter Five, for information on TRIDENT I (C4) missile.
6 HASC, FY 1979 DOD, Part 9, p. 6697.

W78

Figure 3.26 Three Mk-12A reentry vehicles mounted on bus, with MINUTEMAN III shroud, right.

FUNCTION:	Warhead for the Mk-12A multiple independently targetable reentry vehicle (MIRV) deployed on a portion of the MINUTEMAN III ICBM force.
WARHEAD MODIFICATIONS:	none
SPECIFICATIONS:	
Yield:	335 Kt[1] (MM III with Mk-12A carries 3 W78)
Weight:	less than 800 lb
Dimensions:	
Length:	less than 181.3 cm[2]
Diameter:	54.3 cm (Mk-12A base diameter)
Materials:	plutonium as fissile material; probably lithium-6 deuteride for fusion; no IHE[3]
SAFEGUARDS AND ARMING FEATURES:	unknown
DEVELOPMENT:	
Laboratory:	LANL
History:	
Jul 1974	Lab assignment (Phase 3)[4]
FY 1977	development engineering completed[5]
Sep 1979	first production unit
Jan 1980	initial deployment (Phase 5)[6]
FY 1981	W78 reported to be experiencing production delays[7]
Production Period:	FY 1979-FY 1983[8]
DEPLOYMENT:	
Number Planned:	1083 stockpiled as of January 1983;[9] program completed in 1983 with 300 missiles (900 warheads active)[10]
Delivery System:	MINUTEMAN III ICBMs in underground silos
Service:	Air Force

Nuclear Weapons Databook, Volume I **75**

3
W78

Allied User:	none
Location:	Minot AFB, ND; Grand Forks AFB, ND[11]
COMMENTS:	Increased yield of Mk-12A is to offset "the continued Soviet hardening program."[12] W78/Mk-12A was previously planned as baseline warhead on MX missile, with production scheduled to begin in FY 1986.[13] In January 1982 the W87/Mk-21 (formerly ABRV) was designated the baseline RV for the MX, chosen over the W78.[14] W78 is also option as warhead for TRIDENT II missile. Mk-12A has more fuzing options than the Mk-12.[15]

---

1 AW&ST, 16 June 1980, p. 178; some references give 350 Kt, which is within uncertainty limits of warhead yield.
2 Length of Mk-12A reentry vehicle body.
3 SASC, *Strategic Force Modernization Programs*, p. 103; HAC, FY 1983 DOD, Part 4, p. 597.
4 HASC, FY 1980 DOE, p. 137; SASC, FY 1980 DOE, p. 164.
5 ACDA, FY 1980 ACIS, p. 2; ACDA, FY 1982 ACIS, p. 5; ACDA, FY 1983 ACIS, p. 5.
6 JCS, FY 1982, p. 69.
7 SASC, FY 1981 DOE, p. 37.
8 SASC, FY 1981 DOE, p. 194; SASC, FY 1982 DOD, Part 7, p. 3987.
9 SASC, FY 1982 DOD, Part 7, pp. 3986-3987.
10 SASC, FY 1983 DOD, Part 7, p. 4414.
11 HAC, FY 1982 DOD, Part 2, p. 225; SAC, FY 1981 DOD, Part 5, p. 1556.
12 JCS, FY 1981, p. 41.
13 HASC, FY 1982 DOE, p. 38.
14 ACDA, FY 1983 ACIS, pp. 6-7.
15 HAC, FY 1982 DOD, Part 1, p. 192; SAC, FY 1982 DOD, Part 1, p. 522.

W79

Figure 3.27 **M753** 8-inch nuclear artillery projectile in container, raised for insertion of fuse. **Inset, M110** gun firing M753 projectile during test.

FUNCTION: Enhanced radiation warhead for the M753 improved Artillery Fired Atomic Projectile (AFAP) for 8-inch (203mm) artillery.

WARHEAD MODIFICATIONS: W79-0: dual capable fission/ER version, cancelled in favor of ER version; W79-1: currently deployed enhanced radiation version with insertable ER components[1]

SPECIFICATIONS:
Yield:
W79-0: up to 10 Kt[2] selectable yield
W79-1: probably 1 Kt; three yield options ranging from substantially under 1 Kt to about 2 Kt; the lowest yield option will be 50 percent fission and 50 percent fusion, the highest yield option will be 70 or 75 percent fusion.[3] "More yield options" than W33[4]

Weight: approximately 215 lb

Dimensions:
Length: 43 in[5]
Diameter: 8 in

Materials: plutonium and tritium weapon replacing W33 oralloy weapon[6]

SAFEGUARDS AND ARMING FEATURES: Category D PAL built into the warhead section, command disable feature integrated into the M613 projectile storage container;[7] "easier to handle in the field ... modern safety devices not found in the older generation of 8-inch projectiles;"[8] nonviolent explosive destruct system (NEDS) under development for W79 (1974-1977) before program was terminated; one point safe.[9] Fuzing includes "target sensor," electronic programmer, and timing and memory assembly.[10]

DEVELOPMENT:
Laboratory: LLNL[11]

History:
Dec 1973 program study complete (Phase 2)[12]
Jan 1975 Lab assignment (Phase 3)[13]
Jan 1977 President Ford approves Stockpile Memorandum with W79 as ER weapon[14]
Oct 1978 production activities begin on W79 ER warhead[15]
1980 production engineering completed (Phase 4)[16]
Jul 1981 initial deployment (Phase 5)[17]

Production Period: 1981-present (1983)

DEPLOYMENT:
Number Deployed: approximately 120-300 deployed (1983); 800 planned for production[18]

Delivery Systems: dual capable 8-inch howitzers, including the standard M110 and older M115 in Allied use[19]

3
W79

Service:	Army and Marine Corps
Allied User:	Belgium, Greece, Italy, Netherlands, Turkey, West Germany, United Kingdom (current W33 users)
Location:	Warheads will be stored at Seneca Army Depot, New York, and not deployed outside U.S., pending approval of NATO allies; South Korea (?).

COMMENTS: M-753 projectile includes a rocket assist which doubles the range of the present projectile from 18 km to 29 km.[20] The projectile is ballistically similar to a conventional 8-inch high explosive round, thus eliminating the need for a spotting round as in the W33/M422.[21] The round requires no field assembly and includes improved fuzing[22] with a more accurate height of burst.[23] Training rounds include M173 "Type X," M174 "Type W," and M64 explosive ordnance disposal variants.

---

1 ACDA, FY 1981 ACIS, pp. 274-275; ACDA, FY 1982 ACIS, p. 247; SASC, FY 1983 DOD, Part 7, p. 4397.
2 Kaplan, *Scientific American, op. cit.*, p. 48, reports the following yields: Without ER: 5-10 Kt; With ER: 1-2 Kt. The 5-10 Kt may be the W33 yield, however.
3 Fred Kaplan, "The Neutron Bomb," *The Bulletin of the Atomic Scientists*, October 1981, p. 7. See also, George Kistiakowsky, "The Folly of the Neutron Bomb," *Atlantic*, June 1978, p. 9; SASC, FY 1981 DOD, Part 1, p. 411, refers to a "2-3 Kt RB/ER Warhead."
4 DOD, FY 1981 RDA, p. VII-5.
5 Information supplied to the authors by U.S. Army Armament Research and Development Command, Dover, NJ.
6 JCAE, FY 1977 ERDA Authorization Hearing before Joint Committee on Atomic Energy, February-March 1976, Part 3, pp. 1380-82; *Military Applications of Nuclear Technology*, Part 1, p. 26; ACDA, FY 1979 ACIS, p. 153; ACDA, FY 1981 ACIS, p. 275.
7 ACDA, FY 1981 ACIS, p. 274-275.
8 JCS, FY 1982, p. 78; DOD, FY 1983 RDA, p. VII-12.
9 ACDA, FY 1979 ACIS, p. 130.
10 Information supplied to the authors by U.S. Army Armament Research and Development Command, Dover, NJ.
11 Contractors for the nuclear projectile include Motorola Corp., Scottsdale, AZ; Sandia Corp., Livermore, CA and Albuquerque, NM; Chamberlain Manufacturing Corp., Waterloo, IA; and Ferrulmatics, Inc., Patterson, NJ; USA, *Army Weapon Systems*, 80, n.d., p. 24. The electrical system is provided by Sandia Laboratories, Livermore, CA, and fuze design by Harry Diamond Laboratory, U.S. Army.
12 JCAE, FY 1977 ERDA Authorization Hearing before Joint Committee on Atomic Energy, February-March 1976, Part 3, pp. 1380-82; *Military Applications of Nuclear Technology*, Part 1, p. 26; ACDA, FY 1979 ACIS, p. 153; ACDA, FY 1981 ACIS, p. 275.
13 SASC, FY 1979 DOE, p. 35; production activities on the W79 were halted in October 1977 and began again in October 1978; HASC, FY 1980 DOE, p. 100.
14 SASC, FY 1979 DOE, p. 35.
15 HASC, FY 1980 DOE, p. 100.
16 SAC, FY 1981 EWDA, p. 818.
17 JCS, FY 1982, p. 78; DOD, FY 1983 RDA, p. VII-12.
18 Walter Pincus, *Washington Post*, 9 August 1981, p. A8; the number of projectiles deployed is related to the number of 155mm shells also produced; ACDA, FY 1979 ACIS, p. 131.
19 Information supplied to the authors by U.S. Army Armament Research and Development Command, Dover, NJ.
20 SASC, FY 1982 DOE, Part 7, p. 3881; HASC, FY 1982 DOE, p. 451; DOD, FY 1981 RDA, p. VII-5, refers to 18 km as range of present howitzer; JCS, FY 1982, p. 78, refers to 14 km as present range.
21 ACDA, FY 1981 ACIS, pp. 274-275.
22 DOD, FY 1981 RDA, p. VII-5.
23 ACDA, FY 1980 ACIS, p. 153.

W80

Figure 3.28 Top, **Air-launched Cruise Missiles (AGM-86B)** being mounted on B-52 bomber. Missile in top foreground shows opening for **W80** nuclear warhead, shown at bottom.

Figure 3.29 **TOMAHAWK (BGM-109)** Sea-launched Cruise Missile.

FUNCTION:	Common warhead to be used in the strategic Air Force Air-Launched Cruise Missile (ALCM) (AGM-86B) and the Navy TOMAHAWK Sea-Launched Cruise Missile (SLCM) (BGM-109).
WARHEAD MODIFICATIONS:	Mod 0: Sea-Launched Cruise Missile; Mod 1: Air-Launched Cruise Missile

SPECIFICATIONS:

Yield:	selectable yield,[1] circa 200 Kt;[2] 250 Kt also referenced[3]
Weight:	270 lb;[4] total cruise missile weighs 3000 lb;[5] 3144 lb[6]
Dimensions:	
Length:	20 ft 9 in
Diameter:	27.3 in
Materials:	oralloy as fissile material; supergrade plutonium in Mod 0;[7] probably oralloy in Mod 1; tritium;[8] IHE (PBX-9502) as primary HE[9]

3

W80

SAFEGUARDS AND ARMING FEATURES:	weak link/strong link exclusion region for the warhead electrical system.[10] CAT D PAL;[11] Mod 0 will use low ngs (neutron/gram/second) plutonium to produce low intrinsic radiation for personnel protection for use on submarines;[12] coded switch system; unique signal generator		ALCMs planned before change to Advanced Cruise Missiles in early 1983; 3994 SLCM planned, 1000 of which will be nuclear armed.[20]

DEVELOPMENT:

Laboratory: LANL

History:
- 1976 — Lab assignment (Phase 3) for ALCM[13]
- FY 1978 — W80 warhead test program completed[14]
- 1980 — Lab assignment (Phase 3) for SLCM[15]
- 1980 — production engineering (Phase 4) for ALCM[16]
- Sep 1981 — initial deployment (Phase 5) for ALCM
- FY 1985 — production of ALCM ceases with transition to advanced ALCM

Production Period:
- ALCM: 1979-1985[17]
- SLCM: 1983-

DEPLOYMENT:

Number Deployed: approximately 350 (end 1982); ALCM is being produced at a rate of 40 per month;[18] plans are to purchase 4348 ALCMs and Advanced Cruise Missiles for B-52 and B-1 force;[19] 1499

Delivery Systems:
- ALCM: B-52G/H, FB-111, B-1B
- SLCM: sized to fit 21-inch torpedo tubes and general purpose launchers on surface ships and submarines, vertical launching system (VLS) under development

Service:
- ALCM: Air Force
- SLCM: Navy

Allied User: none

Location:
- ALCM: nine bomber bases (see Chapter Four)

COMMENTS: Warhead is essentially the same for each missile (ALCM and SLCM), major differences being warhead-to-missile mounting features and materials used.[21] Basic warhead design is a modification of the B61 bomb.[22] Ground-Launched Cruise Missile (GLCM) will use a different warhead, the W84. The W80 was originally intended as a replacement warhead for the SRAM.[23] It is under consideration as a warhead for the ASALM.[24]

---

1 ACDA, FY 1980 ACIS, p. 27; SASC, FY 1980 DOD, Part 1, p. 334; also refers to "single yield."
2 *Military Balance, 1980-1981*, p. 3.
3 AW&ST, 22 November 1976, p. 15.
4 Kosta Tsipis, "Cruise Missiles," *Scientific American*, February 1977.
5 CRS, Cruise Missiles, (IB 81080), p. 1.
6 Air Force fact sheets.
7 SASC, FY 1980 DOE, p. 190.
8 HASC, FY 1981 DOD, Part 4, Book 2, p. 1708.
9 HASC, FY 1982 DOE, p. 217; ACDA, FY 1983 ACIS, p. 68.
10 DOE, FY 1982 Revised (Reagan) Budget, Reproduced in HAC, FY 1982 EWDA, Part 5, p. 34.
11 SFRC/HIRC, Joint Committee Print, *Analysis of Arms Control Impact Statements Submitted in Connection with the Fiscal Year 1978 Budget Request*, April 1977, p. 96.
12 SASC, FY 1979 DOE, p. 59; SASC, FY 1980 DOE, p. 190.
13 HASC, FY 1980 DOE, p. 137; SASC, FY 1980 DOE, p. 164.
14 ACDA, FY 1980 ACIS, p. 29.
15 SAC, FY 1981 EWDA, p. 818.
16 Ibid.
17 Deliveries were slated to begin in Fiscal Year 1980; ACDA, FY 1980 ACIS, p. 29.
18 AW&ST, 17 January 1983, p. 101.
19 Former plans were to procure 3418 ALCMs for B-52 force. 3780 operational ALCMs will be used in the 4348 total buy; HAC, FY 1983 DOD, Part 4, p. 587.
20 Michael Getler, *Washington Post*, 19 January 1983, p. A15.
21 HASC, FY 1982 DOE, p. 202.
22 HASC, FY 1982 DOE, p. 107.
23 SFRC/HIRC, Joint Committee Print, *Analysis of Arms Control Impact Statements Submitted in Connection with the Fiscal Year 1978 Budget Request*, April 1977, p. 96.
24 ACDA, FY 1980 ACIS, p. 29.

4

4
Roles

Chapter Four
The Role of Nuclear Weapons in U.S. and Allied Military Forces

In 1980, 722 military units were "certified" for nuclear warfare.[1] The units comprised 100,000 specially trained and cleared personnel, with properly wired and inspected weapons. These units are to play a contingency role in the nuclear strategy of the United States.[2] The purpose of this chapter is to explain the nature and magnitude of the nuclear weapons support structure—delivery units, maintenance, and storage. This will shed light on a number of reasons for the large number of weapons in the nuclear stockpile and the diversity of weapons types.

U.S. military forces, which are deployed worldwide, continue to follow a practice of widespread "nuclearization" of military equipment and units begun in the 1950s. The Single Integrated Operational Plan (SIOP), the central nuclear war plan for strategic forces, broadly determines the requirements for roughly 10,000 strategic warheads. A variety of tactical/theater plans account

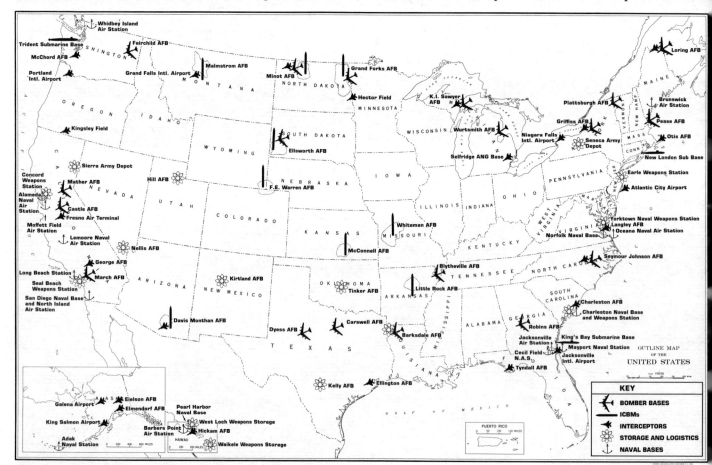

Figure 4.1 Nuclear Weapons Locations in the United States.

---

1 SAC, FY 1981 DOD, Part 3, p. 730; GAO, "Accountability and Control of Warheads in the Custody of the Department of Defense and the Energy Research and Development Administration," PSAD 77-115, 2 June 1977, p. 5, reported 636 nuclear certified units in October 1976.

2 A nuclear certified unit is "a unit or an activity assigned responsibilities for assembling, maintaining, transporting, or storing war reserve nuclear weapons, their assorted components and ancillary equipment"; Defense Nuclear Agency, *Department of Defense Nuclear Weapons Technical Inspection System*, TP 25-1, 1 January 1974, p. 2.

4
Allocations

Table 4.1
Allocation of Nuclear Warheads in the Service Branches (1983)

Warhead	Air Force	Army	Marine Corps	Navy
W25/GENIE	x			
B28 bomb	x			
W31/NIKE-HERCULES/HONEST JOHN		x		
W33/8-inch artillery		x	x	
B43 bomb	x		x	x
W44/ASROC				x
W45/TERRIER				x
W45/MADM		x	x	
W48/155mm artillery		x	x	
W50/PERSHING 1a		x		
B53 bomb	x			
W53/TITAN II	x			
W54/SADM		x	x	x
W55/SUBROC				x
W56/MINUTEMAN II	x			
B57 bomb	x		x	x
B61 bomb	x		x	x
W62/MINUTEMAN III	x			
W68/POSEIDON				x
W69/SRAM	x			
W70/LANCE		x		
W76/TRIDENT I				x
W78/MINUTEMAN III	x			
W79/8-inch artillery		x	x	
W80/ALCM	x			

competitive and aggressive development of new technologies in warheads and delivery systems.

The annual Nuclear Weapons Stockpile Memorandum, approved by the President, determines the number of weapons to be produced and retired. However, the composition of the operational weapons stockpile is primarily influenced by a variety of other plans:

- Annual Unified and Specified Commander-in-Chief[3] requirements as validated by the Joint Chiefs of Staff;
- Annual joint military requirements proposals produced by the Joint Chiefs of Staff, the Joint Strategic Planning Document Nuclear Weapons Annex, Joint Strategic Capabilities Plan, and Joint Planning Assessment Memorandum Nuclear Weapons Annex;
- Annual Secretary of Defense memorandum (Nuclear Weapon Development Guidance) coordinated with the Consolidated Guidance and DOD planning, programming, and budgeting activities; and
- Annual DOD memorandum (Nuclear Weapon Deployment Plan) produced together with the Nuclear Weapons Stockpile Memorandum delineating the allocation of warheads to theater commanders and their storage.

for approximately 11,500 tactical warheads. Roughly 3000 of the tactical warheads are allocated for NATO/European plans, 1000 for U.S. Pacific Command plans, and 2500 for anti-submarine warfare. A few hundred warheads are for strategic defense of the United States, and the remaining 4000 comprise a strategic and tactical reserve. All four services have a wide variety of nuclear weapons (see Table 4.1, Allocation of Nuclear Warheads in the Service Branches).

However, the large number of warheads far exceeds any level of use or destruction which could be contemplated in the plans. This is due to a variety of factors: the duplicative and competitive nuclear weapons missions of the services, the large number of fixed targets designated to be destroyed by strategic forces, the possibility of the use of thousands of battlefield weapons against mobile and non preplanned targets, and the

Military units with nuclear capabilities must pass a certification inspection which determines if they are capable of performing their assigned mission. This inspection is called a Technical Proficiency Inspection in the Army, a Nuclear Weapons Acceptance Inspection (NWAI) in the Navy and Marine Corps, and a Capability Inspection in the Air Force. The certification is not only to ensure knowledge of the unique capabilities of nuclear weapons, but also to indoctrinate the unit as to the safety and control procedures accorded these weapons. The control procedures create enormous additional cost over conventional weapons.

One of the most important ways to prevent the inadvertent or accidental use of nuclear weapons is the Personnel Reliability Program (PRP) (see Table 4.2, Personnel with Nuclear Weapons Duties).[4] The PRP insures the reliability and qualifications of people who have cus-

---

3 The unified commands with nuclear weapons responsibilities include the European Command, Pacific Command, Atlantic Command, Readiness Command, and Central Command (formerly Rapid Deployment Force). The specified commands with nuclear weapons responsibilities are the Strategic Air Command and the North American Aerospace Defense Command.

4 The Personnel Reliability Program is called the Human Reliability Program in the Air Force.

4

Air Force Roles

Table 4.2
Personnel with Nuclear Weapons Duties

	1976[1]	1977[1]	1978	1979	1980
United States					
Military	NA	NA	84,923	84,960	82,920
Civilian[2]	NA	NA	2378	2169	2304
Contractor[3]	NA	NA	29	24	30
U.S. TOTAL	87,415	89,473	87,330	87,153	85,254
Pacific					
Military	NA	NA	5829	6427	4577
Civilian	NA	NA	1	13	2
Contractor	NA	NA	-	-	-
PACIFIC TOTAL	5796	4452	5830	6440	4579
Europe					
Military	NA	NA	23,058	25,558	24,140
Civilian	NA	NA	5	17	25
Contractor	NA	NA	30	30	30
EUROPE TOTAL	22,644	25,063	23,093	25,605	24,195
TOTAL					
Military	NA	NA	113,810	116,945	111,637
Civilian	NA	NA	2384	2199	2331
Contractor	NA	NA	59	54	60
	115,855	118,988	116,253	119,198	114,028

Source: DOD, OSD "Annual Status Report, Nuclear Weapon Personnel Reliability Program," RCS DD-POL(A) 1403, Year Ending 31 December 1980; 31 December 1979; 31 December 1978; RCS DD-COMP(A) 1403, 31 December 1977; 31 December 1976.

1 Breakdown for 1977 and 1976 not available.
2 Federal Civilian Personnel.
3 Contractor Personnel.

tody of, control access to, or have access to nuclear weapons. The investigative and administrative procedures of the PRP also create higher expenses in manning nuclear weapons. For example, it is expensive to train personnel in technical nuclear weapons electronics and maintenance skills.[5] The training costs for each nuclear weapons technician (over the first ten years) is approximately $11,700 for Air Force, $52,300 for Army, and $55,200 for Navy personnel. Training over the second ten year career period costs $22,600 for the Navy and $26,900 for the Army.

Air Force Nuclear Weapons Roles

The Air Force has the dominant position in U.S. strategic and long-range theater nuclear forces, because it controls land-based intercontinental ballistic missiles, bombers, and tactical nuclear fighter bombers. The missile and bomber force comprises the bulk of the strategic capability. Theater bombers constitute the most important long-range regional strike forces.

The kinds of nuclear weapons employed and the missions undertaken by the Air Force nuclear certified units are governed by the regularly revised USAF Program Nuclear Weapons Capabilities and Equipage Document, deriving from JCS, Secretary of Defense, and Presidential guidelines. In October 1976, Air Force capabilities consisted of 74 nuclear certified units; a similar number is estimated to be active today.[6] Generally, the nuclear certified combat unit in the Air Force is a squadron. A squadron consists of 15-24 aircraft (see Table 4.3, Air Force Nuclear Weapons Units), 18 TITAN missiles, or 50 MINUTEMAN missiles. The squadrons are normally subordinate to a wing or group, where the munitions maintenance unit has custody of the nuclear weapons. An exception is in the case of missiles, where the warheads are present in underground silos.

5 SASC, FY 1980 DOD, Part 1, p. 238.
6 GAO, op. cit.

4
Air Force Roles

Figure 4.2 F-16 FALCON, the newest nuclear-capable fighter in the Air Force.

The central maintenance and storage of Air Force nuclear weapons takes place at three bases: Barksdale AFB, Louisiana; Nellis AFB, Nevada; and Kirtland AFB, New Mexico. The warheads are shipped to these bases from the Department of Energy's final assembly plant (PANTEX) at Amarillo, Texas and stored and maintained prior to dispersal to other air bases. Three of five Air Force Air Logistic Centers (ALCs) are also involved in supply and repair of nuclear weapons systems: Ogden ALC, Hill AFB, Utah for missiles; Oklahoma ALC, Tinker AFB, Oklahoma for bombers; and San Antonio ALC, Kelly AFB, Texas for nuclear bombs. Air Force nuclear weapons training takes place at six main bases: Chanute AFB, Illinois; Indian Head, Maryland; Kirtland AFB, New Mexico; Lowry AFB, Colorado; Sheppard AFB, Texas; and Vandenburg AFB, California.[7]

As of 31 December 1982, there were 53,144 Air Force personnel in the PRP involved in nuclear weapons work.[8] Including the above number, in FY 1980, 119,802 military and 16,043 civilian personnel were engaged in strategic weapons work within the Air Force. Many of these personnel were obviously not certified for direct contact work with nuclear weapons (PRP) even though assigned to nuclear weapons units.[9] An additional 3460 military and civilian personnel were engaged directly in theater nuclear forces work.[10]

Air Force strategic offensive forces represent about 90 percent of the total megatonnage delivery capability of U.S. strategic forces.[11] The bomber squadrons of the Strategic Air Command (see Table 4.4, Strategic Bomber Force Basing) each have 14-15 B52 or FB111 aircraft assigned and approximately 150 nuclear weapons. The nuclear weapons include B28, B43, B53, B57, and B61 bombs, SRAM missiles, and, increasingly, ALCMs. The ICBM MINUTEMAN strategic missile squadrons (see Table 4.5, ICBM Deployments) each consist of 50 missile

---

7 AFM 50-5, Volume II.
8 HAC, FY 1983 DOD, Part 3, p. 291; this is a slight reduction from 53,353 as of 31 December 1980.
9 SAC, FY 1980 DOD, Part 3, p. 721.
10 *Ibid.*
11 SAC, FY 1980 DOD, Part 3, p. 722.

4

Army Roles

Table 4.3 Air Force Nuclear Weapons Units	
Squadron Type	**Nuclear Mission / Weapons Type**
Aviation Depot	Receipt and Storage/All weapons
Bomber	Launching Unit/B28, B43, B53, B57, B61, SRAM, ALCM
Fighter Interceptor	Storage and Maintenance, Launching/GENIE
Missile Munitions Maintenance	Storage and Maintenance/ TITAN, MINUTEMAN
Munitions Maintenance	Storage and Maintenance/ Nuclear Bombs, SRAM, ALCM
Munitions Support	Custodial Maintenance and Support/B28, B43, B57, B61
Strategic Missile	Launching Unit/MINUTEMAN II, MINUTEMAN III, TITAN II
Tactical Fighter	Launching Unit/B28, B43, B57, B61

Table 4.4 Strategic Bomber Force Basing (1983)		
Base	**Number[1] / Bomber-Type[2]**	**Nuclear Weapons Type**
Andersen AFB, Guam[3]	14 B-52D[4]	Bombs, SRAM
Barksdale AFB, LA	14 B-52G	Bombs, SRAM, ALCM[5]
Blytheville, AFB, AR	14 B-52G	Bombs, SRAM, ALCM[5]
Carswell AFB, TX[6]	14 B-52D[4]	Bombs, ALCM[5]
Castle AFB, CA	14 B-52G/H	Bombs, SRAM
Dyess AFB, TX[7]	14 B-52H	Bombs, SRAM[8]
Ellsworth AFB, SD	28 B-52H	Bombs, SRAM, ALCM[5]
Fairchild AFB, WA	14 B-52G	Bombs, SRAM, ALCM[5]
Grand Forks AFB, ND	14 B-52H	Bombs, SRAM, ALCM[5],[8]
Griffiss AFB, NY	14 B-52G	Bombs, SRAM, ALCM[5]
K.I. Sawyer AFB, MI	14 B-52H	Bombs, SRAM
Loring AFB, ME	14 B-52G	Bombs, SRAM
March AFB, CA	14 B-52D[4]	Bombs
Mather AFB, CA	14 B-52G	Bombs, SRAM
Minot AFB, ND	14 B-52H	Bombs, SRAM
Pease AFB, NH	30 FB-111	Bombs, SRAM
Plattsburgh AFB, NY	30 FB-111	Bombs, SRAM
Robins AFB, GA[9]	13 B-52G	Bombs, SRAM
Seymour Johnson AFB, NC	14 B-52G	Bombs, SRAM
Wurtsmith AFB, MI	14 B-52G	Bombs, SRAM, ALCM[5]

1 Number is primary active aircraft (PAA) and does not include spares or extras.
2 SASC, FY 1982 DOD, Part 7, p. 4285; DOD, Memorandum for Correspondents, 31 January 1983.
3 Andersen will replace 14 B-52Ds with 14 B-52Gs in late 1983 and add SRAM.
4 Three squadrons of B-52Ds were retired on 1 October 1982. The last two squadrons at Andersen and Carswell will retire in 1983.
5 Bases scheduled for ALCM deployment starting in late 1981.
6 Starting in late 1983, Carswell will receive 20 B-52Hs replacing B-52Ds.
7 Dyess will receive 26 B-1Bs beginning in late 1985.
8 SRAMs moving from Grand Forks to Dyess due to ALCM deployment at Grand Forks.
9 In late 1983, the B-52Gs from Robins will be relocated.

silos, while each full TITAN missile squadron consists of 18 silos. The 1052 active silos (before the beginning of TITAN retirements in October 1982) are located in ten states and spread over approximately 80,000 square miles. Missiles are always prepared to launch within minutes; 30 percent of the bomber force is capable of taking off with nuclear weapons within minutes of any early warning of attack.

Six active and ten Air National Guard fighter interceptor squadrons are also assigned nuclear weapons missions ("strategic defense") with the GENIE air-to-air missile (W25). Four aircraft types (F-106, F-4, F-101, and F-15) are assigned to 28 alert sites (see Table 4.6, Strategic Interception Forces), where at least two armed aircraft are always on 15 minute ground alert. The interceptor force consists of 381 aircraft, 297 operational aircraft, and 84 backup.[12]

In addition to the strategic forces, numerous tactical units of the U.S.-based Tactical Air Command, Pacific Air Force, and United States Air Force Europe (USAFE) are also certified and equipped with nuclear weapons. These tactical fighter wings fly the F-4, F-111, and F-16, and utilize the B28, B43, B57, and B61 bombs. The nuclear equipped units are primarily in Europe. There are thought to be nuclear certified wings in the Pacific area and a large reserve of quickly deployable units in the United States.

Army Nuclear Weapons Roles

Nuclear weapons within the Army represent "a tremendous firepower augmentation of conventiona[l]

---

12 SASC, FY 1981 DOD, Part 2, p. 577.

4
Army Roles

Figure 4.3 Aerial view of **MINUTEMAN** missile launch site.

weapons" with a more intimate integration than in the Air Force. The Army's nuclear systems are predominantly short-range battlefield weapons, unlike the Air Force's long-range pre-targeted weapons. The basic principles of current land warfare doctrine (codified in Field Manual 100-5, *Operations*) regard nuclear weapons as mere additions to normal combat power whether used to directly "destroy enemy forces, to deny an area to enemy movement or to demonstrate national resolve."[13] Although conflict in Europe remains the primary concern of the Army and the political implications of the effects of nuclear warfare are well recognized. The preponderance of short-range Army nuclear weapons and units are only able to fire within friendly territory.

Contingency planning for the tactical use of Army nuclear weapons consists of division and corps plans

Table 4.5
ICBM Deployments (1983)

Base	Missiles
Davis-Monthan AFB, Tucson, AZ	15 TITAN II[1]
Ellsworth AFB, Rapid City, SD	150 MINUTEMAN II
F.E. Warren AFB, Cheyenne, WY	200 MINUTEMAN III[2]
Grand Forks AFB, Grand Forks, ND	150 MINUTEMAN III
Little Rock AFB, Little Rock, AR	17 TITAN II
Malmstrom AFB, Great Falls, MT	150 MINUTEMAN II, 50 MINUTEMAN III[3]
McConnell AFB, Witchita, KS	17 TITAN II
Minot AFB, Minot, ND	150 MINUTEMAN III
Whiteman AFB, Knob Noster, MO	150 MINUTEMAN II

1 Beginning in October 1982, one TITAN missile per month from this base is being retired; as of 1 January 1983, there were 15 deployed.
2 The missile silos at F.E. Warren are spread out in three states: Wyoming, Colorado, and Nebraska.
3 Fifty MINUTEMAN II missiles are going to be replaced with MINUTEMAN III missiles at Malmstrom.

compiled to implement theater (e.g., NATO) tactical objectives. The corps develops plans for the use of "packages" (sets) of nuclear weapons after it has consolidated the "subpackages" from its subordinate division plans. Each package is "a discrete number of nuclear weapons by specific yields and weapon systems for employment in a specified area during a short time period to support a specific division tactical contingency."[14] The package is not a target list, rather it is the number of nuclear weapons deemed necessary for a specific purpose, e.g., halt an attack by a division-size force over hilly terrain, by a tactical commander.

The central storage and maintenance of Army nuclear weapons takes place at two United States depots—Sierra Army Depot in Herlong, California, and Seneca Army Depot in Romulus, New York. These depots receive finished warheads from the Department of Energy assembly plant (PANTEX) at Amarillo, Texas. The warheads are then transferred to field depots and storage sites for subsequent use. "Special Ammunition" ordnance units—"general support" and "direct support"

---

13 Basic sources on Army nuclear weapons doctrine and policy include: U.S. Army, *Operations*, FM 100-5 (20 August 1982); U.S. Army, *Staff Officers Field Manual, Nuclear Weapons Employment Doctrine and Procedures*, FM 101-31-1 (March 1977); U.S. Army, *Tactical Nuclear Operations*, FM 100-30 (Test) (August 1971); U.S. Army, *Operations for Nuclear Capable Units*, FM 100-50 (March 1980); CGSC, "Nuclear and Chemical Operations," *Infantry and Airborne Division and Brigade Operations* (Draft FM) (July 1978), p. 18-1; John P. Rose, *The Evolution of U.S. Army Nuclear Doctrine, 1945-1980* (Boulder, CO: Westview Press, 1981).

14 U.S. Army CGSC, *op. cit.*, p. 18-9.

4
Army Roles

Table 4.6
Strategic Interception Forces (1983)

Main Bases[1]	Aircraft
*Atlantic City AP, Pleasantville, NJ	F-106
Castle AFB, Merced, CA	F-106
*Ellington AFB, Houston, TX	F-101[2]
Elmendorf AFB, Anchorage, AK	F-4
*Fresno Air Terminal, Fresno, CA	F-106**
*Great Falls IAP, Great Falls, MT	F-106
Griffiss AFB, Rome, NY	F-106
*Hector Field, Fargo, ND	F-4
*Jacksonville IAP, Callahan, FL	F-106**
K.I. Sawyer AFB, Gwinn, MI	F-106***
Langley AFB, Hampton, VA	F-15[3]
McChord AFB, Tacoma, WA	F-106***
Minot AFB, Minot, SD	F-106***
*Niagara Falls IAP, Niagara Falls, NY	F-4
*Otis AFB, Falmouth, MA	F-106
*Portland IAP, Portland, OR	F-4
*Selfridge ANGB, Mt. Clemens, MI	F-4

* Air National Guard units and bases.
** Units scheduled to receive F-4D replacements in late 1983.
*** Units scheduled to receive F-15 replacements starting in late 1984.
1 This list does not include 11 alert satellite sites where aircraft are also on full time alert.
2 The F-101 is completing phase out and is being replaced by the F-4.
3 Langley is the first of six bases/fighter interceptor squadrons to receive the F-15 to replace the F-106 starting in 1982; HAC, FY 1983 DOD, Part 5, p. 548.

Figure 4.4 Two GENIE (AIR-2A) rockets mounted under Air Force F-101.

(assigned to or in support of a unit)—maintain custody of all warheads until they are transferred to the using delivery units. Nuclear weapons supply and maintenance support—from the Theater Army Area Command, Corps Support Command, or Nuclear Weapons Support Command, to the nuclear capable unit—includes a supply of the basic components, periodic exchange of limited life components (e.g., those containing tritium), and any maintenance which the receiving unit cannot or is not authorized to perform. Army Nuclear weapons training takes place at a number of bases, including Indian Head, Maryland; Kirtland AFB, New Mexico; Fort Sill, Oklahoma; Aberdeen Proving Grounds, Maryland; Fort Bliss, Texas; Fort Belvoir, Virginia; and Redstone Arsenal, Alabama.

In October 1976, 275 units in the Army were nuclear certified,[15] including air defense, artillery, atomic demolitions, and ordnance. Table 4.7 lists the types of units with nuclear missions in the Army. As of 31 December 1982, there were 16,733 Army personnel in the PRP involved in nuclear weapons work.[16] The total number of Army personnel involved in nuclear weapons work is not known. However, with far over 200 certified units and 110 nuclear weapons storage sites worldwide,[17] the number is probably much larger than the number of people involved in the PRP. There are 4780 Army personnel involved in nuclear weapons security, 566 in the United States and 4214 overseas.[18]

The longest-range nuclear weapon system in the Army (750 km) is the PERSHING 1a (W50) missile deployed in West Germany (108 U.S. launchers). These are also the only Army weapons on stand-by alert at all times with nuclear warheads aimed at pre-determined Warsaw Pact targets. The shorter-range LANCE (W70) missile is also deployed with both U.S. corps in West Germany. Six battalions, with six launchers each, provide the general support to U.S. ground forces in Central Europe. Both the PERSHING and LANCE missiles have reloads with nuclear warheads available. Army nuclear capable artillery consists of a variety of 155mm and 8-inch guns (see Chapter Nine) and is widely deployed in Infantry, Armored, or Mechanized divisions and Armored Cavalry regiments (see Table 4.8, Allocation of Nuclear Weapons in Army Units). The use of non-nuclear 105mm and 175mm guns has been virtually eliminated in the past ten years. Almost all Army artillery is now nuclear capable. An Army division has both gun sizes assigned to it, with similar ranges and nuclear capabilites.

---

15 GAO, *op. cit.*; the number today is thought to be similar.
16 HAC, FY 1983 DOD, Part 3, p. 291; this is a large reduction from 24,420 personnel as of 31 December 1980.
17 SAC/SASC, FY 1983 Joint Mil Con, pp. 314-315.
18 SAC, FY 1979 DOD, Part 3, pp. 72-75; SAC, FY 1979 DOD, Part 2, pp. 146-148.

4
Marine Corps Roles

Figure 4.5 PERSHING 1a (MGM-31A/B) platoon in launch position.

"Defensive" nuclear systems used by the Army consist of the NIKE-HERCULES surface-to-air missile system certified for nuclear warheads (W31) and Atomic Demolition Munitions (nuclear land mines) (W54 and W45). A large number of both weapons are deployed, particularly in West Germany.

Marine Corps Nuclear Weapons Roles

The Marine Corps nuclear weapons roles are similar to those of the Army, but because of the high mobility requirements of the Marines, the heavier weapons (LANCE and PERSHING) are not assigned to them.[19] However, the Marines have their own air force which provides 'organic' (internal) nuclear weapons support.

In peacetime, the Marine Corps does not have custody of its own nuclear weapons but would receive them from the Navy during crisis or hostilities.[20] A structure exists for the receipt, supply, and maintenance of these warheads. For air delivered weapons, the Marine Air Wing has a Marine Wing Weapons Unit responsible for nuclear weapons. For Marine ground forces, the Nuclear Ordnance Platoon of the Marine Divisions prepares for the receipt, storage, and assembly of nuclear artillery and atomic demolition munitions. Within the operational structure (combat units), the Marine Amphibious Unit (MAU)—a composite ground and air force combat team—has a weapons shop which is also responsible for nuclear weapons supply and maintenance.

Operationally configured Marine units are carried aboard Navy amphibious ships. The following ships are also certified to carry nuclear weapons for the Marines:

- Amphibious Assault Ships (LPH),
- Amphibious Transport Docks (LPD),
- Dock Landing Ships (LSD), and
- Tank Landing Ships (LST).

---

19 The basic source of Marine Corps nuclear doctrine is *Staff Officers' Field Manual, Nuclear Weapons Employment Doctrine and Procedures*, FM 11-4, March 1977.

20 Information provided by the Department of Defense.

4

Marine Corps Roles

Figure 4.6 M109 155mm howitzer.

Marine CH-46 and CH-53 helicopters are also authorized to transfer nuclear weapons between ships and land.

The Marine Corps presently flies two nuclear capable aircraft: the A-4 SKYHAWK and the A-6 INTRUDER.[21]

They are certified to carry the B43, B57, and B61 bombs, flying from Navy aircraft carriers or land bases. The A-4, a light attack aircraft, will be replaced with the AV-8B HARRIER II, a vertical/short takeoff and landing (V/STOL) aircraft, which, unlike the AV-8A first generation V/STOL, will be nuclear capable. The A-6, the primary long-range bomber, will remain in the inventory through the 1980s. The new F/A-18 under development will more than double the Marine Corps nuclear capable inventory when it enters service during FY 1982-1983. It will replace the current Marine Corps F-4 force which is not nuclear certified.[22]

The ground forces are equipped with dual capable 155mm and 8-inch artillery also used in the Army. Marine Corps policy is to certify all nuclear capable artillery units to fire nuclear weapons.[23] The 155mm howitzers—older towed M-114s, self-propelled M-109s, and newer towed M-198s—fire the W48 and will be compatible with the W82, which is under development and already adopted by the Marines. The only 8-inch gun in use by the Marines, the self-propelled M-110, fires both the W33 and the W79 nuclear projectiles.

In 1975, a reorganization of artillery in the Marine Corps resulted in the shift from 105mm (non nuclear)

Table 4.7
Army Nuclear Weapons Units[1]

U.S. Unit Type		Nuclear Mission
Air Defense Artillery:	Battery	NIKE-HERCULES basic firing unit with nine launchers
	Detachment	NIKE-HERCULES custodial unit supporting allied battalion/squadron[2]
	Team	NIKE-HERCULES custodial unit supporting allied firing battery[2]
Atomic Demolitions:[3]	Company	Corps level general support ADM unit
	Platoon	Division/Regiment direct support ADM unit
	Team	ADM basic firing unit providing direct support to maneuver units and allied forces
Field Artillery:	Battalion	Artillery, LANCE, or PERSHING unit
	Battery	Artillery, LANCE, or PERSHING basic firing unit[4]
	Detachment	Artillery, HONEST JOHN, LANCE, or PERSHING custodial unit supporting allied battalions[2]
Ordnance:	Company	Direct and General unit support providing maintenance and/or storage of nuclear weapons[2]

1 Other units with nuclear support missions include security (military police), maintenance (ordnance), storage (ordnance), transportation (organic and external [assigned or in support of]), explosive ordnance disposal, command and control, and planning.
2 Custodial units are subordinate to "U.S. Army Artillery Groups" which provide command and control of custodial units and weapons.
3 Atomic demolitions units are officially designated "engineer (atomic demolition munitions)."
4 Artillery batteries have 6-12 guns; LANCE battery has six launchers; PERSHING battery has nine launchers.

---

21 JCS, FY 1982, p. 79.
22 JCS, FY 1981, p. 49.
23 Maj. Roger A. Jacobs, 'Artillery's Nuclear Mission,' *Marine Corps Gazette*, April 1982, p. 24

4
Navy Roles

Table 4.8
Allocation of Nuclear Weapons in Army Units

Headquarters Unit	Nuclear Unit/Weapon
Theater Army	Field Artillery Brigade/PERSHING 1a
	Army Air Defense Command/NIKE-HERCULES
Corps	Corps Artillery/LANCE, 8-inch artillery
	Engineer Brigade/ADMs
Division	Division Artillery/155mm and 8-inch Artillery
	Engineer Battalion/ADMs
Armored Cavalry Regiment	Howitzer Battery/155mm Artillery
	Engineer Company/ADMs
Special Forces Group	Engineer Team/ADM

Figure 4.7 The Marine Corps' newest fighter, the **AV-8B HARRIER**, which is nuclear capable.

artillery in the division organization to 155mm artillery. With the beginning of deployment in 1981 of the new nuclear-certified M-198 155mm gun, the number and range of Marine dual-capable artillery will increase by more than 300 and 60 percent, respectively.[24] The new 155mm gun will replace the existing non-nuclear 105mm howitzers as the direct support weapon in two of the three Marine Divisions[25] and will replace all other older towed 155mm guns in Marine Corps artillery. Marine engineers and commandos are also equipped with the Medium and Special Atomic Demolition Munitions (MADM and SADM).[26]

Navy Nuclear Weapons Roles

The Navy has the greatest diversity of nuclear weapons responsibilities, including: strategic warfare (missile firing submarines), tactical/theater land attack warfare, defensive anti-air warfare, and anti-submarine warfare. Its nuclear capability is assigned to surface ships, submarines, and ship and land based aircraft (see Table 4.9, Navy Nuclear Weapons Units). The strategic weapons of the Navy are on average the newest weapons in the strategic nuclear arsenal. The tactical nuclear weapons, on the other hand, are some of the oldest. The Navy has not introduced a new theater nuclear weapon for 16 years.[27]

Most of the nuclear certified units in the Navy are ships. The number of certified units is counted by the quarterly average. In October 1976, there were 287 certified units in the Navy.[28] That number has probably remained stable, but will rise in the next few years as new nuclear weapons are introduced into the Navy. As of 31 December 1982, there were 34,871 Navy (and

---

24 JCS, FY 1982, p. 97; JCS, FY 1981, p. 49.
25 SASC, FY 1980 DOD, Part 2, p. 490.
26 U.S. Navy, *Loading and Underway Replenishment of Nuclear Weapons*, NWP 14-1, Rev. A (November 1979) p. 2-25.

27 JCS, FY 1982, p. 32.
28 GAO, *op. cit.*

4

Navy Roles

Table 4.9
Navy Nuclear Weapons Units

Flying Squadron / Ship Type / Support Units[1]	Nuclear Weapons Function / Weapon Types
AIRCRAFT[2]	
Air Anti-Submarine (VS)	S-3 ASW Unit / B57
Attack Squadron (VA)	A-6 and A-7 Attack Unit / B43, B57, B61
Fighter Squadron (VF)	F-4 and F/A-18 Fighter Unit / B43, B57, B61
Helicopter Anti-Submarine (HS)	SH-3 ASW Unit / B57
Patrol Squadron (VP)	Land Based P-3 ASW Unit / B57
COMBAT SHIPS	
Aircraft Carrier (CV)	Storage, Supply, and Maintenance of air-launched weapons / B43, B57, B61
Attack Submarine (SSN)	Launching Unit / SUBROC
Cruiser (CG/CGN)	Launching Unit / ASROC, TERRIER
Destroyer (DD/DDG)	Launching Unit / ASROC, TERRIER
Frigate (FF/FFG)	Launching Unit / ASROC
Strategic Submarine (SSBN)	Launching Unit / POSEIDON C3, TRIDENT I C4
SUPPORT SHIPS[3]	
Ammunition Ship (AE)	Underway Replenishment / ASROC, TERRIER, Marine Corps Warheads
Destroyer Tender (AD)	Supply, Maintenance, and Storage for surface ships / ASROC, TERRIER
Fast Combat Support Ship (AOE)	Underway Replenishment and Depot / ASROC, TERRIER, Marine Corps Warheads
Replenishment Oiler (AOR)[4]	Limited Underway Replenishment / ASROC, TERRIER
Submarine Tender (AS)	Supply, Maintenance, and Storage for submarines / POSEIDON C3, TRIDENT I C4, SUBROC
SHORE SUPPORT UNIT	
Naval Weapons Station	Central Supply and Maintenance, Storage, Assembly / All Warheads
Naval Magazine	Fleet Supply and Maintenance, Storage, Assembly / All Warheads
Naval Air Facility	Airfield Storage and Maintenance / B43, B57, B61
Naval Aviation Weapons Facility	Custodial Storage and Maintenance / B57

1 Nuclear capable ships are identified in U.S. Navy, *Loading and Underway Replenishment of Nuclear Weapons*, NWP 14-1, Rev. A (November 1979).
2 Helicopter Combat Support Squadrons (HC) and Helicopter Mine Countermeasures Squadrons (HM) with H-46 and H-53 helicopters are also certified to transfer nuclear weapons between ships, and between ships and shore.
3 A number of amphibious ships (for Marine Corps support) also carry nuclear weapons (see Marine Corps above).
4 One Oiler (AO-51, USS *Ashtabula*) is also certified for nuclear weapons transport; *Ibid*.

Marine Corps) personnel in the PRP involved in nuclear weapons work.[29] Inclusive of that number, in FY 1980, 20,580 military and 2512 civilian personnel were engaged in strategic weapons work within the Navy.[30] An additional 971 military and 186 civilian Naval personnel were responsible solely for theater nuclear work.[31] There were also 6269 Navy and Marine Corps personnel involved in nuclear weapons security (in 1979).[32] Of this number, 3767 were assigned to bases in the continental United States, 466 to bases in Europe, and 2036 to bases in the Pacific.

U.S. Navy ships are assigned a nuclear weapons capability by the Chief of Naval Operations in accordance with contingencies plans and validated Navy requirements for defense of ships, battle groups, anti-submarine warfare, and land attack theater missions in support of ground forces and naval objectives. The capabilities of ships assigned nuclear weapons missions are detailed in a regularly revised classified report known as the Naval Atomic Planning and Support Capabilities (NAPSAC) Report. Combat, logistical, and amphibious ships are assigned tasks relating to nuclear weapons while they are at sea, and receive support from the shore establishment of training and operational bases, depots, and supply centers (Table 4.10, Naval Bases for Nuclear Armed Ships and Submarines, describes the operational shore establishment).

---

29 HAC, FY 1983 DOD, Part 3, p. 291; this is an increase from 33,683 as of 31 December 1980.
30 SAC, FY 1980 DOD, Part 3, p. 550.
31 SASC, FY 1980 DOD, Part 6, p. 1581.
32 SAC, FY 1979 DOD, Part 3, pp. 72-75; SAC, FY 1979 DOD, Part 2, pp. 146-148.

4
Navy Roles

Figure 4.8 U.S.S. *Virginia* (CGN-38) cruiser.

Figure 4.9 U.S.S. *Ohio* (SSBN-726), the Navy's first TRIDENT submarine.

Nuclear weapons supply for the Navy is centered at five major depots in the United States: Naval Weapons Station, Concord, California; Naval Weapons Station, Seal Beach, California; Naval Base, Pearl Harbor, Hawaii; Naval Weapons Station, Charleston, South Carolina; and Naval Weapons Station, Yorktown, Virginia. These depots receive warheads from the final assembly plant (PANTEX) of the Department of Energy at Amarillo, Texas and subsequently transfer them to sub-depots, ships, airfields, and custodial units. Two other Naval support bases—Naval Ordnance Station, Indian Head, Maryland and Naval Detachment, Army Ammunition Plant, McAlester, Oklahoma—provide technical support for naval nuclear weapons. Nuclear weapons training is centered at the Atlantic Fleet Nuclear Weapons Training Group in Norfolk, Virginia and the Pacific Fleet Group in North Island, California.

The strategic force consists of 32 fleet ballistic missile submarines, 31 POSEIDON submarines (19 with 16 POSEIDON C3 missiles and 12 with 16 TRIDENT C4), and one new TRIDENT submarine (with 24 TRIDENT C4). More than half of the submarine force is on patrol at any one time, with the rest at six major bases. A large

Nuclear Weapons Databook, Volume I **93**

4
Allied Roles

Table 4.10
Naval Bases for Nuclear Armed Ships and Submarines (1983)

Base	Supported Unit
Alameda, CA	Aircraft carriers
Apra Harbor, Guam	Strategic and attack submarines, surface ships
Bangor, WA	Strategic and attack submarines
Charleston, SC	Strategic submarines, surface ships
Gaeta/Naples, Italy	Surface ships
Groton, CT	Strategic and attack submarines
Holy Loch, Scotland, UK	Strategic submarines
Kings Bay, GA	Strategic submarines
La Maddalena, Italy	Attack submarines
Long Beach, CA	Surface ships
Mayport, FL	Aircraft carriers, surface ships
Newport, RI	Surface ships
Norfolk, VA	Surface ships, attack submarines
Pearl Harbor, HI	Strategic and attack submarines, surface ships
San Diego, CA	Attack submarines, surface ships
Subic Bay, Philippines	Surface ships

percentage of the Navy aircraft carriers, cruisers, destroyers, frigates, and attack submarines are equipped with nuclear capable weapons systems and are supplied with nuclear warheads during operations (see Chapter Eight). The surface ships either have launchers for the nuclear TERRIER (W45) surface-to-air system or ASROC (W44) anti-submarine rocket; attack submarines have the capability of firing the SUBROC (W55) anti-submarine rocket. However, the preponderance of the tactical nuclear weapons capability is in Naval aviation, both land and aircraft carrier based. Six aircraft types— A-4, A-7, F-4, F/A-18, S-3, and P-3—and one (SH-3) helicopter are certified to carry the B43, B57, and B61 nuclear bombs for land attack and anti-submarine warfare.

Allied Nuclear Weapons Roles

Nuclear warheads for NATO countries are provided by the U.S. under Programs of Cooperation (POC)— bilateral agreements between the U.S. and NATO countries involving the transfer and certification of nuclear capable delivery vehicles or the deployment of nuclear warheads on foreign soil for support of foreign forces.[33] The U.S. unit which maintains control of nuclear weapons for use by allied units is called a custodial unit. All three services have custodial units. The United States maintains POCs with nine nations: Belgium, Canada, Greece, Italy, The Netherlands, South Korea, Turkey, The United Kingdom, and West Germany (see Table 4.11, Allied Nuclear Capabilites) and has nuclear weapons deployed in each of those countries.

There are over 600 allied dual capable tactical fighters and medium bombers available for nuclear duties.[34] Allied aircraft certified for nuclear weapons duty include the F-4 PHANTOM in Greek and Turkish units, the F-100 in Turkey, and the F-104 STARFIGHTER in Belgian, Dutch, Greek, Italian, and West German units. The new F-16 and the multi-national TORNADO now being introduced are planned for nuclear certification in Belgian, Dutch,[35] Italian, and West German units. Other countries are currently seeking suitable replacements capable of nuclear certification for some of their older aircraft. They are also considering the F-18 in addition to the planes already mentioned above.

Nuclear bombs known to be in allied use include the B43, B57, and B61. The B23RE is also thought to be in limited use. Some allied NEPTUNE and P-3 ORION maritime patrol and anti-submarine aircraft are also nuclear certified. British and Dutch forces have both B57 depth bombs stored in the United Kingdom. Canadian CF-101 interceptor aircraft, part of the North American Aerospace Defense Command (NORAD), are also equipped with nuclear armed GENIE (W25) air-to-air rockets at their bases in Canada.

Three Army missile systems (PERSHING, LANCE, and HONEST JOHN) are currently nuclear armed in allied military formations. The PERSHING 1a missile system supplied solely with nuclear warheads (W50), with two "wings" each equipped with 36 launchers and missiles, is utilized by the West German Air Force for medium-range nuclear support. Five allied armies utilize the nuclear armed dual capable short-range LANCE missile system (W70), while Greece and Turkey are still armed with the older and obsolete HONEST JOHN rocket (W31). These battlefield missiles are deployed at

---

33 SASC, FY 1980 DOD, Part 6, p. 3426.
34 JCS, FY 1982, p. 76, and previous JCS reports stated 400 allied dual capable aircraft; JCS, FY 1984, p. 19, reported 600.

35 The final decision by the Dutch government as to whether the new F-16s will be nuclear certified has not been made.

4
Allied Roles

Table 4.11
Allied Nuclear Capabilities (1983)

	Belgium	Canada	Greece	Italy	Netherlands	South Korea	Turkey	United Kingdom	West Germany
W25		x							
B28[1]				unknown					
W31(HJ)			x			x	x		
W31(NH)	x		x	x	x	x			x
W33	x		x	x	x	x	x	x	x
B43[1]				unknown					
W45(MADM)					x			x	x
W48	x		x	x	x	x	x	x	x
W50									x
B57(Bomb)[1]				unknown					
B57(ASW)					x			x	
B61[1]	x		x	x	x		x		x
W70	x			x	x			x	x

1 B28RE, B43, B57, and B61 nuclear bombs are deployed to Europe; AFM 50-5, Volume II, p. 3-87.

the Corps level for general nuclear support of military operations and comprise the most capable and longest-range nuclear delivery means of the NATO ground forces. Seventy LANCE and 26 HONEST JOHN launchers are estimated to be deployed in non-U.S. NATO military formations, supported by approximately 400 warheads. The South Korean Army also uses the HONEST JOHN rocket and it is possible that the United States maintains warheads for those rockets in South Korea.

Five countries currently have nuclear warheads (W31) for their NIKE-HERCULES surface-to-air air defense missile launchers. It is estimated that more than 700 warheads are available for nuclear air defense (including U.S. Army systems in West Germany).[36] Although the nuclear NIKE-HERCULES air defense missile system is obsolescent and is currently being replaced by the conventional Improved HAWK missile (or being reduced in the nuclear role as plans are laid for introduction of the future conventional PATRIOT system), numerous allied batteries remain nuclear armed with nuclear warheads on a high level of alert. The NIKE-HERCULES also has a surface-to-surface capability.

Seven allied armies are supplied with nuclear warheads for their artillery. The warhead supply includes both the current 155mm (W48) and 8-inch (W33) versions. A large number and wide variety of guns are certified for nuclear missions (see Chapter Nine). Nuclear artillery is typically deployed at Corps level, although there are a number of certified units and guns at the division and even brigade level. An estimated 1700 155mm guns and 400 8-inch guns are available for nuclear missions in Europe.[37] The two most common guns are the standard American designs, the self-propelled M-109 (155mm) and M-110 (8-inch). Seven additional 155mm gun types in the armies of allied countries are also certified for nuclear weapons use: the older

---

36 Although the NIKE-HERCULES is assigned to the Army in the U.S. military, the allied NIKE units all fall under the Air Force.

37 This number includes U.S. Army guns deployed in Europe.

4
Allied Roles

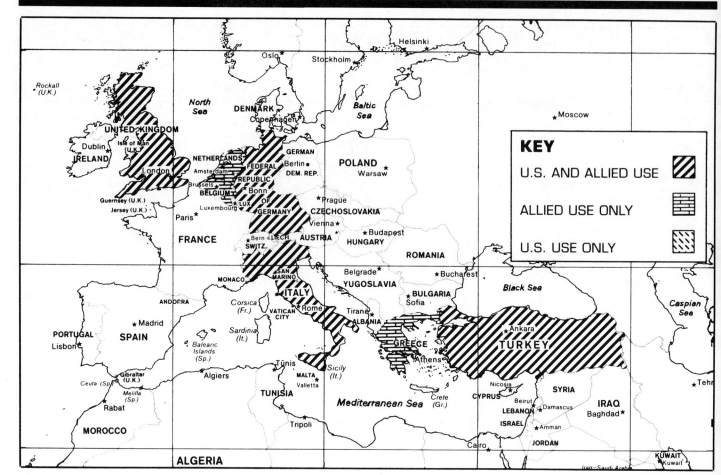

Figure 4.10 Deployment of U.S./NATO nuclear weapons in Europe.

American designed M-44, M-53, and M-114 guns, the newer towed M-198, and two European designed guns, the FH-70 (towed) and the SP-70 (self-propelled). Two older American designed 8-inch gun types are also utilized with nuclear weapons: the M-55 and the M-115. With the current nuclear warheads, the nuclear artillery of both calibers average 17-18 kilometers in range.

Atomic demolitions are also allocated to allied forces. Little is known of the nature of the agreements for the supply of atomic demolition munitions, but it is known that at least West Germany, the Netherlands, and Britain have special engineer units trained and certified for the use of the Medium Atomic Demolition Munition (W54). Since the ADM is a nuclear weapon without a delivery system, per se, the procedures for the sharing of ADM tasks remains unclear.

4
Allied Roles

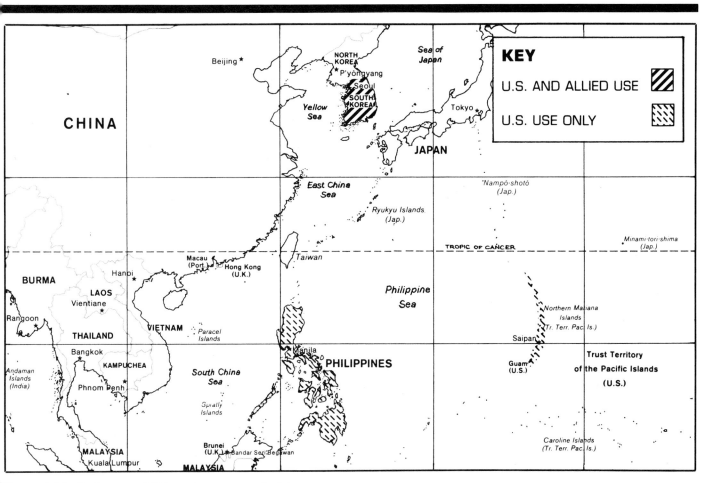

Figure 4.11 Deployment of U.S. nuclear weapons in Asia.

5
Strategic Forces

Chapter Five
Strategic Forces

U.S. strategic nuclear weapons are delivered by three principal means: land-based intercontinental ballistic missiles (ICBMs), submarine-launched ballistic missiles (SLBMs), and bombers. These three strategic systems are referred to by the Department of Defense as the "Triad."[1] A fourth element—Sea-Launched Cruise Missiles (SLCM)—will be added to strategic forces in 1984 (see Chapter Six) as part of a so-called "Strategic Reserve Force." One "defensive" strategic weapon—the GENIE air-to-air missile (W25)—is also deployed for bomber interception, and development of a nuclear armed anti-ballistic missile system continues. U.S. Strategic Forces in 1983 comprise about 10,000 operational nuclear warheads. Approximately $30 billion per year[2] is spent on these forces and some 140,000 military personnel are involved.[3]

The strategic warheads are either in ballistic missile reentry vehicles, missiles, or bombs. Once outside the earth's atmosphere, ballistic missiles (e.g., ICBMs and SLBMs) continue on a trajectory governed by gravity until reentry through the earth's atmosphere. The term ballistic derives from the reentry vehicle's free-fall trajectory after the rocket's boost phase and separation. There are currently five types of reentry vehicles deployed on ballistic missiles, with warheads ranging in yield from 40 Kt to 9 Mt. The ballistic missiles currently deployed all use reentry vehicles which are themselves ballistic, but future RVs may be maneuverable (and consequently non-ballistic).

Strategic ballistic missiles are multi-staged rockets with intercontinental ranges of well over 5000 miles. The demonstrated accuracy of land-based ICBMs is close to one-tenth of a nautical mile (200 m) and improving. The shape and construction of the RV is chosen to minimize drag upon atmosphere reentry, thus maintaining accuracy under varying weather conditions and rendering the high-speed RV difficult to defend against. The transit time of the missiles over intercontinental range is about 30 minutes.[4]

Each of the elements of the Triad has relative advantages and disadvantages in terms of reliability, accuracy, safety, and responsiveness (see Table 5.1). The land-based ICBM, however, has three characteristics, other than its high accuracy, that make it superior to the other strategic delivery systems: a large percentage of missiles prepared for immediate launch, a high probability of missile survival under nuclear attack, and a more secure communications link with command authorities.

The sea-launched systems (SLBMs and SLCMs) have the advantages of reduced vulnerability as long as the launching submarine is travelling quietly and invisibly under the ocean surface. SLBMs and SLCMs are also very flexible in deployment and movement. The submarine has a more tenuous communication link with the national command authorities, particularly under wartime conditions. The SLBM is also not as accurate as the ICBM because of the uncertainty of the submarine's location, orientation, and velocity, although with the deployment of the TRIDENT II missile starting in 1989 submarine missiles will begin to approach land-based missile accuracy.

Bombers, which carry nuclear warheads in gravity bombs, cruise missiles, or short-range air-to-surface missiles (SRAM), can be recalled after launch and en-route retargeting is possible. They are thus the most responsive to changing political or tactical circumstances. The short-range attack missile (SRAM) and cruise missiles launched from the aircraft also provide bombers with a standoff delivery capability and thus passive survivability. The bomber force is also highly accurate in targeting.

Land-Based Missiles

The land-based missile force currently consists of 1000 MINUTEMAN missiles (450 MINUTEMAN II and 550 MINUTEMAN III) armed with 2100 warheads, and 49 TITAN II missiles[5] armed with 49 warheads. Four

---

1 Triad is the Latin term for a union or group of three.
2 This is calculated based on the FY 1984 budget request, including Strategic Forces, Research and Development, and Department of Energy warhead expenses.
3 John M. Collins, *U.S.-Soviet Military Balance: Concepts and Capabilities, 1960-1980* (NY: McGraw-Hill, 1980), p. 249.
4 The launch to target time for a missile with a range of 6500 nm is about 36 minutes for a "minimum energy trajectory," neglecting effects of the earth's atmosphere. The launch speed is 14,500 nm/hr, the maximum height is 1300 km (701.5 nm), the speed at the top of the trajectory is 11,400 nm/hr and the angle between the trajectory and ground at impact is 18 degrees. See Abram Chayes and Jerome B. Weisner, eds., *ABM* (New York: New American Library, 1969), p. 278.
5 As of 1 January 1983; TITAN II missiles are being retired at a rate of about one per month; JCS, FY 1984, p. 13.

5
Strategic Forces

Table 5.1
Features of "Legs" of the Strategic TRIAD

ICBMs
- full target coverage
- high degree of accuracy (depending on model)
- assured ballistic penetration
- rapid retargeting capability
- constant survivable command and control
- highest degree of reliability (98%)[1]
- highest degree of alert (90%+)
- hardened silos
- post attack survivability
- quickest reaction time
- low operating cost ($330 m/yr)[2]

SLBMs
- highest degree of survivability (60% of forces at sea)
- assured ballistic penetration
- tenuous communications link
- high degree of reliability
- ability to withhold from initial attack
- invulnerable to detection or attack[3]

Bombers
- survivability of forces on alert (30%)[4]
- recallable after takeoff
- flexible targeting to include mobile targets, targets of opportunity, and multiple targets separated by long distances
- highest degree of accuracy
- vulnerable to air defenses
- ability to withhold from initial attack

1 SASC, FY 1981 DOD, Part 2, p. 549.
2 SASC, FY 1982 DOD, Part 7, p. 4002.
3 SASC, FY 1982 DOD, Part 3, p. 163.
4 SASC, FY 1982 DOD, Part 7, p. 3799.

Table 5.2
Strategic Nuclear Forces (1971-1981)[1]

	1971	1972	1973	1974	1975	1976	1977	1978	1979	1980	1981
B-52	412	402	422	422	420	419	417	344	343	343	344
FB-111	66	60	71	72	69	68	66	66	65	63	62
MINUTEMAN	990	955	970	999	1010	1094	1162	1180	1170	1167	1164
TITAN II	58	57	57	57	57	58	57	57	57	56	56
SRAM	—	227	651	1149	1451	1431	1415	1408	1396	1383	1374
ALCM	—	—	—	—	—	—	—	—	—	—	14
HOUND DOG	340	338	329	327	308	288	249	—	—	—	—

1 SAC, "The Development of Strategic Air Command: 1946-1981," 1 July 1982.

5

Land-Based Missiles

RV/warhead types are deployed on land-based missiles: the Mk-6/W53 on TITAN with 9 Mt yield, the Mk-11C/W56 on MINUTEMAN II with 1.2 Mt yield, the Mk-12/W62 on 250 MINUTEMAN IIIs with 170 Kt yield, and the Mk-12A/W78 on 300 MINUTEMAN IIIs with 335 Kt yield. The MINUTEMAN III missiles have multiple independently-targetable warheads (MIRVs). The remainder of the land-based missile force has single warheads.

The land-based missile force is deployed in hardened underground launching silos, primarily in the western United States (see Chapter Four). The first ICBM, the liquid fuel ATLAS, was deployed in above ground launching sites starting in 1959 and had an accuracy of about one mile. Eight years after the first ATLAS ICBM deployment numerous upgrades of the missile had been deployed, and the new heavy TITAN I and solid fuel MINUTEMAN I were added to the land-based force. Almost as quickly as the new missiles were deployed, however, they were phased out. By 1970, the missile force was stabilized at 1054 ICBMs, already with second generation TITAN and MINUTEMAN missiles. The first third generation missile—the MINUTEMAN III—with the new multiple independently-targeted reentry vehicle began testing in 1968. Between 1970 and 1975, 550 MINUTEMAN IIIs were deployed, replacing the same number of older single warhead MINUTEMANs.

Although the number and type of land-based missiles deployed has not changed for a decade, their military effectiveness has been continuously improved. The original warheads (W62) on a portion of the MINUTEMAN III force (300) missiles were upgraded to a higher yield variant (W78). The accuracy of MINUTEMAN II, MINUTEMAN III, and TITAN II missiles was improved through the deployment of new and more accurate guidance systems, and targeting and retargeting options were improved through the development of newer, faster, and more responsive systems.

Work on a completely new missile design—the MX—was already well underway when MINUTEMAN III was first deployed. Ten years later, the MX missile, now designated "PEACEKEEPER," is still under development and planned for a late-1986 initial deployment. The MX missile will carry 10 warheads, more than three times that of the MIRVed MINUTEMAN III, and it will be twice as accurate. Although deployment has been bogged down in controversy over strategic, environmental, cost, and political issues, a massive investment is still going into the new missile. The deployment plans are to retrofit 100 MX missiles into MINUTEMAN

Table 5.3
Strategic Nuclear Forces (1983)

Delivery System	No.[1]	Warheads/Launcher	Total Warheads	Yield (Mt)	Total Mt
MINUTEMAN II	450	1	450	1.2	540.0
MINUTEMAN III	550	3	1650		
Mk-12	(250)		(750)	.17	127.5
Mk-12A	(300)		(900)	.335	301.5
TITAN II	49	1	49	9.0	441.0
Total ICBMs	1049		2149		1410.0
POSEIDON	304	10	3040	.05	152.0
TRIDENT	240	8	1920	.1	192.0
Total SLBMs	544		4960		344.0
B-52G/H	241		2096		
ALCM	(14)	20[2]	(280)	11.08	155.1
Non-ALCM	(227)	8	(1816)	8.68	1970.3
B-52D	31	4	124	8.00	248.0
FB-111	60	6	360	4.68	280.8
Total Bombers	328		2580		2654.2
GRAND TOTAL	1921		9689		4408.2

1 As of 1 January 1983.

2 Weapons include 12 ALCM, 4 SRAM, 4 bombs.

5
Sea-Based Systems

underground launching silos, in lieu of placing the missiles in one of the previously chosen "survivable" basing modes (see MX Basing). The MX will carry a new reentry vehicle, the Mk-21 (Advanced Ballistic Reentry Vehicle), and a new warhead, the W87. The W87 warhead has an initial yield of 300 Kt, but can accommodate a change in fissile material and provide an upgraded yield of 475 Kt. A single warhead ICBM is also being developed, for an early-1990s IOC, as an eventual MINUTEMAN and MX replacement.

The Air Force is removing and dismantling approximately one TITAN II missile per month as part of a five year retirement program begun in October 1982. During FY 1983-1984, 50 single warhead MINUTEMAN II missiles will also be replaced in their silos with triple warhead MINUTEMAN IIIs. The remainder of the MINUTEMAN II force (400) will then begin retirement in 1986 as the PEACEKEEPER/MX enters the inventory. Six hundred MINUTEMAN IIIs will probably remain deployed through the 1990s.

Sea-Based Systems[6]

The present force of strategic ballistic missile submarines includes two TRIDENT and 31 POSEIDON submarines capable of firing 520 submarine-launched ballistic missiles (SLBMs) and carrying approximately 4960 warheads (see Table 5.4).[7] The POSEIDON submarines, constructed between 1960 and 1967, have 16 launch tubes for either POSEIDON C3 or TRIDENT I C4 ballistic missiles. Twelve submarines carry the TRIDENT I and 19 carry the POSEIDON C3. The new TRIDENT submarine has 24 launch tubes and carries the TRIDENT I C4; it will begin carrying the TRIDENT II D5 in 1989. Numerous improvements have taken place since the original POLARIS submarines and missiles were deployed in 1960. Besides new missiles and newly designed submarines, there have been significant improvements involving the latest communications, computing, quieting, and electronics equipment. The latest class of submarine, the OHIO, commonly referred to as TRIDENT, is now beginning to enter the SLBM force (see Tables 5.19 and 5.6).

The missile inventory consists of 304 POSEIDON C3 launchers in 19 POSEIDON submarines, 192 TRIDENT I C4 launchers in 12 POSEIDON submarines, and 48 TRIDENT I C4 launchers in two TRIDENT submarines.[8] Two RV/warhead types are deployed on the submarine missiles: the Mk-3/W68 on POSEIDON with 40-50 Kt yield, and the Mk-4/W76 on TRIDENT I with 100 Kt yield. Both missiles are MIRVed, with the POSEIDON capable of carrying 14 RVs, but averaging about 10, and the TRIDENT I carrying eight RVs.

There are 20 TRIDENT submarines in the current Five Year Defense Plan (Reagan Administration as of FY 1984). Although still undetermined, estimates are that 20-25 TRIDENT submarines will be built.[9] Ten TRIDENT submarines are authorized for construction through FY 1983. The shipbuilding program proposed by the Department of Defense for Fiscal Year 1983 includes funds for construction of two more TRIDENT submarines and one sub per year thereafter (see Table 5.6).[10]

The TRIDENT program, first called the Undersea Long-Range Missile System (ULMS), began as a follow-on to the POLARIS and POSEIDON fleet in the late 1960s. It was envisioned as a more survivable system capable of launching missiles at intercontinental ranges from quieter submarines. The need for the eventual modernization of POSEIDON was based upon the projected 20-25 year service life for deployed submarines (this has since been extended to 30 years). Development of a new missile (TRIDENT I C4) and submarine (OHIO class) was approved on 14 September 1971 by the Secretary of Defense.

The new TRIDENT submarines, the largest submarines ever built by the U.S., are more than twice the size of the present POSEIDON submarines. They are designed to operate at greater speeds and to emit less noise than the POSEIDONs. A new refit, maintenance and overhaul cycle, and the longer range of the missiles permit basing in the United States and operations off the protected U.S. coast. The increased patrol area still allows targeting throughout the Soviet Union.

Development of the TRIDENT I and TRIDENT II missiles has always been part of the TRIDENT program. The size of the TRIDENT I missile was limited to allow its deployment in smaller POSEIDON submarines. The

Table 5.4
Strategic Missile Submarines (1983)

Type (class)	Active	Building	Missiles
POSEIDON (*Lafayette, Madison*)	19	—	16 POSEIDON C3
POSEIDON (*Franklin*)	12	—	16 TRIDENT C4
TRIDENT (*Ohio*)	2	9	24 TRIDENT C4

---

6 An excellent history is *Fleet Ballistic Missile System: Polaris to Trident* (R.A. Fuhrman, President, Lockheed Missiles and Space Company, Inc., AIAA von Karman Lecture for 1978) (Washington, D.C.: AIAA, February 1978).
7 ACDA, FY 1982 ACIS, p. 76; ACDA, FY 1983 ACIS, p. 33.
8 Ibid.
9 Ibid; HASC, FY 1983 DOD, Part 4, p. 93; SASC, FY 1983 DOD, Part 7, p. 4179, mentions a 20 boat force.
10 ACDA, FY 1982 ACIS, p. 78; ACDA, FY 1983 ACIS, p. 36.

5
Strategic Submarines

Table 5.5
Strategic Submarine Chronology

Date	Event
Sep 1955	Sea basing of ballistic missile system considered in "Killian Report"
Mar 1956	Fleet Ballistic Missile submarine and surface combatant development program authorized
Apr 1956	Lockheed awarded contract to determine feasibility of submarine missile development
Dec 1956	Navy authorized to proceed with development of small, solid propellent POLARIS missile
Jan 1958	First POLARIS test flight; construction begun of first three POLARIS submarines
Apr 1959	First full successful POLARIS vehicle flight test
Dec 1959	U.S.S. *George Washington*, first POLARIS submarine, commissioned
Sep 1960	Development of POLARIS A3 approved
Nov 1960	U.S.S. *George Washington* leaves on first operational patrol with 16 POLARIS A1
Oct 1961	First launch of POLARIS A2 missile
Dec 1961	Last production model of POLARIS A1 delivered
May 1962	U.S.S. *Ethan Allen* successfully fires a POLARIS missile with a nuclear warhead
May 1962	U.S.S. *Lafayette* launched, first of new POSEIDON class
Aug 1962	First flight test of POLARIS A3 missile
Sep 1964	U.S.S. *Daniel Webster* goes on first patrol with POLARIS A3
Oct 1965	Last POLARIS A1 patrol
Oct 1965	Development of POSEIDON missile begins
Dec 1966	Production of POSEIDON C3 approved
Dec 1966	Electric Boat awarded contract for STRAT-X studies
Apr 1967	Last POSEIDON submarine, U.S.S. *Will Rogers*, commissioned
Aug 1967	STRAT-X system designated ULMS (Undersea Long-Range Missile System)
Jul 1968	Last production-line POLARIS A3 missile delivered
Aug 1968	First flight test of POSEIDON C3
Feb 1969	Program begun to convert 31 SSBNs from POLARIS to POSEIDON
Aug 1969	Production of POSEIDON C3 begins
Oct 1969	ULMS program formally established
Jun 1970	Development flight testing of POSEIDON C3 completed
Mar 1971	U.S.S. *James Madison* makes initial POSEIDON operational patrol
Dec 1971	TRIDENT missile advanced development begins
May 1972	ULMS renamed TRIDENT
Feb 1973	Bangor, WA selected as initial TRIDENT base
1973	TRIDENT C4 enters engineering development
Nov 1973	Funds for first TRIDENT submarine authorized
1974	Production of POSEIDON C3 completed
Mar 1974	First test flight of TRIDENT C4
Jun 1974	Last POLARIS A2 patrol
Jul 1974	Electric Boat receives contract for lead TRIDENT submarine
Nov 1975	Final flight test of TRIDENT C4
Apr 1976	Keel laid for *Ohio* (SSBN-726), first TRIDENT submarine
Jan 1977	First full scale production flight of TRIDENT C4, production begins
Feb 1978	Completed conversion of 31 POLARIS to POSEIDON
Apr 1979	Launch of U.S.S. *Ohio*
Jul 1980	UK announces decision to purchase TRIDENT system
Jul 1981	Delivery of U.S.S. *Ohio* by Electric Boat to Navy
Nov 1981	Commissioning of *Ohio*
Jan 1982	First test firing of TRIDENT I C4 from *Ohio*
Apr 1982	Deployment of *Ohio* on operational sea trials
May 1982	UK announces decision to acquire TRIDENT II rather than TRIDENT I
Mar 1983	Retrofit program of 12 POSEIDON SSBNs backfitted with TRIDENT I C4 missiles completed
Apr 1983	Planned deployment of the second TRIDENT, U.S.S. *Michigan*
End 1986	Six TRIDENTs scheduled for deployment
late 1988	First TRIDENT II deployed
1993	First POSEIDON hull scheduled for retirement
1993-1999	POSEIDON submarines scheduled for retirement

Sources: USN, Strategic Systems Project Office, "FBM Facts Polaris, Poseidon, Trident," 1978; General Dynamics Corporation, Electric Boat Division, "General Dynamics Trident Progam Milestones," April 1980; ACDA, FY 1982 ACIS, p. 85; SASC, FY 1983 DOD, Part 7, p. 4516.

5
Strategic Bomber Force

Table 5.6
Strategic Submarine Forces (1979-1990)[1]

	FY 79	FY 80	FY 81	FY 82	FY 83	FY 84	FY 85	FY 86	FY 87	FY 88	FY 89	FY 90
TRIDENT SSBNs with TRIDENT I SLBMs	0	0	1	1	2	4	5	7	8	8	8	8
TRIDENT SSBNs with TRIDENT II SLBMs	0	0	0	0	0	0	0	0	0	0	1	2
POSEIDON SSBNs with TRIDENT I SLBMs	0	5	11	12	12	12	12	12	12	12	12	12
POSEIDON SSBNs with POSEIDON SLBMs	31	26	20	19	19	19	19	19	19	19	19	19

Source: ACDA, FY 1983 ACIS, p. 38; SASC, FY 1981 DOD, Part 6, p. 3469.

1 Cumulative Forces of launched (but not operational) submarines.

TRIDENT I missile, first deployed in FY 1979, increased the range over the POSEIDON and doubled the yield (from 50 to 100 Kt), but had similar accuracy. The TRIDENT II, planned for initial deployment on the ninth TRIDENT submarine scheduled for late 1989, will have increased range over the TRIDENT I and employ a much higher yield warhead. However, its most significant feature is its accuracy, which approaches the capability of land-based missiles. The deployment of TRIDENT II missiles (with 10 warheads each) in at least 20 new submarines will cost over $30 billion. The high yield variant of the W87 MX warhead (see MX Warhead), with a yield of 475 Kt, on a modification of the Advanced Ballistic Reentry Vehicle (designated the Mk-5), has been chosen as the developmental baseline for the TRIDENT II. A maneuvering reentry vehicle (designated the Mk-600) is also being considered for the missile.

In 1978, U..S. SSBNs had completed 1723 alert patrols of approximately 104,000 patrol days.[11] By 1981, 2043 patrols had been completed.[12] Approximately 60 percent of the submarine force is on patrol at any one time, and this percentage will increase as more TRIDENTs are deployed.

Strategic Bomber Force

The Srategic Air Command operates over 400 B-52 and FB-111 bombers, of which 272 B-52s and 56 FB-111s are in the active force[13] with the others used for training and backup. Thirty percent of the bomber force is always kept on 15-minute alert.[14] This percentage could be increased to 50% during an emergency.[15] Some bombers could be launched in as little as three minutes.[16] The bomber force carries five different types of nuclear bombs: the B28, B43, B53, B57, and B61. These bombs have various weights, yields, accuracies, and delivery profiles (see Chapters Three and Seven). Two missile systems are also carried: the Short-Range Attack Missile (SRAM) with its W69 warhead in the 170-200 Kt range and the new Air-Launched Cruise Missile (ALCM) with a 200 Kt warhead (W80).

The B-52 heavy bomber was first developed during the Korean War to create an intercontinental capability which was not possible with the B-29 or B-47 bombers it replaced. The B-52's airframe is old. This makes it difficult for the bomber to meet the current requirements of penetrating Soviet defenses or flying to targets at low altitudes. The usefulness and reliability of the B-52 in this role is undeniably decreasing, but the airframe is still reliable as a high altitude cruise missile platform. In fact, recent Air Force studies have confirmed the serviceability of the airframe until the year 2000. As the ability of B-52s to penetrate Soviet defense is diminishing, its role is also changing to the less demanding stand-off role for cruise missile carriage. Beginning in late 1981, B-52Gs were being deployed as cruise missile carriers. By the 1990s, B-52s will no longer have a penetration role.

Shortly after the B-52 was deployed in the 1950s, a modernization program began to evaluate and upgrade the capability and effectiveness of the B-52 force. Numerous modifications to the B-52 over the last 20-25 years have increased aircraft weight and drag.[17] Table 5.22 describes the extensive nature of the modernization program.

---

1 SASC, FY 1980 DOD, Part 5, p. 2498.
2 SASC, FY 1983 DOD, Part 7, p. 4515.
3 JCS, FY 1984, p. 13.
4 DOD, FY 1979 Annual Report, p. 44.

15 *Ibid.* p. 57.
16 *Military Applications of Nuclear Technology*, Part 1, p. 7.
17 SASC, FY 1982 DOD, Part 7, p. 3790.

5

Ballistic Missile Reentry Vehicles

Table 5.7
Strategic Bomber Force Loadings

Average Weapons Per Plane

Aircraft	Number[1]	ALCM	SRAM	Bombs
B-52G	151	12	4	4
B-52H	90	—	4	4
B-52D	31	—	—	4
FB-111	56	—	4[2]	2[3]

Capable Load

Aircraft	Number	ALCM	SRAM	Bombs
B-52G	151	12	20	12[4]/4
B-52H	90	20[5]	20	12[6]/4
B-52D	31	—	—	4
FB-111	56	—	6	6[7]

1 The number of bombers reduces every year as a result of attrition; these are operational bombers as of 1 January 1983; SASC, FY 1983 DOD, Part 7, p. 4556; JCS, FY 1984, p. 13.
2 SAC, FY 1981 DOD, Part 5, p. 1629.
3 *Ibid.*
4 Maximum number of B61s possible; SASC, FY 1982 DOD, Part 7, p. 4329.
5 Planned.
6 Maximum number of B61s possible; SASC, FY 1982 DOD, Part 7, p. 4329.
7 *Ibid.*

Table 5.8
Bomber Forces Funding (1970-1980)
($ millions)

Year	RDT&E*	Procurement*	Operations*
1970	131.1	288.2	476.1**
1971	73.7	473.7	435.5**
1972	6.8	724.5	496.1**
1973	50.0	544.1	583.6
1974	14.8	388.0	573.5
1975	67.8	284.6	637.1
1976/7T	70.8	181.1	780.5
1977	154.8	217.0	634.7
1978	395.3	269.2	639.8
1979	537.7	493.8	686.8
1980	266.3	1137.1	709.4
TOTAL	1769.1	5001.3	6654.2

* Then year dollars.
** Includes South East Asia funding.
Source: SASC, FY 1980 DOD, Part 1, p. 398. Table excludes B-1 funding, but includes bomber/bomber weapons funding.

The bomber force is more capable of destroying hardened military targets than MINUTEMAN.[18] This capability will be enhanced with the continued deployment of some 4000 ALCMs (and its Advanced Cruise Missile replacement), the new B83 bomb (planned for the mid-1980s), and a replacement for the SRAM, the Advanced Strategic Air-Launched Missile (see Chapter Six). Although the bombers are capable of carrying 24 nuclear weapons, the location of high priority targets within the Soviet Union makes it unlikely that they could go to 24 different places.[19] The widespread deployment of new long-range ALCMs and SRAM replacements will increase target flexibility and allow for great distances between targets.[20] The new weapons incorporate further technological advances in many areas, including lightweight materials, miniaturized electronics, modern warhead design, and advanced guidance systems. The new bomb can be delivered at low level and at supersonic speeds.

In addition to changes in the weapons load, the bomber force itself is in the midst of significant changes and upgrading. Older B-52Ds, the last of the gravity bomb carriers armed with four high yield nuclear bombs, began retirement in FY 1983 and will be completely withdrawn in FY 1984.[21] The B-52G, which was the first ALCM launcher starting in 1982, will continue to be modified to carry ALCMs on two large external pylons mounted under the wings. Plans to modify these bombers for internal ALCM carriage have been cancelled. The B-52H force will be modified to carry ALCMs both internally and externally, starting deployment in 1985.[22] The B-52Gs will convert entirely to standoff bombers as B-1Bs are deployed starting in 1985-1986. The B-52Gs will begin retirement with deployment of the Advanced Technology Bomber (ATB) ("Stealth") in the early 1990s. The B-52H force will continue as an ALCM carrier well into the 1990s. After a long, difficult development program, a new manned bomber, the B-1B, will be deployed in FY 1985 and will carry cruise missiles and gravity bombs. Deployment of 100 B-1Bs and 100-150 ATBs is planned. The FB-111s will be transferred to the tactical inventory as ATBs are deployed.[23]

Ballistic Missile Reentry Vehicles

The reentry vehicle (RV) on a ballistic missile carries the nuclear warhead. Reentry vehicles are designed to minimize the environmental factors, such as wind and atmospheric density, which accompany the missile's reentry to the earth's atmosphere. Both missile speed and accuracy are only slightly reduced. Assuming perfect ballistic trajectories, aside from reentry effects, RVs

---

18 SASC, FY 1982 DOD, Part 7, p. 4010; ACDA, FY 1979 ACIS, p. 3.
19 *Military Applications of Nuclear Technology*, Part 1, p. 7.
20 ACDA, FY 1983 ACIS, p. 67.
21 SASC, FY 1983 DOD, Part 7, p. 4589.
22 ACDA, FY 1983 ACIS, p. 67.
23 DOD, FY 1984 Annual Report, pp. 222-224.

Table 5.9
U.S. Ballistic Missile Reentry Vehicles

RV	Warhead	Missile System	RVs/Missile
Deployed (1983)			
Mk-3	W68	POSEIDON (C3)	up to 14 (MIRV) (avg 10)[1]
Mk-4	W76	TRIDENT I (C4)	up to 10 (MIRV) (avg 8)
Mk-6	W53	TITAN II	1
Mk-11C	W56	MINUTEMAN II	1
Mk-12	W62	MINUTEMAN III	2-3 (MIRV)
Mk-12A	W78	MINUTEMAN III	3 (MIRV)
Retired			
Mk-1	W38,W49	ATLAS/THOR/JUPITER	1
Mk-2	W38,W49	TITAN I	1
Mk-1 (Navy)	W47	POLARIS (A1/A2)	1
Mk-2 (Navy)	W58	POLARIS (A3)	3 (MRV)[2]
Mk-5	W59	MINUTEMAN I	1
Mk-11	W59	MINUTEMAN I	1
Cancelled Programs			
Mk-17	—	MINUTEMAN III/POSEIDON	1
MK-18	—	MINUTEMAN III	multiple unguided RVs
Mk-19	—	MINUTEMAN III	MIRV/MaRV
Mk-20	—	MINUTEMAN III	MIRV/MaRV

1 JCS, FY 1984, p. 16; prior to FY 1984, the average loading was nine; the C3 missile has been tested with 14 warheads. Since there is a tradeoff between throwweight and range, actual loadings are less than the maximum depending on target and submarine locations.

2 Not independently targetable.

can achieve overall theoretical accuracies of better than 250 feet CEP.

All ballistic missile systems may carry one or several reentry vehicles which may be independently targeted. If the missile system carries several RVs that are *not* independently-targetable, the system is referred to simply as a multiple reentry vehicle (MRV) system. In a multiple independently-targetable reentry vehicle system (or MIRVed system), the separate reentry vehicles are carried on a "bus" which releases the RVs one by one after making preselected changes in speed and orientation so as to direct each RV to separate targets. These RVs will land inside a "footprint" of perhaps 100 miles by 300 miles. Missiles with multiple RVs have less targeting flexibility than single warhead missiles, because the other RVs on the bus are not completely independent in arrival time or location of the RV first released.

The newest RV, the Mk-12A, has been retrofitted on 300 MINUTEMAN III missiles, replacing the Mk-12.[24] The Mk-12A, designed in the mid-sixties, retained approximately the same dimensions, aerodynamic properties, and radar cross section of older RVs, and its weight was only slightly greater than the Mk-12. The Mk-12A incorporated a larger yield warhead, increased accuracy, and an improved arming and fuzing system over the Mk-12.[25] The Air Force justification for this new RV was that it was necessary to compensate for continual Soviet hardening of its strategic targets.

The Mk-12A RV was originally intended to be deployed on all 550 MINUTEMAN IIIs. However, because of additional weight over the Mk-12 and the resultant decrease in range, some Mk-12 equipped MM IIIs were retained in order to reach all targets.[26] Until January 1982, the Mk-12A was also the baseline RV for the PEACEKEEPER/MX missile, but has now been

24 1083 Mk-12A RVs are planned for procurement for MM III forces; SASC, FY 1982 DOD, Part 7, pp. 3986-87.

25 ACDA, FY 1979 ACIS, p. 2.
26 HAC, FY 1980 DOD, Part 2, p. 499.

5

Ballistic Missile Reentry Vehicles

Table 5.10
RV Developments

Mk-80	MX/TRIDENT II/ MINUTEMAN III	Lightweight oralloy warhead/RV, cancelled
Mk-81	MX/TRIDENT II/ MINUTEMAN III	Plutonium warhead/ RV, cancelled
Mk-500 EVADER	TRIDENT	Maneuvering RV
Mk-5/ Mk-21/ ABRV	MX/TRIDENT II/ Small ICBM	Highly accurate RV
AMaRV	MX/TRIDENT II/ Small ICBM	Highly accurate maneuvering RV
PGRV	MX/TRIDENT II	Highly accurate terminally guided RV
Mk-600	TRIDENT II	Highly accurate, terminal homing maneuvering RV

- Concealment: reduction of radar cross section, new shapes, and materials,
- Countermeasures: radar blackout, saturation, replica decoys, traffic decoys, active ECM,
- Evasion: maneuvering,
- Speed: increased,
- Accuracy: terminal guidance, post maneuver accuracy.

Maneuvering RVs could also have the purpose of attacking mobile targets, such as ships or mobile missiles. It is generally assumed that with a long-range ballistic missile it is difficult to observe continuously and accurately the position of a mobile target. But once launched, a MaRV equipped missile together with an

shelved in favor of the more accurate Mk-21 (formerly ABRV) with a new warhead, the W87.

The U.S. ballistic missile RVs currently deployed are shown in Table 5.9. All currently deployed missile RVs are ballistic. Future RVs, however, including possible reentry vehicles for the PEACEKEEPER/MX and TRIDENT II D5 missile systems, may be nonballistic or Maneuvering Reentry Vehicles (MaRVs). For accuracy or evasion, the MaRV will be able to correct its flight path after reentry.

The Advanced Strategic Missile Systems (ASMS) program is continuing DOD research and development of RVs and subsystems for ICBMs, IRBMs, and SLBMs. The ASMS program (its name was changed from the Advanced Ballistic Reentry Systems (ABRES) program in FY 1982) includes advanced development for ballistic missile systems, subsystems, and reentry and penetration aids for existing and future weapons. These reentry vehicles are discussed later in this section.

The ASMS program, started in 1962, generally focuses on post-ballistic phases of the trajectory. The feasibility of large MaRVs was demonstrated by flight tests in the mid-1960s.[27] In the 1970s, flight tests examined advanced design concepts and more severe maneuvering environments, including:.[28]

Table 5.11
RV Chronology[1]

Mar 1963	First ABRES test launch atop ATLAS missile
May 1963	Advanced Ballistic Reentry Systems (ABRES) program started
Feb 1964	First ATHENA rocket launched to test subscale ballistic reentry vehicles
Aug 1966	First Maneuvering Ballistic Reentry Vehicle (MBRV-1) launched
Sep 1973	Advanced Nosetip Test program to develop Multiple Small RVs for MINUTEMAN tested
Dec 1974	Mk-12A development contract signed with GE
Mar 1975	First Mk-500 flight test
May 1975	Second Mk-500 flight test
Aug 1975	Third Mk-500 flight test
Jan 1976	Fifth Mk-500 flight test
Jan 1976	AMaRV concept review conducted
Apr 1976	Mk-12A critical design review conducted
FY 1976	PGRV program initiated
Jan 1977	First Mk-12A flight
Jul 1977	AMaRV prototype construction began
FY 1978	First ABRV flight
FY 1979	Mk-12A completed development
FY 1980	First two AMaRV flight tests conducted
FY 1980	IOC of Mk-12A on MM III
FY 1981	Third and final AMaRV flight test
Mar 1981	93 MM III fitted with Mk-12A[2]
Jan 1982	ABRV chosen for MX
1983	Deployment of Mk-12A on 300 MM III completed

1 ACDA, FY 1979 ACIS, p. 1; information was also received from USAF Space Division.
2 HAC, FY 1982 DOD, Part 2, p. 225.

---

27 ACDA, FY 1981 ACIS, p. 17; ACDA, FY 1982 ACIS, p. 18. 28 Ibid.

5
Ballistic Missile Reentry Vehicles

Table 5.12
ABRES/ASMS Costs:[1]

FY	Total Appropriation ($ million)
1977 & prior	1757.5[2]
1978 & prior	1855.5[3]
1979 & prior	1961.2
1980	95.4
1981 & prior	2153.3[4]
1981	103.8
1982	99.6[5]
1983	52.3

Mk-12A COSTS:

FY	Total Appropriation ($ million)
1977 & prior	152.7
1980 & prior	402.5[6]
1981 & prior	626.6[7]
1982	56.4[8]

1 ACDA, FY 1981 ACIS, p. 23; and DOD, *Program Acquisition Costs by Weapon System*, FY 1982, 15 January 1981.
2 ACDA, FY 1979 ACIS, p. 48.
3 HAC, FY 1980 DOD, Part 6, p. 677.
4 ACDA, FY 1983 ACIS, p. 12.
5 In FY 1982, Congress appropriated $100 million even though DOD requested $50 million.
6 ACDA, FY 1982 ACIS, p. 24.
7 *Ibid.*, p. 11.
8 *Ibid.*

autonomous sensor could reach the target area and maneuver in order to attack a non-fixed target.

The goal of the current MaRV development program is to establish whether ballistic missiles can reliably fly nonballistic reentry trajectories in order to evade ballistic missile defenses and improve accuracy. The two MaRV efforts under development include the Advanced Maneuvering Reentry Vehicle (AMaRV) program initiated in FY 1976 and the Precision Guided Reentry Vehicle (PGRV). The PGRV is a longer term effort involving technology developed in AMaRV with terminal sensors.[29]

Funding for maneuvering RVs over the past five years has been approximately $100 million per year.[30] Current research emphasis is on nosetip ablation/erosion studies, tests, materials development, maneuvering subsystems, decoys, and other penetration aids.[31] The ASMS program contracts with approximately 40 corporations and consultants and extensively uses government laboratories.[32] A new Air Force study, "Strategic Missile Systems 2000," begun during FY 1982, will determine the most promising ballistic concepts, technologies, and areas of development for the future.[33]

Mk-21/Mk-5 (Advanced Ballistic Reentry Vehicle) (ABRV)

The ABRV program began in 1975 to demonstrate the maximum accuracy achievable with small ballistic reentry vehicles. An ABRV was originally envisioned for the MX and, in January 1982, was chosen as the RV for the MX carrying the 300 Kt W87 warhead.[34] The ABRV has also been selected as the baseline RV for the TRIDENT II (designated Mk-5) and for the Small ICBM.[35] The development program has sought to optimize yield-to-weight ratios, improve packaging, and incorporate a new, improved (interactive) fuze.[36] The choice of the W87 warhead for the Mk-21 results in a number of new characteristics—use of less nuclear material, the ability to increase low yield later by adding more materials, and incorporation of IHE.[37] The weight of the ABRV, however, is reportedly greater than the Mk-12A. This restricts its range.[38]

The current R&D program includes flight testing, data analysis, and development.[39] Flight testing for the ABRV, which began in 1978, has been used to demonstrate the use of shapes, materials, fuzing, heat shielding, cooling, and composite structuring.

The size of the ABRV permits a larger number of warheads to be carried by MX and TRIDENT II, but the increased weight and SALT II adherence would restrict the number.[40] There are three warheads compatible with the ABRV: the current warhead; the versatile 300 Kt light-weight warhead (W87) (which can be upgraded to 475 Kt);[41] the 500-600 Kt CALMENDRO warhead; and the 800+ Kt MUNSTER warhead.[42] The W78 has also been considered for the ABRV.

Advanced Maneuverable Reentry Vehicle (AMaRV)

The AMaRV R&D program, a follow-on to the earlier Mk-500, aims to develop a more accurate maneuvering RV capable of evading enemy terminal offenses with

---

29 ACDA, FY 1981 ACIS, p. 17.
30 SASC, FY 1982 DOD, Part 7, p. 3995.
31 DOD, *Program Acquisition Costs by Weapon System*, FY 1982, 15 January 1981, p. 155.
32 *Ibid.*
33 AF, FY 1983 RDTE Statement, 2 March 1982; *Air Force Magazine*, February 1982, p. 21; ACDA, FY 1983 ACIS, p. 8.
34 See MX and W87 warhead.
35 AW&ST, 17 January 1983, p. 26.
36 ACDA, FY 1981 ACIS, p. 16.
37 SASC, FY 1983 DOD, Part 7, p. 4179.
38 AW&ST, 4 May 1981, p. 52.
39 ACDA, FY 1982 ACIS, p. 18.
40 ACDA, FY 1981 ACIS, p. 16.
41 AW&ST, 17 January 1983, p. 26.
42 AW&ST, 9 March 1981, p. 25.

5

Ballistic Missile Reentry Vehicles

advanced interceptor missiles. The AMaRV program includes the development of a high-altitude maneuver capability, with an inertial measurement system on the reentry vehicle[43] affording accuracy equal to or better than the ballistic RV it replaces.

AMaRV completed a concept review in January 1976 and a contract was awarded in September 1976 for the development and flight testing of two AMaRVs.[44] These two flight tests occurred in FY 1980. A third and final AMaRV test took place in FY 1981.

The AMaRV program has now replaced the Mk-500 EVADER by utilizing advances in technology and by adding new angles of attack, speed, acceleration, and guidance features. The AMaRV program is evaluating a laser gyro, stellar navigation updating, and a new inertial platform.

Mk-500 EVADER Maneuverable RV (MaRV)

The Mk-500 development program began in the late 1960s as an evasive maneuvering endoatmospheric RV for the TRIDENT missile. The program, which enables the U.S. to respond to potential changes in Soviet ABM defenses, is essentially designed to maneuver against terminal defensive missiles. The Mk-500 lacks a terminal maneuvering guidance and thus is less accurate than the current generation SLBM ballistic RVs. This first generation MaRV is essentially ready for deployment on TRIDENT I, even at the price of degraded accuracy, should a rapid change in Soviet ABM capabilities occur.[45]

The program has remained an advanced test program conducted by Lockheed, the prime contractor, and General Electric, the principal sub-contractor.[46] Flight testing on MINUTEMAN boosters began in 1975 and two flight tests on TRIDENT I missiles have been conducted. The DOD goal is to obtain an acquisition readiness during FY 1983. Presently there are no firm plans to produce the RV.[47] The Mk-500 Advanced Development Program has fully demonstrated the feasibility of MaRV and compatability with the TRIDENT I.[48] The Readiness Maintenance Program maintains a capability to deploy a maneuvering RV with little delay and low risk.[49] Included in this program is design and testing of fire control software, parts, and guidance system.

Precision Guided Reentry Vehicle (PGRV)

The Precision Guided Reentry Vehicle (PGRV) program began in FY 1976 with completion of a system design study. Development of PGRVs is a long term effort utilizing AMaRV technology and adding terminal sensors. The terminal sensor would allow for corrections in guidance in the final phase of flight by providing relative position and velocity updates for the RV guidance system as the RV approaches its target.[50]

A maneuvering PGRV is under development which could, given expected technological developments, be accurate enough to significantly increase the U.S. ability to destroy the Soviet ICBM force. Alternatively, a PGRV with a low yield warhead could permit "greater targeting flexibility with lower collateral damage, but without a hard target capability."[51] A PGRV, designated Mk-600 with terminal homing guidance, has been adopted as an alternative warhead for the TRIDENT II. Deployment of the Mk-600 PGRV is thought possible by the 1990s.[52]

---

43 Currently only ballistic missile boosters, not RVs, have inertial navigation systems.
44 ACDA, FY 1979 ACIS, p. 53.
45 ACDA, FY 1979 ACIS, p. 54.
46 Other sub-contractors include Litton, Rockwell Autonetics, Batelle, Bell, Northrop, and Hamilton Standard.
47 ACDA, FY 1983 ACIS, p. 40.
48 SASC, FY 1981 DOD, Part 2, p. 612.
49 SASC, FY 1981 DOD, Part 2, p. 612; ACDA, FY 1981 ACIS, p. 20.
50 ACDA, FY 1981 ACIS, p. 18.
51 ACDA, FY 1979 ACIS, p. 54.
52 AW&ST, 8 March 1982, p. 27.

5
TITAN II

Land-Based Missile Systems
TITAN II (LGM-25C)

Figure 5.1 **TITAN II (LGM-25C)** missile.

DESCRIPTION: Largest Air Force Intercontinental Ballistic Missile (ICBM), with single high yield warhead, and the only ICBM remaining which is liquid fueled.

CONTRACTORS: Martin Marietta Corp.
Denver, CO
(prime)
General Electric
Philadelphia, PA
(RV)
Delco Electronics
(guidance)
Aerojet General
Sacramento, CA
(propulsion)
IBM
(guidance)

SPECIFICATIONS:

Length: 103 ft (31.3 m)

Diameter: 120 in (305 cm)

Stages: 2

Weight at Launch: 327,000 lb (149,700 kg)

Fuel: liquid[1]

Propulsion:
Stage 1: two Aerojet LR87-AJ-5s, 98,000 kg thrust each
Stage 2: Aerojet LR91-AJ-5, 45,500 kg thrust

Speed: 24,000+ km/h at burn-out

Guidance: inertial gimballed; original guidance system replaced FY 1979-FY 1981[2]

Throwweight/Payload: 7500 lb;[3] 8275 lb[4]

Range: 6296 nm;[5] 7250 nm;[6] 8100 nm;[7] 4000 km;[8] 15,000 km[9]

Ceiling: about 700 miles

DUAL CAPABLE: no

NUCLEAR WARHEADS: one W53/Mk-6 reentry vehicle with penetration aids; 9 Mt (see W53)

DEPLOYMENT:
Launch Platform: fixed site underground hardened silo

Silo Hardening: 300 psi[10]

Number Deployed: 49 missiles deployed (Jan 1983); retirement program began October 1982 with one missile per month dismantled.[11] 65 warheads in stockpile prior to start of withdrawal[12]

Nuclear Weapons Databook, Volume I **111**

5

TITAN II

Location:

Wing	Base	Missiles[13]
308 SMW	Little Rock AFB, AR	17
381 SMW	McConneell AFB, KS	17
390 SMW	Davis-Monthan AFB, AZ	15

HISTORY:

IOC:	8 June 1963[14]
Dec 1963	TITAN II achieves full operational capabilities[15]
1980	54 TITAN IIs deployed prior to silo accidents in 1978 and 1980 at Rock, Kansas and Damascus, Arkansas
Oct 1981	Reagan strategic program calls for early retirement of TITAN missile force
Oct 1982	first TITAN missiles at Davis-Monthan AFB, AZ, begin retirement[16]
late 1986	10 TITANs remain on alert at MX IOC[17]
FY 1983-1987	TITAN force deactivated[18]

TARGETING:

Types:	wide area soft military targets industry and urban areas "hundreds" of "high-yield aggregate Desired Ground Zero (DGZs) which contain more than one primary DGZ" (target)[19]
Selection Capability:	two target selection capability
Retargeting:	missile silos are also launch centers
Accuracy/CEP:	0.7-0.8 nm;[20] 0.5 nm[21]

COST:

Annual Operations:	$330 m (FY 1980)[22] $345 m (FY 1982)[23]

FY	Number Procured	Total Appropriation ($ million)
1981 & prior	-	1785.7[24]
1982	-	266.8[25]

COMMENTS: TITAN warhead is the largest in the U.S. land-based inventory and provides the most destructive soft target (population center, industry, etc.) capability in the U.S. strategic forces. Early retirement will result in FY 1982-1987 savings of $500 million.[26]

---

1 Fuel is mixture of 50 percent hydrazine (N_2H_4) by weight and unsymmetrical dimethylhydrazine ($2H_2(CH_3)_2$); oxidizer is nitrogen tetroxide (N_2O_4); SAC, Fact Sheet, "TITAN II," August 1981.
2 JCS, FY 1980, p. 67.
3 *Military Balance*, 1980-1981, p. 88; Heritage Foundation, *SALT Handbook*, p. 75.
4 John Collins, *op. cit.*; Paul H. Nitze, Prepared Statement Before Senate Foreign Relations Committee, 12 July 1979, Revised 13 January 1981, Reprinted in *Congressional Record*, 20 July 1972, p. S10078.
5 *Military Balance*, 1975-1976, p. 71.
6 John Collins, *op. cit.*, p. 446.
7 *Military Balance*, 1980-1981.
8 *Flight International*, 30 May 1981, p. 1637.
9 *Jane's Weapons Systems*.
10 HAC, FY 1982 Mil Con, Part 6, p. 275.
11 49 missiles were deployed as of 1 January 1983; JCS, FY 1984, p. 13.
12 AF, "Missile Procurement Justification," FY 1981 (January 1980), p. 184.
13 Before the decision to retire the TITAN force, the silo at Damascus, Arkansas (site of the accident in September 1980) was not planned for rebuilding. A second site in Kansas destroyed by an oxidizer leak in 1978 had been reported as being repaired, but it is now likely that the weapons will be retired in lieu of repair. See, for instance, Walter Pincus, *Washington Post*, 23 May 1981, p. A-11.
14 SASC, FY 1977 DOD, Part 1, p. 393.
15 ACDA, FY 1983 ACIS, p. 2.
16 HASC, FY 1983 DOD, Part 2, p. 163; *New York Times*, 12 November 1982, p. A16; the first wing will phase out over a two year period; the other two wings will follow over a three year period.
17 SASC, FY 1983 DOD, Part 7, p. 4159.
18 AF RDTE Statement, FY 1983, p. II-3; Walter Pincus, *Washington Post*, 24 September 198 ; DOD, FY 1984 Annual Report, p. 221.
19 SASC, FY 1982 DOD, Part 7, p. 3841.
20 A CEP of 0.7 nm is given by Paul H. Nitze, *op. cit.*, and by UN Secretary General ("General and Complete Disarmament," A/35/392, 12 September 1980). John Collins, *op. cit.*, assumes 0.8 nm. These estimates are probably low since, according to the Joint Chiefs of Staff (FY 1982, p. 69), "A new more accurate guidance system has been installed in the TITAN missile."
21 Colin S. Gray, "The Future of Land-Based Missile Forces" (London: IISS, Adelphi No. 14), p. 32.
22 Annual MINUTEMAN/TITAN operating costs, including military personnel; SASC, FY 1982 DOD, Part 7, p. 4002.
23 *Ibid.*, p. 4337.
24 Does not include initial procurement costs; ACDA, FY 1983 ACIS, p. 11.
25 System safety, reliability, and maintainability modifications planned; SASC, FY 1982 DOD, Part 7, p. 3829.
26 *Ibid.*, p. 4232.

5
MINUTEMAN II

MINUTEMAN II (LGM-30F)

Figure 5.2 **MINUTEMAN II (LGM-30F)** missile.

DESCRIPTION:	Air Force three-stage, solid fuel, single warhead ICBM.
CONTRACTORS:	Boeing Aerospace Co. Seattle, WA (prime, assembly and test) AVCO Systems Wilmington, MA (reentry vehicle) GTE Sylvania Needham Heights, MA (ground electronics) Autonetics Division, Rockwell International Anaheim, CA (guidance) Aerojet General Sacramento, CA (propulsion) Thiokol Chemical Corp. Brigham City, UT (propulsion) Hercules Inc. Wilmington, DE (propulsion) Tracor Inc. Austin, TX (penetration aids)
	TRW Systems Redondo Beach, CA (technical direction)
SPECIFICATIONS:	(same as MINUTEMAN III except for top stage)
Length:	57 ft 6 in (691.2 in)[1]
Diameter:	67.2 in; 72 in[2]
Stages:	3
Weight at Launch:	73,000 lb; 70,000 lb[3]
Propulsion:	three solid-propellant rocket engines
Speed:	15,000+ mph; 24,000+ km/h at burn-out
Guidance:	inertial gimballed NS-17 guidance and control system
Throwweight/ Payload:	1000-1500 lb;[4] 2500 lb;[5] 1625 lb[6]
Range:	6080 nm;[7] 8000 nm;[8] 6500 nm;[9] 7000 mi[10]
Ceiling:	about 700 miles
DUAL CAPABLE:	no
NUCLEAR WARHEADS:	one W56 warhead/Mk-11C reentry vehicle;[11] 1.2 Mt (see W56)
DEPLOYMENT:	
Launch Platform:	fixed 25 m deep underground hardened silo with missile suspension, shock isolated floor, debris collection system and EMP protection
Silo Hardening:	1200-2200 psi

Nuclear Weapons Databook, Volume I   **113**

5

MINUTEMAN II

Number Deployed: 450 active (1983).[12] Fifty MM II missiles will be replaced with MM IIIs at Malmstrom AFB, MT, starting in FY 1983.[13]

Location:

Wing	Base	Missiles
I/ 341 SMW	Malmstrom AFB, MT	150
II/ 44 SMW	Ellsworth AFB, SD	150
IV/351 SMW	Whiteman AFB, MO	150[14]

HISTORY:
IOC: 1966[15] (see Table 5.13 for MINUTEMAN chronology)

TARGETING:
Types: moderately hard targets; soft large-area military and industrial installations requiring high yield but less than pinpoint accuracy; isolated targets[16]

Selection Capability: eight target selection in missile computer, one set designated default primary

Retargeting: re-programming of target data in missile computer could take 36 hours. Command Data Buffer System, which permits retargeting in 25 minutes, is not being installed in MINUTEMAN II.[17]

Accuracy/CEP: 0.2-0.34 nm[18]

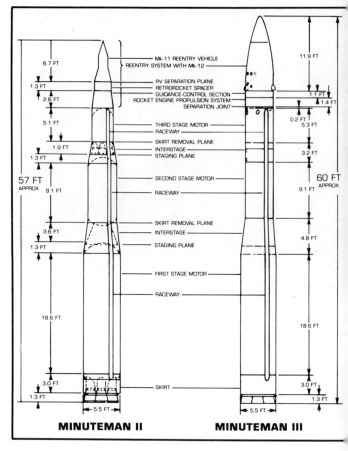

Figure 5.3 Comparison of MINUTEMAN II and MINUTEMAN III.

COST:
Annual operations: $330 m (FY 1980)[19]
$345 m (FY 1982)[20]

5
MINUTEMAN II

FY	Number Procured	Total Appropriation ($ million)
1979 & prior	620[21]	unknown
1980	-	144.5
1981	-	196.0
1982	-	140.7

COMMENTS: MM II has the capability to carry chaff and the Mk-1A penetration aids canister.[22] Missile is showing signs of deterioration and by mid-1980s missiles will have to be replaced or overhauled.[23]

---

1 *The World's Missile Systems*, 6th Ed., p. 286.
2 *Ibid*.
3 *Ibid*.
4 *Military Balance*, 1980-1981, p. 88; Heritage Foundation, *SALT Handbook*, p. 75.
5 William Schneider, Jr. and Francis P. Hoeber, *Arms, Men and Military Budgets: Issues for Fiscal Year 1977* (New York: Crane, Russak, 1978), p. 27.
6 John Collins, *op. cit.*; Paul H. Nitze, *op. cit.*, assumes 1600 lb.
7 *The World's Missile Systems*, 6th Ed., p. 269.
8 John Collins, *op. cit.*, p. 446.
9 Heritage Foundation, *SALT Handbook*, p. 75.
10 *The World's Missile Systems*, 6th Ed., p. 269.
11 Mk-11C is an evolution of the Mk-11 RV originally deployed on MINUTEMAN I. Mk-11 had no penetration aids and two target selection capability. Mk-11A reportedly had a different warhead yield. Mk-11B designed for MM II had an eight target selection capability and incorporated penetration aids. Mk-11C was improved and hardened against nuclear weapons effects.
12 Colin S. Gray, "The Future of Land-Based Missile Forces," *op. cit.*, p. 33.
13 JCS, FY 1983, p. 72.
14 Eight MINUTEMAN II missiles at Whiteman AFB are armed with Emergency Rocket Communications System (ERCS) transmitters rather than warheads.
15 DOD, FY 1981 RDA, p. II-14.
16 JCS, FY 1984, p. 15.
17 Colin S. Gray, *op. cit.*
18 The UN Secretary General ("General and Complete Disarmament," A/35/392, 12 September 1980); AW&ST, 16 June 1980, p. 178; Paul H. Nitze, *op. cit.*, assume a CEP of 0.2 nm; John Collins, *op. cit.*, assumes a CEP of 0.34 nm; Colin Gray, *op. cit.*, assumes 0.3 nm.
19 Annual MINUTEMAN/TITAN operations cost, including military personnel; SASC, FY 1982 DOD, Part 7, p. 4002.
20 *Ibid.*, p. 4337.
21 An additional 48 missiles were procured during R&D.
22 SAC, FY 1981 DOD, Part 5, p. 1522.
23 SASC, FY 1981 DOD, Part 2, p. 508.

5
MINUTEMAN III

MINUTEMAN III (LGM-30G)

Figure 5.4 MINUTEMAN III (LGM-30G) missile in silo.

DESCRIPTION:	Air Force three-stage, solid fuel MIRVed ICBM with improved rocket motor, new reentry system and new guidance.
CONTRACTORS:	Boeing Aerospace Co. Seattle, WA; Ogden UT (prime, assembly and test) General Electric Philadelphia, PA (reentry vehicle) GTE Sylvania Needham Heights, MA (ground electronics) Autonetics Division, Rockwell International Anaheim, CA (guidance) Aerojet General Sacramento, CA (2 stage propulsion) Thiokol Chemical Corp. Brigham City, UT (1 and 3 stage propulsion) TRW Systems Redondo Beach, CA (technical direction) Bell Aerospace Textron Buffalo, NY (post boost RV control)
SPECIFICATIONS:	(same as Minuteman II except for third stage)
Length:	59 ft 11 in (718.8 in)[1]
Diameter: Stage I: Stage II: Stage III:	66 in 52 in 52 in
Stages:	3
Weight at Launch:	77,900 lb; 76,000 lb[2]
Propulsion:	three solid propellant rocke engines plus post boost systen
Speed:	15,000+mph; 24,000+ km/h a burn-out (Mach 19.7)
Guidance:	all inertial gimballed, im proved NS-20 (INS-20) ha been deployed on all 550 MM IIIs.[3]
Throwweight/ Payload:	2400 lb,[4] 1500-2000 lb,[5] 1975 lb, 2000 lb;[7] 2300 lb[8]
Range:[9]	8000 nm,[10] 6950 nm[11]
Ceiling:	about 700 mi
DUAL CAPABLE:	no
NUCLEAR WARHEADS:	2 or 3 MIRV/missile;[12] missile carries W62/Mk-12 warhead with 170 Kt or W78/Mk-12A warhead with 335-350 Kt

5
MINUTEMAN III

Figure 5.5 **MINUTEMAN III** launch sequence from Vandenburg Air Force Base, California.

DEPLOYMENT:

Launch Platform: fixed 25+ m underground hardened silo with missile suspension, shock isolated floor, debris collection system and EMP protection; silo upgrade program to protect launch facilities completed in January 1980[13]

Silo Hardening: approximately 2000 psi[14]

Number Deployed: 550 (250 with Mk-12, 300 with Mk-12A) missiles active (1983); 867 missile production deliveries, 152 flight tests[15]

Nuclear Weapons Databook, Volume I **117**

5

MINUTEMAN III

Location:

Wing	Base	Missiles
I/ 341 SMW	Malmstrom AFB, MT	50
III/ 91 SMW	Minot AFB, ND	150
V/ 90 SMW	F.E. Warren AFB, WY	200
VI/321 SMW	Grand Forks AFB, ND	150

A number of silos at Vandenburg AFB, CA can also be used to launch missiles in the event of war.[16]

HISTORY:
IOC: Jun 1970 (see Table 5.13 for MINUTEMAN chronology)
1980-1983: W62/Mk-12 RVs on 300 MM IIIs replaced by W78/Mk-12A[17]
1983-1985: Fifty MM II at Malmstrom replaced with MM III[18]

TARGETING:
Types:
  Mk-12A: hardened target system across the entire spectrum
  Mk-12: hard targets

Selection Capability: three target selection capability in missile computer, one for each set of MIRVs, one set designated default primary

Retargeting: Command Data Buffer allows infinite retargeting of missiles in 25 minutes and retargeting of the entire force in 10 hours.

Accuracy/CEP: 0.12 nm (Mk-12A);[19] 600 ft;[20] 0.12 nm (Mk-12);[21] 900-1000 ft (INS-20)[22]

Table 5.13
MINUTEMAN Chronology

Oct 1958	Boeing chosen as MINUTEMAN missile contractor
Feb 1961	First MINUTEMAN prototype launch
Apr 1962	First production MM I completed
Oct 1962	First MM I goes on alert
Dec 1962	IOC of MM I (20 missiles)
Feb 1963	First complete operational MM I squadron active at Malmstrom AFB, MT
Sep 1964	First flight test of MM II
Oct 1965	First MM II deployed in underground silos
Feb 1966	First salvo firing of MM from Vandenburg AFB, CA
1966	Development of MM III begins
Apr 1967	450th MM II deployed, and 1000th MM goes on strategic alert
Aug 1968	First flight test of MM III
Dec 1970	First squadron of MM III active at Minot, ND
Mar 1973	MM II completes replacement of MM I
Feb 1975	Command Data Buffer/Upgrade Silo Modification completed at first MM III base
Jul 1975	Last MM III deployed at Malmstrom AFB, MT
Jul 1976	MM III tested with INS-20 guidance
Sep 1978	Guidance improvement/software modification completed at all MM III wings
Nov 1978	Production of MM III completed
Jul 1979	IOC of INS-20 guidance on MM III
Jan 1980	Silo Upgrade Program completed
1983-1984	50 MM III replace 50 MM II at Malmstrom

5
MINUTEMAN III

COST:
Unit Cost: $4.622 m (FY 1976) (flyaway)
                        $7.875 m (program costs)
                        $4.842 m (FY 1977) (flyaway)

Annual Operations: $330 m (FY 1980)[23]
                               $345 m (FY 1982)[24]

FY	Number Procured	Total Appropriation ($ million)
1979 & prior	794[25]	12,800.0[26]
1981 & prior	794	12,586.2[27]
1980	-	144.5
1981	-	196.0
1982	-	185.7

COMMENTS: Airborne Launch Control System (ALCS) provides a backup launch control capability to underground launch control centers for 200 of 550 MM IIIs.[28] The Mk-12 reentry vehicle has the capability to carry chaff.[29]

---

1 *The World's Missile Systems*, 6th Ed., p. 288.
2 *Ibid.*
3 ACDA, FY 1982 ACIS, p. 3; ACDA, FY 1983 ACIS, p. 3. The improved software is "predicted" to improve operational accuracy but more tests are required to confirm estimates.
4 2400 lb for 3 reentry vehicles. General Accounting Office, "The MX Weapon System: Issues and Challenges," 17 February 1981, p. 34.
5 GAO, *op. cit.*, p. 88.
6 John Collins, *op. cit.*, p. 446.
7 Colin S. Gray, *op. cit.*, p. 132.
8 Heritage Foundation, *SALT Handbook*, p. 75.
9 Range is for MM III/Mk-12; missiles with Mk-12A RVs are heavier and thus have a shorter range; HAC, FY 1980 DOD, Part 2, p. 499.
10 John Collins, *op. cit.*, p. 446; *The World's Missile Systems*, 6th Ed., p. 288.
11 *The World's Missile Systems*, 6th Ed., p. 269.
12 Three RVs per missile is the nominal loading. Some may be deployed with less; DOD, FY 1981 RDA, p. 11-14; SAC, FY 1980 DOD, Part 1, p. 1407; SAC, FY 1981 DOD, Part 5, p. 1522; payload varies for mission; ACDA, FY 1979 ACIS, p. 1; it is technically possible to use seven RVs on MM III, and this has been demonstrated; SASC, FY 1980 DOD, Part 1, p. 389.
13 ACDA, FY 1983 ACIS, p. 3.
14 HAC, FY 1982 DOD, Part 9, p. 112.
15 AW&ST, 19 April 1982, p. 65.
16 *Air Force Times*, 28 July 1980.
17 AW&ST, 9 March 1981, p. 25. Current plans, however, do not call for the Mk-12A to be deployed on the remaining 250 MM III missiles; ACDA, FY 1983 ACIS, p. 3. SAC has asked the USAF to continue production of Mk-12A beyond FY 1982 and to deploy W78/Mk-12A on at least 200 of the remaining 250 MINUTEMAN IIIs.
18 Fifty MM II missiles at Malmstrom AFB, Montana, will be replaced with MM IIIs starting in FY 1983, thus adding 100 warheads; JCS, FY 1983, p. 72; SASC, FY 1983 DOD, Part 7, p. 4159.
19 Paul H. Nitze, *op. cit.*, and AW&ST, 16 June 1980, p. 178. The UN Secretary General, *op. cit.*, assumed a CEP of 300 m for both the MINUTEMAN III Mk-12 (initial deployment-1970) and the MINUTEMAN III Mk-12A (initial deployment-1979); NS-20 guidance estimated to be .15 nm with Mk-12; see also Colin S. Gray, *op. cit.*, p. 33, fn. 8.
20 *Military Balance*, 1980-81, p. 3.
21 Paul H. Nitze, *op. cit.*
22 AW&ST, 22 March 1982, p. 18.
23 Annual MINUTEMAN/TITAN operating cost including military personnel; SASC, FY 1982 DOD, Part 7, pp. 3992, 4002.
24 SASC, FY 1982 DOD, Part 7, p. 4337.
25 An additional 44 missiles were procured for R&D; *U.S. Missile Data Book*, 1980, 4th Ed., p. 2-52.
26 Prior to FY 1981, $12.8 billion had been appropriated for procurement of MINUTEMAN ICBMs and spares; ACDA, FY 1982 ACIS, p. 4-5.
27 "Minuteman Squadrons"; ACDA, FY 1983 ACIS, p. 11.
28 ACDA, FY 1982 ACIS, pp. 2, 4.
29 SAC, FY 1980 DOD, Part 1, p. 1407.

5

PEACEKEEPER/MX

PEACEKEEPER/MX Missile System

The PEACEKEEPER/MX missile is a completely new weapon system under development incorporating advanced components and technology in its missile booster, guidance control system, post boost vehicle, reentry system, and warhead. The guidance system improvements, larger post boost vehicle (more warheads), new warhead design,[1] and greater range and accuracy make the MX a significant improvement over the present MINUTEMAN III.

The MX program formally began in 1971 and entered advanced development in 1974. Initially, there was considerable study of basing modes, concentrating on "multiple aim point systems" where each missile would have a large number of launching points (see MX Basing). Although a large missile was quickly chosen to maximize payload capability, the program became bogged down in political controversy surrounding the kind of basing mode, environmental concerns, arms control implications, and cost.

The Reagan Strategic Program, announced in October 1981, determined that MX development would continue and that at least 100 operational missiles would be deployed.[2] Forty MX missiles are to be deployed initially in converted MINUTEMAN silos, even though a plan to "superharden" them to 5000 psi strength was cancelled.[3] Full-scale engineering development contracts for all components of the missile were concluded by FY 1982, and MX development is reportedly on schedule. The first flight test is scheduled for 1983, a full scale production decision is projected for mid-1983, and an IOC is planned for late 1986.

The Strategic Program also presented a number of basing options in place of multiple protective structures and other mobile land-based options. On 22 November 1982, the DOD announced selection of a "closely spaced basing" (CSB) or "dense pack" mode, which would involve placing 100 MX missiles in "vertical shelters" so as to avoid a calculated single attack to destroy all the missiles. Congress, however, decided in December 1982 to restrict expenditure of MX funds until a permanent basing mode was approved, and asked the President to submit a report on MX alternatives. The President then established a Commission on Strategic Forces to examine such alternatives. The Commission was guided in part by the DOD's requirement to deploy a missile in a basing mode or combination of modes "resistant to future Soviet threats resulting from further technological advances in missile accuracy and proliferation of missile warheads."[4]

The Presidential Commission on Strategic Forces reported its recommendations in April 1983:[5]

- immediate deployment of 100 MX missiles in existing MINUTEMAN silos, replacing older MINUTEMAN and TITAN II missiles,
- research to resolve uncertainties regarding silo hardness,
- investigation of different types of land-based vehicles and launchers, including hardened vehicles, and
- engineering design of a new single-warhead ICBM.

---

1 Before the W87 was chosen, the warheads generally thought to be under consideration for MX were of higher yield than the W78/Mk-12A currently on a portion of the MINUTEMAN force. The W87, however, has a yield approximately equal to the W78.
2 AF/RD, "MX Development and Deployment Plan," 8 February 1982; the previous plan had been to deploy a force of 200 operational missiles.
3 HAC, FY 1982 DOD, Part 9, p. 112.
4 AF/RD, op. cit., p. 2.
5 "Report of the President's Commission on Strategic Forces," April 1983.

5
MX

MX Missile

Figure 5.6 Full scale mock-up of **MX** missile.

DESCRIPTION:	Air Force large payload, solid fuel, cold launch,[1] four stage ICBM under development.
PRIME CONTRACTORS:	(see Table 5.15 for list of "associate" prime contractors)
SPECIFICATIONS:	
Length:	71 ft (21.6 m)
Diameter:	92 in (233 cm)
Stages:	4
Weight at Launch:	193,000 lb (87,500 kg)
Fuel:	three solid propellant booster motors, storable liquid hypergolic propellant in the fourth stage, post boost vehicle[2]
Guidance:	inertial floating ball (advanced inertial reference sphere)[3]
Throwweight/ Payload:	7900 lb;[4] 7200 lb[5]
Range:	5800 nm;[6] 7000+ nm (13,000+ km)
DUAL CAPABLE:	No
NUCLEAR WARHEADS:[7]	W87 on the Mk-21 (formerly Advanced Ballistic Reentry Vehicle (ABRV)); 10 MIRV/ missile baseline;[8] there is room for 12 RVs on the MX bus without stacking;[9] 300 Kt range (see W87)
DEPLOYMENT:	(see MX Basing)
Silo Hardening:	circa 2200 psi
Number Planned:	226 missiles for 100 MX system; 339 missiles for 200 MX system[10]
HISTORY:[11]	
IOC:	December 1986[12] (see Table 5.16, MX Chronology)
TARGETING:	
Types:	all hardened targets, including "superhard" control centers; W87 allows targeting of "fourth generation ICBM silos" and "very hard leadership bunkers."[13]
Selection Capability:	five fuzing modes remotely selectable via targeting instructions[14]
Retargeting:	"automatic retargeting" capability including a capability to "reprogram target information to compensate for missiles that malfunction or are destroyed by an enemy attack"[15]
Accuracy/CEP:	less than 400 ft[16]

Nuclear Weapons Databook, Volume I   **121**

5
MX

Table 5.14
MX Missile System Costs

System Costs ($ billion):

Total Acquisition:	GAO (78)[1]	DOD (82) (FY 78 BY)[2]	(TY)[3]	ACDA (83)[4]	DOD (83)[5] (FY 82)
Development:	6.8	6.7	10.6	4.3	9.8
Procurement:	13.0	12.9	26.3	-	8.6
Construction:	9.5[6]	9.0	19.1	.1	.9
Operations (to 2000):	4.9	-	-	-	-
TOTAL:	34.2	28.6	56.0	4.4	19.3

Annual Cost:

FY	Number Procured	Total Appropriation ($ million)
(1977 & prior)	-	159.4[7]
(1978 & prior)	-	293.8[8]
1980	-	732.4
(1981 & prior)	-	2451.6[9]
1981	-	1605.1
1982	-	1994.1[10]
1983	(5)	4773.6[11]
1984	27	6636.3[12]

Annual Operations Cost: 448.0 million[13]

The Air Force has consistently claimed that the MX system will cost no more than $33-34 billion, including 200 missiles in the MPS basing mode to the year 2000, omitting operations and maintenance costs; a number of other sources, however, estimated MX costs for the same system at $55.6 billion, including operations and maintenance through the year 2000.[14] Some unofficial estimates are more than twice this amount.

New Reagan Administration figures for a 200 MX (100 operational) system are approximately $27 billion.[15] However, since silo basing is itself $19.3 billion, and the additional DOD quoted costs of long term basing is $10-30 billion, system costs appear $30-50 billion. Since most previous estimates have been based upon a particular basing mode, comparison with other estimates is impossible.

1 GAO, "The MX Weapon System: Issues and Challenges," 17 February 1981, p. 4. These costs exclude the DOE costs to develop, acquire, and maintain warheads for MX, and impact assistance funds to the areas where MX will be deployed.
2 SASC, FY 1982 DOD, Part 7, p. 3970.
3 Ibid.
4 Sunk Costs; ACDA, FY 1983 ACIS, p. 12.
5 This estimate includes 100 operational missiles and silo basing costs of 40 MX in MINUTEMAN silos.
6 MPS basing mode.
7 ACDA, FY 1979 ACIS, p. 15.
8 SASC, FY 1980 DOD, Part 5, p. 2488.
9 Sunk Costs; ACDA, FY 1983 ACIS, p. 12.
10 Ibid.
11 Ibid.
12 DOD, FY 1984 Annual Report, p. 221.
13 200 missiles in 4600 MPS system; ACDA, FY 1981 ACIS, p. 15.
14 See Council on Economic Priorities, *Misguided Expenditure: An Analysis of the Proposed MX Missile System* (NY, 1981), pp. 115-126; and GAO, "The MX Weapon System—A Program with Cost and Schedule Uncertainties," 29 February 1980.
15 AW&ST, 12 October 1981, p. 18.

COST: (See Table 5.14)

Missile Costs Only: 226 missiles (FY 1982): $4700 m
339 missiles (FY 1982): $6900 m[17]

1 AF/RD, op. cit.; The missile is initially boosted from its protective storage cannister by pressure created by gas generators. As the missile clears the cannister, the first stage main engines ignite, beginning the powered flight phase.
2 DOD, FY 1981 RDA, p. VI-3; SASC, FY 1982 DOD, Part 7, p. 3945.
3 Floating ball suspended in a fluid, less than a foot in diameter; SASC, FY 1982 DOD, Part 7, p. 3951. AIRS provides the flight computer with information on missile movement during flight.
4 SASC, FY 1981 DOD, Part 2, p. 539; GAO indicates 7000 lb for 10 reentry vehicles (Mk-12A); GAO, "The MX Weapon System: Issues and Challenges," 17 February 1981, pp. 16, 34.
5 Heritage Foundation, *SALT Handbook*, p. 75.
6 SASC, FY 1982 DOD, Part 7, p. 4004.
7 The W78/Mk-12A Warhead/RV was the baseline warhead for the MX until January 1982 when the W87/ABRV was chosen; ACDA, FY 1983 ACIS, p. 6.
8 ACDA, FY 1982 ACIS, p. 16.
9 AF/RD, op. cit., p. 7; SASC, FY 1983 DOD, Part 7, p. 4495, states that 12 ABRV warheads is possible "without significant performance limitations."
10 DOD News Release, 31 December 1981; HAC, FY 1982 DOD, Part 9, p. 264.
11 Air Force Systems Command, "Brief Facts About M-X," Andrews Air Force Base, MD.
12 Original IOC was July 1986; HAC, FY 1982 DOD, Part 2, p. 225; new IOC of Dec 1986 occurred during Reagan Administration.
13 SASC, FY 1983 DOD, Part 7, p. 4173.
14 AVCO Systems Division, "Advanced Ballistic Reentry Vehicle," Fact Sheet, n.d. (circa 1981).
15 AF/RD, op. cit., p. 8.
16 AW&ST, 22 March 1982, p. 18.
17 HAC, FY 1982 DOD, Part 9, p. 264.

5

MX

Figure 5.7 MX prototype vehicle launch validation test at Mercury, Nevada, November 1982. MX missile loaded into cannister (top left); completion of loading phase (top right); MX ejected from cannister in "cold launch" method (bottom left); prototype missile clearing cannister in flight (bottom right).

5

MX Contractors

Table 5.15
Major MX Contractors*

Company	Location	Component
Aerojet Strategic Propulsion Co.	Sacramento, CA	propulsion, Stage II
Avco Corp., Systems Division	Wilmington, MA	reentry vehicle
Boeing Aerospace Co.	Seattle, WA; Las Vegas, NV	transporter; basing
Charles Stark Draper Laboratory, Inc.	Cambridge, MA	technical support for guidance and control
Dynamics Research Corp.	Wilmington, MA	inertial measurement study
Economics Technology Assoc.	Los Angeles, CA	system development
Ertech Western	Long Beach, CA	siting studies
Fugro National	Long Beach, CA	siting
General Electric Co.	Philadelphia, PA	reentry vehicle backup
GTE Sylvania, Inc., Strategic Division	Needham Heights, MA; Westboro, MA	command and control
Henningson, Durham and Richardson	Santa Barbara, CA	environmental studies
Hercules, Inc., Aerospace Div.	Magna and Bacchus, UT	stage III
Honeywell, Inc., Avionics Div.	St. Petersburg, FL; Clearwater, FL;	guidance and control
Logicon, Inc.	San Pedro, CA	targeting
Martin Marietta Corp.	Denver, CO	assembly, test and support, cannister
Northrop Corp., Electronics Division	Hawthorne, CA	AIRS, inertial measured unit
Physics International	San Leandro, CA	engineering support
Northrop Corporation Precision Products Division	Norwood, MA	gyro
Ralph M. Parsons Co.	Pasadena, CA	basing
Person, Brinkerhoff, Quade & Douglas	New York, NY	design hardened protective structure
Rockwell International, Autonetics Division	Anaheim, CA	flight computer, guidance
Rocketdyne Division	Canoga Park, CA	stage IV
Sandia Corporation	Albuquerque, NM	arming and fuzing system
Science Applications	San Diego, CA	analysis support, development study
SofTech, Inc.	Waltham, MA	software compiler
Systems Science & Software	San Diego, CA	nuclear hardness
TASC	Reading, MA	guidance/control study
TRW, Inc.	Redondo Beach, CA	integration and targeting
Thiokol Corp., Wasatch & Elkton Divisions	Brigham City, UT; Elkton, MD	stage I and ordnance
UltraSystems, Inc.	Irvine, CA	logistics support
University of Houston	Houston, TX	maintenance/management study
Westinghouse	Sunnyvale, CA	canister

* Prime contractors as of July 1982, information provided by MX Program Office; Adapted from CEP, *Misguided Expenditures, op. cit.*, pp. 173–219, and HAC, FY 1980 DOD, Part 2, pp. 455–456.

5
MX Warhead

MX Warhead and Reentry Vehicle

On 29 January 1982, DOD made a final decision on the MX warhead. DOD chose a new warhead, the W87, to be mated with the new Mk-21 Advanced Ballistic Reentry Vehicle (ABRV).[1] The fourth stage of the MX Missile carries the reentry vehicle "bus" and computer systems which release the W87/Mk-21 RVs in their intricate spacing and deployment maneuvers, allowing them to continue accurately to their targets.

The W87 warhead was originally thought to be of higher yield than the original baseline warhead for the MX, the W78.[2] It has since been revealed that the W87's yield is approximately the same as that of the W78.[3] Two new features of the W87 were important to its selection. First, the W87 uses less fissile material than the W78 through a more efficient design. Second, the yield of the W87 can be increased by changing the mix of fissionable materials.

Two other factors went into choosing the W87/Mk-21 rather than the W78/Mk-12A: the DOD believes (1) that each leg of the Triad in the strategic force should "have at least two [types of] warheads," and (2) that the ABRV is more accurate.[4] In distinguishing the Mk-21 from the Mk-12A, the DOD states that the Mk-21 is "a more militarily effective weapon in the context of accuracy, hardness and overall military efficiency."[5]

Two other higher yield nuclear warhead designs were earlier considered for MX, but were rejected: the 500-600 Kt CALMENDRO and the 800 Kt MUNSTER.[6] The CALMENDRO was developed at LANL, was moved to LLNL[7] where it entered Phase 3 (Development Engineering) in FY 1982,[8] and was reportedly favored to replace the W78. However, both of these high yield variants were dropped from consideration and development then focused on the W87/Mk-21. The W87 is expected to be ready in time for a 1986 IOC on the first 10 MX missiles.[9]

Table 5.16
MX Chronology

1963	Air Force begins study of the "Improved Capability Missile" (WS120A)
1967	Air Force introduces concept for a mobile land-based missile to be shifted among silos
Nov 1971	SAC forwards requirement for new ICBM
May 1974	Advanced development of MX missile begins
Feb 1975	Secretary of Defense Schlesinger rejects air-mobile concepts for MX in favor of multiple protective shelter and buried trench.
Mar 1976	DSARC I approves large missile with emphasis on trench
Jun 1979	full-scale development of MX missile in MPS basing mode authorized by President Carter
Sep 1979	Presidential decision on horizontal MPS basing and proceeding with full scale development
Oct 1981	Reagan Strategic Program cancelling MPS basing and restructuring of MX program results in delay from July 1986 to late 1986[1]
Jan 1982	W87/ABRV chosen as warhead/RV for MX
Nov 1982	Closely Spaced Basing announced as latest preferred basing mode
Dec 1982	Congress requests report on MX alternatives
early 1983	First flight tests scheduled
Apr 1983	Report recommends MX in silos, with small missile follow-on
Jul 1983	DSARC III
1984	MPS construction slated to start
Dec 1986	Initial operational capability planned[2]
FY 1990	Full operational capability

1 SASC, FY 1983 DOD, Part 7, p. 4484.
2 AF/RD, "MX Development and Deployment Plan," 8 February 1982, p. 6; previous DOD estimates were "mid-1986," see for instance, JCS, FY 1982, p. 70.

---

1 ACDA, FY 1983 ACIS, p. 7; DOE, FY 1982 Supplemental Request to the Congress, Atomic Energy Defense Activities, March 1982, p. 5.
2 AW&ST, 9 March 1981, p. 51, identified the yields as 500 Kt for the CALMENDRO and 800 Kt for the MUNSTER; New York Times, 10 October 1981, p. 26, later identified the ABRV yield as 600 Kt, but it has also been referred to as 500 Kt in AW&ST, 4 May 1981, p. 51.
3 AW&ST, 9 March 1981, p. 25; AW&ST, 22 March 1982, p. 18.
4 HAC, FY 1982 EWDA, Part 5, pp. 34, 180.
5 SASC, Strategic Force Modernization Programs, pp. 102-103.
6 AF/RD, "MX Development and Deployment Plan," 8 February 1982, p. 7.
7 AW&ST, 22 March 1982, p. 18.
8 HAC, FY 1983 DOD, Part 4, p. 595.
9 New York Times, 10 October 1981, p. 26.

5
W87

Figure 5.8 MX bus with four Mk-21 reentry vehicles mounted.

FUNCTION:	Warhead on the Mk-21 (formerly Advanced Ballistic Reentry Vehicle (ABRV)), for the MX and Mk-5 RV for the TRIDENT II missile.
WARHEAD MODIFICATIONS:	none known
SPECIFICATIONS:	
Yield:	300 Kt upon deployment for MX,[1] upgradable to 475 Kt[2]
Weight:	unknown[3]
Dimensions (Mk-21):	
Base Diameter:	21.8 in
Nose Radius:	1.4 in
Overall Length:	68.9 in
Half Angle:	8.2°
Materials:	contains oralloy;[5] uses less materials than W78/Mk-12A;[6] has feature to increase yield by adding additional oralloy at later date; contains IHE[7]
SAFEGUARDS AND ARMING FEATURES:	primary inertial (interactive) path length fuze with microprocessor immune to jamming;[8] secondary dual mode radar with microprocessor for airburst, surface/proximity fuzing; five modes: high altitude fuze, airburst, low airburst, surface/proximity burst and surfact/contact burst[9]
DEVELOPMENT:	
Laboratory:	LLNL[10]
History:	
IOC:	1986
FY 1983	Lab assignment (Phase 3)
(1986)	initial deployment (Phase 5)
Production Period:	1985-?
DEPLOYMENT:	
Number Planned:	1050;[11] 1055[12]
Delivery System:	MX missile, possibly TRIDENT II and small missile

5
W87

Service: Air Force; possibly Navy

Allied User: no

Location: see MX Basing

COMMENTS: Additional program cost of larger warhead over previously planned W78/Mk-12A was estimated at $1.2 billion (in FY 1982 dollars).[13] Given a reduction in cost of fissile materials of some $500 million with the ABRV, the actual additional cost of the ABRV is estimated at $881 million.[14]

---

1 AW&ST, 9 March 1981, p. 25, identified the CALMENDRO, one of the two earlier competing warheads for the ABRV/MX, as having a yield of 500 Kt; New York Times, 10 October 1981, p. 26, identified the ABRV yield as 600 Kt; AW&ST, 22 March 1982, p. 18, identified the chosen warhead as 300 Kt.
2 AW&ST, 17 January 1983, p. 26.
3 The Mk-21/W87 is heavier than the Mk-12A, but the Air Force states that this will not affect its operational requirements; SASC, FY 1983 DOD, Part 7, p. 4174.
4 AVCO Systems Division, "Advanced Ballistic Reentry Vehicle," Fact Sheet, n.d. (circa 1981).
5 AW&ST, 22 March 1982, p. 19; HAC, FY 1983 DOD, Part 4, p. 597.
6 SASC, FY 1980 DOD, Part 5, p. 2496.

7 SASC, *Strategic Force Modernization Programs*, p. 103; SASC, FY 1983 DOD, Part 7, p. 4179.
8 SASC, FY 1983 DOD, Part 7, p. 4179.
9 AVCO Systems Division, *op. cit.*
10 The CALMENDRO warhead was developed at LANL, but tested and engineered at LLNL. The MUNSTER warhead was developed at LANL.
11 SASC, FY 1983 DOD, Part 7, p. 4487.
12 SAC, FY 1983 DOD, Part 2, p. 84.
13 SASC, FY 1982 DOD, Part 7, p. 4000.
14 SASC, FY 1983 DOD, Part 7, p. 4487.

5
MX Basing

MX Basing

Since 1965, the Air Force has studied over forty different options (see Table 5.17) for mobile basing of an advanced ICBM. Options have included trains, trucks, air cushion vehicles, and aircraft, as well as deployment in waterways, in proliferated shelters, or underground buried trenches.

Although basing proposals over the years have varied widely, they have essentially included four major kinds:

- Land Basing: including "multiple protective shelters," either mobile or semi-fixed; silo basing, with and without BMD adjunct; road or off-road mobile,
- Air Mobile,
- Sea Basing: SUM, surface ship or TRIDENT II, and
- Defense: BMD of existing silo, launch under attack, defense of MPS.

The Reagan Administration has found it as difficult as previous Administrations to choose an acceptable basing mode for the new MX missile. In October 1981 it cancelled the Multiple Protective Shelter (MPS) basing scheme chosen by the Carter Administration and announced that initial deployment would be in 40 MIN-

Table 5.17
MX Basing Options[1]

Covered Trench	Unmanned TELs traveling randomly in trench covered with camouflage.	Hard Rock Silo/ Deep Underground	Silo launchers built in granite outcroppings in SW United States.
Hybrid Trench	Unmanned TELs in shallow buried tunnels with hardened firing points.	Hard Tunnel	Missile stored in very deep hardened tunnels able to withstand direct hit and then digout on launch command.
Dash to Shelter	TELs at center of radial road or rail network, dashing to hardened shelters on warning.	Launch Under Attack	ICBM force capable of launch from early warning.
Pool	Transporters deposit water-tight encapsulated missiles in opaque water pools, serving as shelters.	Shallow Underwater Missile	Encapsulated missiles fastened to small submarines patrolling off US coast.
Sandy Silo	Buried encapsulated missiles in 2000 ft deep holes covered with sand. Pressurized water would fluidize the sand and capsules would float to surface for launch.	HYDRA	Waterproof missiles anchored to offshore sea bed.
		ORCA	Encapsulated missiles anchored to offshore sea bed.
Dedicated Rail	Randomly moving unmanned nuclear hardened trains carrying missiles on grid network.	Ship Ocean	Missiles on special ships moving in oceans.
		Road Mobile MX/ MINUTEMAN	Truck launched missiles dispersed on warning.
Public Railroads	Missiles on special cars randomly moving on public railroads.	MINUTEMAN MPS	Expand MM fields by adding new silos.
Off-Road Mobile	Fleet of off-road mobile TELs scattered over uninhabited areas of SW United States.	BMD	Deploy ABMs in MX or MINUTEMAN fields.
Wide Body Jet	Ground alert missile launching 747 or C-5 class air craft.	Mesa Basing	Horizontal tunnels on south side of mesas.
Short Takeoff and Landing	Ground alert missile launching STOLs, possible with network of new small airfields.	Grasshopper	VTOL aircraft with new small missile.
		Great Lakes	Small submarines or barges in Great Lakes or inland waterways.

1 HAC, FY 1982 DOD, Part 2, pp. 254-255; SASC, FY 1982 DOD, Part 6, pp. 3745-3747; DOD, "ICBM Basing Options: A Summary of Major Studies to Define a Survivable Basing Concept for ICBMs," December 1980.

5
MX Basing

Figure 5.9 One-eighth scale MX silo test model after being subjected to a large TNT blast.

UTEMAN silos.[1] The silos, however, would not be further hardened although the option of increased hardening remained available.[2] Four alternatives for long-term basing of MX were introduced:

- Deep Basing: deployment in survivable locations deep underground,
- Continuous Patrol Aircraft: deployment aboard long-endurance aircraft that could launch MX,
- Ballistic Missile Defense: active defense of missiles in present ICBM silos, and
- Deceptive Basing: deceptive basing with Ballistic Missile Defenses.

The decision on a long-term basing mode was originally required by 1 December 1982, as directed by Congress.[3]

Following the 22 November 1982 decision by DOD to deploy MX in a Closely Spaced Basing configuration near F.E. Warren AFB, Cheyenne, Wyoming, Congress requested that the Administration reexamine MX deployment and alternatives. The Presidential Commission on Strategic Forces, which was formed in response to Congress, made the following recommendations relevant to MX basing in April 1983:[4]

- Putting the MX missile into production while scaling down, at least initially, the original deployment plan to 100 missiles,
- Deploying 100 MX missiles in underground silos now used for older MINUTEMAN missiles. This option would allow the first MX missiles to be fielded roughly one year earlier than with any mobile basing,
- Accelerating research, development and testing—although not necessarily deployment—of an antiballistic missile (ABM) defense system,
- Beginning engineering design of a missile smaller and lighter than MX that could be produced in large numbers in the early 1990s, be mobile or fixed, and be relatively invulnerable (see Small Missile),
- Accelerating development of the advanced (D5) version of the submarine-based TRIDENT missile, and

Figure 5.10 Reinforcing steel bar skeleton used in the MX silo model test.

---

1 The MX, by design, has always been compatible with silo basing but Congress specified in the FY 1977 authorization bill that none of the program's funds be spent on silo basing; ACDA, FY 1979 ACIS, p. 12.
2 AW&ST, 11 January 1982, p. 20; AF/RD, op. cit., p. 9.
3 HAC, FY 1983 DOD, Part 9, p. 723.
4 "Report of the President's Commission on Strategic Forces," April 1983. These were essentially the options discussed much earlier in the Administration; see, for instance, Michael Getler and Lou Cannon, Washington Post, 7 June 1981, pp. A1, A16.

5

MX Basing

- Examining uncertainties regarding silo or shelter hardness and different types of land-based vehicles and launchers, for later possibility of shifting MX or deploying small missiles.

Multiple Protective Structure (MPS) Basing

In August 1979, President Carter announced the selection of a "race track" basing mode, involving 200 tracks of about 25 miles circumference, each equipped with 23 shelters and one MX missile. The missile, its capsule, and a transporter-erector-launcher (TEL) would be docked at one of the horizontal shelters, under cover of a "shield vehicle" which would visit each of the shelters in turn. The shield vehicle would contain either the TEL or a decoy simulator, so that electronic surveillance or observation would not enable one to determine which shelter contained the missile. One option was for the TEL to have the ability to move on warning. Thus, if warned of an ICBM launch against the MX complex, some or all of the 200 TELs would race from the shelters where they had been hidden to other shelters. The shelters would be clustered in valleys in Utah and Nevada. The Air Force favored a variation of this basing mode which called for 200 MX missiles to be shuttled at random among the 4600 shelters spaced out over 5000 square miles. Testimony by U.S. Secretary of Defense Brown and by others emphasized that the capability to move on warning would avoid any vulnerability which might arise if the location of the missiles became known.

Figure 5.11 MINUTEMAN I (LGM-30B) missile launched from a C-5A transport during 1974 test of air-launched ICBM concept.

DOD believed MX could remain survivable against 10,000 RVs.[5]

Critics argue that if the basing mode were limited to 4600 shelters, the Soviet Union, in the absence of SALT II constraints, could have sufficient warheads to defeat this system. Estimates of the required expenditure for this system range from $30-60 billion and more, of which only about $5 billion was for procurement of the missiles themselves.

Deep Basing

Deep Basing (DB) includes several possibilities for deployment, ranging from deep underground silo tunnels for individual missiles to underground "citadels" for several missiles. The greatest asset of this basing mode is that missiles at great depth can survive a nuclear attack.

Although many DB systems have been described, perhaps the best known is the Mesa Concept. This is a system of interconnected deep tunnels 2000 to 3000 feet below the surface of a mesa or similar geological formation that would provide attenuation of weapons effects. Stored within the complex would be the MX missiles and all the necessary equipment, communications, and personnel to operate and maintain them for post-attack "dig-out." Such a system could supposedly provide a secure survivable reserve force.

The primary operational problem seen with deep-based MX was poor reaction time, which would make it available only for launch after an attack. A system with predug portals could provide quick-response capability, but survival of the portals would be difficult to achieve. "Dig-out" capability would increase survivability, but at the sacrifice of quick response. Other significant problems were arms control verification, environmental impact, and cost.

Continuous Patrol Aircraft

The "Continuous Patrol Aircraft" (CPA) concept called for deployment of MX on large, "long-loiter-time," fuel-efficient aircraft. A portion of the MX force would be constantly aloft, with another portion maintained on alert at ground bases. During brief periods of high alert, more aircraft could be kept aloft where they could survive for limited periods of time.

Survivability under this concept would derive from the difficulty in attacking the airborne portion of the MX force. The precise location of CPAs would be kept secret and many would be expected to survive any preemptive

---

5 AW&ST, 4 May 1981, p. 49.

5

MX Basing

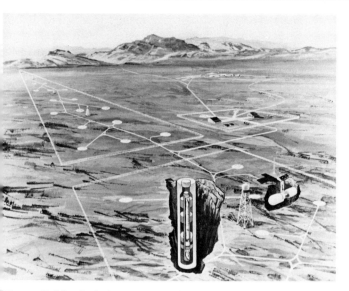

Figure 5.12 Artist's concept of **Closely Spaced Basing** for MX missile.

strike. The primary problem, however, is that CPA would be vulnerable to destruction on the ground by the same means that currently are claimed to threaten U.S. strategic bombers and missile submarines in port.

Ballistic Missile Defense

The feasibility and benefits of a "Ballistic Missile Defense" (BMD)—the concept of actively defending deployed missiles with anti-missile missiles—is being closely examined by DOD. The concept is claimed to be attractive because of technological advances that have been made since deployment of the first ABM system in the early 1970s. Design and development work is presently being focused upon BMD approaches that are compatible with MX basing in silos (the initial 100 plus small missiles and others that could be added in the future), as well as defended deceptive basing of MX or small missiles in some yet to be determined mode. Since silo emplacement is the basing mode for MX, the BMD program is oriented toward plans to increase silo hardness, extend post-attack endurance, and integrate a viable command and control capability concurrent with BMD/silo deployment.

The initial BMD system capability is in the form of a low altitude defense system (LoADs) (called SENTRY) that has for several years been undergoing advanced development for potential use with a variety of MX basing concepts. This system could be used alone or in conjunction with a high-altitude system to enhance the survivability of MX and other ICBMs. The effectiveness of this type of BMD system would be multiplied if deceptive basing of MX missiles would be employed. Then a relatively small number of hardened silo locations would need to be defended by BMD interceptors.

If the BMD option is selected, modifications to the ABM Treaty would be required. The treaty places restrictions on BMD deployments. The possibility of changes in the treaty was anticipated in 1972 when it was agreed that a comprehensive joint review of the treaty would occur every five years. Either side also could unilaterally withdraw from the treaty with six months formal notice.

Closely Spaced Basing/Dense Pack[6]

In November 1982, the last of the formal mobile basing modes for MX—Closely Spaced Basing (CSB)—was recommended as the "permanent" mode. CSB is a new basing mode for land-based ICBMs that would compensate for increasing missile accuracy by using hardness and concentration. CSB involves deploying 100 MX missiles in superhard capsules, spaced 1800 feet apart in a column, which would maximize the phenomenon of "fratricide." Hardness would prevent destruction by an airburst, but at the same time concentration would take advantage of the effects of many incoming ground burst warheads to enhance missile survivability. Fratricide would occur when explosions of incoming warheads attacking closely spaced silos would deflect or destroy other warheads and severely affect their accuracy to destroy hardened silos. The distance between capsules would be small enough to create fratricide but would also be great enough (and capsules hard enough) to ensure that multiple capsules could not be destroyed by one warhead.

The technical claims for CSB were widely disputed. While the Air Force has been reported to believe that 50 to 70 percent of the force could survive an attack in CSB, some analysts believe that only a few missiles would survive.

---

6 DOD, "MX/CSB System," November 1981; CBO, "Contribution of MX to the Strategic Force Modernization Program," n.d. (1982).

5

Small Missile

Small Missile

The inability to find an acceptable basing mode for the PEACEKEEPER/MX has directed greater attention to the concept of a small, highly dispersed, single warhead, land-based ICBM. The Presidential Commission on Strategic Forces recommended engineering design of a single warhead ICBM, weighing about 15 tons, deployed in either hardened silos or shelters or hardened mobile launchers. The Air Force is now developing conceptual designs for such a missile, which it plans to deploy in the early 1990s.[1] The deployment of thousands of these small missiles in hardened silos or on mobile launchers, which will disperse the targets and increase survival in a nuclear attack, is now receiving widespread support as a follow-on to the MX and MINUTEMAN programs. The increased penetration potential of such a small missile is also advanced in its favor. There are, however, significant arms control implications in developing a system that violates the 2250 strategic launcher limit in SALT II. At least three small ICBM alternatives have been suggested: MIDGETMAN, SICBM, and long-range PERSHING II (designated PERSHING III). Both fixed and mobile (so-called "ARMADILLO") deployments have been suggested.

The "MIDGETMAN" missile was the original small missile proposal. It is about 50 feet long, has a range of 7000 miles, and weighs twenty to thirty thousand pounds. Three to four thousand missiles would be deployed in blast-resistant silos spaced about a mile apart. With a total deployment area of 4500 square miles, the small vertical shelters would be highly survivable. The major negative features of MIDGETMAN are high cost and potential technological problems with guidance.[2]

The Small ICBM (SICBM),[3] an outgrowth of a study by Boeing Aerospace, was also offered as an alternative to MX. Encapsulated in a canister similar to the MX and dormant for up to one year without servicing, the

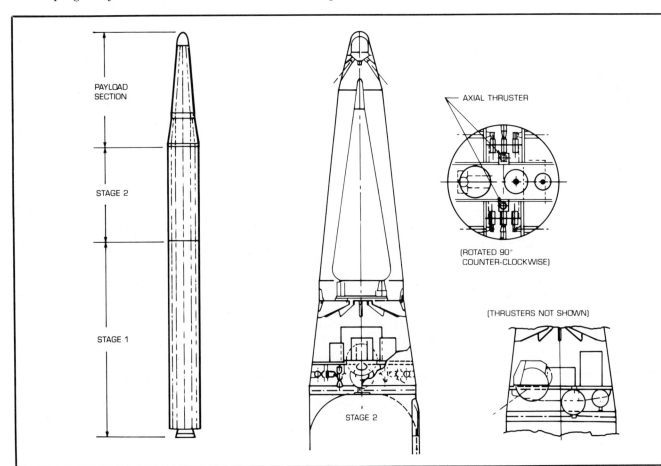

Figure 5.13 Artist's concept of **Small ICBM** prototype.

---

1 SASC, FY 1983 DOD, Part 7, p. 4500.
2 DOD, "ICBM Basing Options: A Summary of Major Studies to Define a Survivable Basing Concept for ICBMs," December 1980, pp. 42-43.
3 AW&ST, 4 May 1981, pp. 49-51.

5
Small Missile

SICBM is envisioned as a small, highly survivable, land-based ICBM system. A number of conceptual designs have been suggested, predicated on a 38.2 foot long baseline design weighing 30,410 pounds.[4] These designs include a 37.8 foot missile weighing 28,840 pounds, a 36.6 foot missile weighing 27,280 pounds, and a 35.4 foot missile weighing 23,700 pounds.[5]

The small missile alternative identified in the Presidential Commission report of April 1983 was similar to the SICBM:

- three stage,
- 25,000-35,000 pound (depending on guidance system),
- circa 1000 pound throwweight,
- 38 foot length, 42 inch diameter, and
- CEP 1-1.8 times MX.

The original SICBM proposal was for some 3350 missiles to be deployed in silos on existing military reservations throughout the western U.S. or at MINUTEMAN missile fields. Each canister would be placed initially in a vertical silo hardened to between 7000 and 8000 psi. The silos would be spaced 1500 to 2000 feet apart. Also suggested was road mobile basing, where a SICBM would move on public roads on a TEL. The warheads would either be joined to the missile and be in continuous movement on roads, or the missile would be dispersed from storage prior to launch (Beehive Basing).

A third suggestion for a small missile has been a modification of the PERSHING II missile (designated PERSHING III), with a third (and fourth) stage added for intercontinental range. The Presidential Commission examined an extended range version of the missile with an approximately 8000 mile range.[6] In material submitted with its report, the Presidential Commission identified PERSHING III as follows:

- four stage,
- circa 25,000 pound,
- circa 1000 pound throwweight,
- 43 foot length, 40 inch diameter, and
- CEP 1-1.8 times MX.

Both the PERSHING III and Boeing Small ICBM proposals have been suggested for use in both mobile and fixed deployments. Air and helicopter transportation and launching systems have also been promoted. Two designs have been reported for mobile TEL vehicles for these missiles. Boeing has promoted a tractor and trailer, 60 feet long, 9 feet high, weighing 67,552 pounds, which would carry the missile in its canister on a launch pallet which would be raised into a vertical position for launch.[7] A second concept, reportedly under development by General Dynamics, is "ARMADILLO," a specially armored carrier and launcher for a 38 foot missile. General Dynamics contends that ARMADILLO's thick armored shell, low silhouette, and ability to anchor itself to the ground to strengthen itself against the blast and winds of a nuclear attack will give it a high degree of survivability.[8]

Each of the small missiles would carry one nuclear warhead. Both the W78/Mk-12A and the W87/Mk-21 (ABRV) have been suggested as possibilities. A new 500 Kt warhead, designated the Advanced Mobile ICBM warhead (or possibly the high yield variant of the W87), has also been mentioned.[9] The light weight TRIDENT I C4 W76/Mk-4 has been suggested for the PERSHING III. If the PERSHING II was adapted as a small ICBM, it would need a new warhead because its present warhead (W85) has a low yield. Both small missile alternatives are now given 1992 IOC projections. Boeing claimed in 1981 that its SICBM could begin deployment by the end of 1986 instead of MX,[10] but as of early 1983, it projected a 1989-1990 IOC. Martin Marietta, however, claimed that the use of its PERSHING II would allow deployment by the planned 1986 MX IOC date. The Presidential Commission's support for the concept of a small missile, however, envisioned an "early 1990s" IOC as an augmentation of MX, rather than a replacement.

---

4 The original SICBM suggested by Boeing in 1981 weighed only 22,000 lb; AW&ST, 4 May 1981, pp. 49-51.
5 Ibid.
6 Walter Pincus and Lou Cannon, Washington Post, 15 February 1983, p. 7.
7 AW&ST, 21 February 1983, p. 14.
8 Leslie H. Gelb, New York Times, 8 February 1983, p. 1.
9 Time, 21 February 1983, p. 18.
10 Walter Pincus, Washington Post, 29 June 1981, pp. A1, A3.

5
POSEIDON Submarine

Sea-Based Missile Systems
POSEIDON Submarine

Figure 5.14 U.S.S. Sam Rayburn (SSBN-635) with POSEIDON missile hatches open.

DESCRIPTION: New POSEIDON submarine class (616 class), consisting of three classes of converted POLARIS boats, of nuclear powered strategic weapons launchers fitted with 16 tubes for POSEIDON C3 or TRIDENT I C4 submarine-launched ballistic missiles (SLBMs).

CONTRACTORS: Electric Boat Division, General Dynamics Corp. Groton, CT (13 submarines) Mare Island Naval Shipyard Vallejo, CA (6 submarines) Newport News, VA (10 submarines) Portsmouth Naval Shipyard Portsmouth, NH (2 submarines) (See Table 5.18 for list of major subcontractors)

SPECIFICATIONS:[1] POLARIS submarines converted to POSEIDON included 3 classes: LAFAYETTE, MADISON and FRANKLIN SSBNs

Length:	425 ft
Diameter:	33 ft
Displacement:	7250 t (surface), 8250 t (submerged)
Draught:	31 ft 6 in
Propulsion:	water-cooled pressurized (S5W) nuclear reactor
Speed:	20 knots (surface), circa 30 knots (submerged)
Crew:	145 personnel (147 berths)
Armament:	4 21-inch forward torpedo tubes
MISSILE SYSTEM:	POSEIDON C3 or TRIDENT I C4; gas steam generator launch system
Number:	16 missile tubes, each with POSEIDON C3 or TRIDENT I C4 missiles
Nuclear Warheads:	
POSEIDON:	W68/Mk-3 MIRV, with 10 warheads (average)
TRIDENT:	W76/Mk-4 MIRV, with 8 warheads

134 Nuclear Weapons Databook, Volume I

5
POSEIDON Submarine

Warheads per Submarine:
   POSEIDON: 144-160; current average force loading is 10 warheads per missile

   TRIDENT: 192; current force loading is 8 warheads per missile

Fire Control System: Mk-88[2]

Navigation System: 2 Mk-2 Mod-6 Ships Inertial Navigation System (SINS) and Satellite Receivers[3]

DEPLOYMENT:

Cycle: 55% at-sea availability based on a 32 day refit period, 68 day patrol period, and a 6 year interval between 16 month long overhauls[4]

Homeport: Groton, CT; Charleston, SC, and Kings Bay, GA; submarines operate out of Holyloch, U.K., a forward deployment location. Kings Bay, GA, is being developed as the east coast TRIDENT base and to support 12 POSEIDON submarines.[5]

Patrol Areas: North Atlantic, Mediterranean Sea[6]

HISTORY:

IOC: First LAFAYETTE class sub (USS *Lafayette*) commissioned 1963 (see Table 5.5 for POSEIDON chronology)

COST: The first POSEIDON submarines cost $109 million each.[7] Total POSEIDON program cost (31 submarines and 619 missiles; construction and support equipment) for FY 1966-FY 1980 was $4847 million.[8]

Annual Operations: $1039 m (FY 1980)[9]
$1627 m (FY 1982)[10]

COMMENTS: Originally designed 20 year service life span of POSEIDON SSBNs has been extended to 30 years.[11] First hull is planned for retirement in FY 1993 and last hull in 1999.[12]

---

1 See various annual issues of *Jane's Fighting Ships*, 1975-76 to present.
2 Fire control system performs target calculations, insertion of data into the guidance system, test and checkout launch order, and sequence control.
3 U.S. Navy, Strategic Systems Project Office, "Polaris & Poseidon FBM Facts," 1970, p. 6; "FBM Facts: Polaris, Poseidon, Trident," 1978, p. 11.
4 SASC, FY 1980 DOD, Part 1, p. 327.
5 ACDA, FY 1982 ACIS, p. 77.
6 ACDA, FY 1983 ACIS, p. 32; SASC, FY 1982 DOD, Part 7, p. 4024.
7 *Ships and Aircraft of the U.S. Fleet*, 11th Ed., p. 20.
8 DOD, Selected Acquisition Report, 30 June 1975.
9 SASC, FY 1982 DOD, Part 7, p. 3992, p. 4002.
10 *Ibid.*, p. 4337.
11 ACDA, FY 1982 ACIS, p. 88.
12 SASC, *Strategic Force Modernization Programs*, p. 169.

5
POSEIDON C3 Missile

POSEIDON C3 Missile System (UGM-73A)

Figure 5.15 POSEIDON C3 (UGM-73A) missile.

DESCRIPTION:	Two-stage, solid propellant MIRVed SLBM with improved accuracy and larger payload than POLARIS A3.
CONTRACTORS:	Lockheed Missiles and Space Co. Sunnyvale, CA (prime) (See Table 5.18 for list of major subcontractors)

SPECIFICATIONS:[1]

Length:	34 ft 1 in (409 in)
Diameter:	74 in
Stages:	2 (1st and 2nd stages, glass fiber)

Table 5.18
Major POSEIDON Subcontractors

Subcontractor	Work Contracted
Aerojet-General Corp. Sacramento, CA	missile propulsion
Autonetics Div., Rockwell International Anaheim, CA	navigation
Bell Telephone Labs Whippany, NJ	communications
General Electric Co., Ordnance Systems Pittsfield, MA	fire control/ missile guidance
General Electric Corp. Lynn, MA	propulsion
Hercules, Inc. Wilmington, DE	missile propulsion
Honeywell Minneapolis, MN	missile guidance
Hughes Aircraft Co. Culver City, CA	fire control/ missile guidance
Interstate Electronics Corp. Anaheim, CA	instrumentation
ITT Labs Nutley, NJ	communications
MIT Cambridge, MA	missile guidance
Northrop Corp. Anaheim, CA	missile checkout
Raytheon Co. Lexington, MA	missile guidance
RCA, Princeton Labs Princeton, NJ	communications
Sperry Systems Great Neck, NY	navigation
Sylvania Electric Products Co. Buffalo, NY	communications
Thiokol Chemical Brigham City, UT	1st stage propulsion
Vitro Labs Silver Spring, MD	weapons system coordination
Western Electric Corp. Pittsburgh, PA	propulsion
Westinghouse Electric Corp. Sunnyvale, CA	missile launching

5

POSEIDON C3 Missile

Weight at Launch:	64,000+ lb	**HISTORY**:	(see Table 5.5 for POSEIDON chronology)
Propulsion:	solid fuel (1st stage, composite)	IOC:	31 March 1971, in USS *James Madison* (SSBN-627)
Speed:	Mach 10+		
Guidance:	all inertial	1969-1977	31 SSBNs converted from POLARIS to POSEIDON[13] missiles
Throwweight/Payload:	2000 lb;[2] 3300 lb;[3] 3000 lb[4]	**TARGETING**:	
Range:	2500 nm;[5] 3200 mi;[6] 10 RVs: 3230 mi; 14 RVs: 2485 mi	Types:	mostly soft targets, military airfields, bases, command and communications installations
DUAL CAPABLE:	no	Selection Capability:	four target
NUCLEAR WARHEADS:	6-14 W68/Mk-3 MIRV/missile;[7] 10 is average;[8] number of RVs was announced as being upgraded from 9 to 14 in October 1980;[9] 40-50 Kt,[10] with penetration aids (see W68)	Retargeting:	unknown
		Accuracy/CEP:	0.25 nm; 0.3 nm[14]
		COST:	$2.8 m (unit cost) (FY 1975)

FY	Number Procured	Total Appropriation ($ million)
1979 & prior	619[15]	3487.7[16]
1980	-	23.8
1981 & prior	-	2609.1[17]
1981	-	26.2[18]
1982	-	18.7
1983	-	10.5

DEPLOYMENT:
Launch Platform: LAFAYETTE class, JAMES MADISON class, and BENJAMIN FRANKLIN class SSBNs, designed for launching from submerged submarines

Number Deployed: 619 operational missiles procured;[11] as of 1983, 19 submarines, 304 missiles, and some 3040 warheads were deployed[12]

---

1 *World's Missile Systems*, 5th Ed., p. 123; USN, Strategic Systems Project Office, "FBM Facts: Polaris, Poseidon, Trident," 1978.
2 *Military Balance*, 1980-1981, p. 88.
3 Paul H. Nitze, *op. cit.*
4 *U.S. Missile Data Book, op. cit.*
5 *Jane's Weapons Systems*.
6 *The World's Missile Systems*, 6th Ed., p. 328.
7 HAC, FY 1980 DOD, Part 7, p. 630. The C3 missile has been tested with 14 warheads. Since there is a tradeoff between throwweight and range, actual loadings are less than the maximum, depending on target and submarine station locations. For average loading, the 1981 SIPRI Yearbook assumes 10/missile. Paul H. Nitze, *op. cit.*, indicates 8-10 RVs per C3 missile and uses an average of nine.
8 JCS, FY 1984, p. 16; with withdrawal of POLARIS, warheads on POSEIDON missiles were selectively increased; HASC, 4 March 1982, Statement of VADM Walters, p. 5; HAC, FY 1982 DOD, Part 7, p. 544; HASC, FY 1982 DOD, Part 3, p. 156.
9 *New York Times*, 30 October 1980, p. A23.
10 *Military Balance*, 1980-1981, p. 88.
11 *U.S. Missile Data Book*, 1980, 4th Ed., pp. 2-72 - 2-73; DOD, Selected Acquisition Report, 30 June 1975.
12 The first POSEIDON submarine was backfitted with TRIDENT I C4 missiles in October 1979. At time of writing, C4 missiles have been backfitted onto 12 of 31 POSEIDON SSBNs. (The 12th and last was backfitted in FY 1982.) See also C4 missile system.
13 ACDA, FY 1982 ACIS, p. 85.
14 Colin S. Gray, *op. cit.*, p. 32.
15 Ibid.
16 *U.S. Missile Data Book, op. cit.*, p. 2-74.
17 Excluding development costs; ACDA, FY 1983 ACIS, p. 48.
18 POSEIDON (C3) missiles are no longer in production. Funding continues to support the weapons system; ACDA, FY1982 ACIS, p. 85.

5
TRIDENT Submarine

TRIDENT Submarine

Figure 5.16 U.S.S. *Ohio* **(SSBN-726)**, the first TRIDENT submarine.

DESCRIPTION:	TRIDENT submarine, designated as the OHIO-class, is the newest and largest of the nuclear powered submarine strategic weapons launchers, fitted with 24 tubes for TRIDENT I C4 or TRIDENT II D5 submarine-launched ballistic missiles.
CONTRACTORS:	Electric Boat Division, General Dynamics Groton, CT; Quonset Point, RI
SPECIFICATIONS:[1]	
Length:	560 ft
Diameter:	42 ft
Displacement:	16,600 t (surface), 18,750 t (submerged)
Draught:	36.5 ft
Propulsion:	water-cooled pressurized (S8G) nuclear reactor, 60,000 horsepower[2]
Speed:	20+ knots (submerged)[3]
Crew:	154 personnel[4] (164 berths)
Stores:	90 days
Armament:	4 21-inch torpedo tubes amidships (Mk-48 torpedoes)

5
TRIDENT Submarine

Figure 5.17 Missile hatches open on **TRIDENT** submarine.

Figure 5.18 Interior of missile compartment, showing vertical launch tubes for **TRIDENT** missiles.

MISSILE SYSTEM: TRIDENT I C4; TRIDENT II D5 starting with SSBN-734, the ninth TRIDENT submarine

Number: 24 missile tubes,[5] each currently with TRIDENT I C4 missiles; ninth TRIDENT submarine will initially have TRIDENT II D5 deployed; first eight TRIDENT submarines will be retrofitted[6]

Nuclear Warheads: W76/Mk-4 MIRV, with 8 warheads, each with yield of 100 Kt

Warheads per Submarine: 192

Fire Control System: Mk-98

Navigation System: 2 Mk-12 Mod-7 Ships Inertial Navigation Systems (SINS), electrostatically supported gyro, satellite receiver

DEPLOYMENT:

Number Planned: 20 submarines are planned by 1998[7] (see Table 5.19); up to 1983 the estimate was 15 submarines[8]

Cycle: 66 percent at-sea availability based on a 25-day refit period, 70-day patrol period, and a 9-year interval between 12-month long overhauls.[9] TRIDENT increases at-sea patrol time of SLBM force by 21 percent.[10]

Homeport: Plans are to deploy the first 10 TRIDENT submarines in the Pacific from a new base at Bangor, WA.[11] Kings Bay, GA, has been chosen as the site for the Atlantic coast base.[12]

5

TRIDENT Submarine

Table 5.19
TRIDENT Submarine Construction[1]

Submarine	FY Authorized	Original Contract Delivery Date	Estimated Delivery	Commissioning Date[2]
SSBN-726 (Ohio)	74	Apr 1979	Dec 1981	Nov 1981[3]
SSBN-727 (Michigan)	75	Apr 1980	Sep 1982	Sep 1982
SSBN-728 (Florida)	75	Dec 1980	Sep 1983	Apr 1983
SSBN-729 (Georgia)	76	Aug 1981	May 1984	—
SSBN-730 (Rhode Island)	77	Apr 1982	Jan 1985	—
SSBN-731 (Alabama)	78	Dec 1982	Sep 1985	—
SSBN-732	78	Aug 1983	May 1986	—
SSBN-733	80	May 1986	Jan 1987	—
SSBN-734[4]	81	Dec 1988	Dec 1988	—
SSBN-735	83	Aug 1989	Aug 1989	—
SSBN-736	83	Apr 1990	Apr 1990	—
SSBN-737	84	Dec 1990	Dec 1990	—

1 SASC, FY 1980 DOD, Part 1, p. 323; SASC, FY 1981 DOD, Part 2, p. 561; HASC, FY 1982 DOD, Part 3, p. 158; SASC, FY 1982 DOD, Part 7, p. 4081.
2 By the end of 1986, 6 TRIDENTs are planned for deployment; 8 were previously planned; ADCA, FY 1983 ACIS, p. 37; ACDA, FY 1982 ACIS, p. 77.
3 U.S.S. Ohio commissioned 11 November 1981.
4 First TRIDENT submarine to be initially equipped with TRIDENT II D5; GAO, "Information Regarding Trident II (D-5) Missile Configured Trident Submarine Costs and Schedule" (MASAD-82-47), 3 September 1982.

HISTORY:
IOC: November 1981, commissioning of USS Ohio, first TRIDENT submarine (see Table 5.5 for TRIDENT chronology)

COST:
Program Cost: $31,731 m[13]
$14,085.2 m (Dec 1982) (TRIDENT II submarines only)

Annual Operations: $663 m (15 SSBNs) (FY 1980)

FY	Number Procured	Total Appropriation ($ million)
1977 & prior	5	5405.3[14]
1978 & prior	7	7352.8[15]
1979 & prior	7	7930.5[16]
1980	1	1501.1
1981 & prior	9	10,656.5[17]
1981	1	1218.9
1982	0[18]	

5
TRIDENT Submarine

COMMENTS: TRIDENT C4 eliminates the need for overseas basing and increases its patrol areas 10 to 20 times.[19] The TRIDENT submarine reduces acoustic observables, improves defensive systems, and decreases dependence on outside electronic navigational aids compared with POLARIS and POSEIDON submarines.[20]

---

1. See various issues of *Jane's Fighting Ships*, 1975-76 to present; SASC, FY 1977 DOD, Part 12, p. 6573; USN, Strategic Systems Project Office, "FBM Facts: Polaris, Poseidon, Trident," 1978.
2. Michael Getler, *Washington Post*, 4 October 1981, p. A26.
3. *Ships and Aircraft of the U.S. Fleet*, 11th Ed., p. 18.
4. Each submarine will be manned by two crews who will conduct alternate patrols consisting of a 25-day refit period followed by a 70-day at-sea period; ACDA, FY 1979 ACIS, p. 29.
5. DOD has reportedly considered dedication of one ballistic missile launcher on each TRIDENT submarine to a small communications satellite with booster to replace Defense communications spacecraft destroyed in wartime; AW&ST, 13 April 1981, p. 15.
6. DOD, FY 1983 Annual Report, p. 222.
7. HAC, FY 1983 DOD, Part 4, p. 599.
8. DOD, Selected Acquisition Report, 30 June 1982.
9. ACDA, FY 1982 ACIS, p. 78; SASC, FY 1980 DOD, Part 5, p. 327.
10. JCS, FY 1981, p. 43.
11. ACDA, FY 1982 ACIS, p. 77.
12. *Ibid*.
13. Estimate of 15 submarine TRIDENT force; SASC, FY 1982 DOD, Part 7, p. 4002.
14. ACDA, FY 1979 ACIS, p. 32.
15. HASC, FY 1979 Mil Con, p. 53; ACDA, FY 1980 ACIS, p. 49.
16. ACDA, FY 1981 ACIS, p. 85.
17. ACDA, FY 1983 ACIS, p. 47.
18. No TRIDENT submarines were funded in FY 1982.
19. HAC, FY 1982 EWDA, Part 7, p. 312.
20. DOD, FY 1981 RDA, p. VI-6.

5
TRIDENT I C4 Missile

TRIDENT I C4 Missile System

Figure 5.19 TRIDENT I C4 (UGM-93A) missile.

DESCRIPTION:	Three-stage, solid propellant, MIRVed SLBM with greater range than POSEIDON C3.
CONTRACTORS:	Lockheed Missiles and Space Co. Sunnyvale, CA (prime/missile/RV) GE/Raytheon/MIT (guidance) Hercules Inc. Wilmington, DE (propulsion) Thiokol (propulsion)

SPECIFICATIONS:

Length:	34 ft 1 in (10.4 m)
Diameter:	74 in (1.9 m)
Stages:	3 (Kevlar fiber materials)
Weight at Launch:	greater than 65,000 lb
Fuel:	advanced, more efficient solid plus post boost system
Guidance:	stellar-aided inertial[1] digital computer
Throwweight/Payload:	2900 lb;[2] 3000+ lb[3]
Range:	4230 nm at full payload;[4] 7400 km at full payload;[5] greater with fewer RVs

DUAL CAPABLE: no

NUCLEAR WARHEADS: 8 W76/Mk-4 MIRV/missile;[6] 100 Kt (see W76)

Future Possibility: TRIDENT I C4 missiles have also been tested with the Mk-500 EVADER MaRV.[7]

DEPLOYMENT:

Launch Platform:	First 8 OHIO class SSBNs and 12 converted POSEIDON SSBNs[8]
Number Planned:	740 missiles (Dec 1982); 712 missiles reported to be in original procurement program;[9] 327 (302 operational, 25 development) planned in FY 1982;[10] program reduced by 60 missiles in FY 1984[11]
Location:	Longer range of TRIDENT I C4 over POSEIDON and POLARIS missiles eliminates the need for overseas basing of submarines carrying this missile.

HISTORY:

IOC: 20 October 1979 (First POSEIDON SSBN backfitted with TRIDENT I C4) (see Table 5. for TRIDENT chronology)

142 Nuclear Weapons Databook, Volume I

5
TRIDENT I C4 Missile

Figure 5.20 TRIDENT I C4 missile being loaded into U.S.S. *Ohio*.

TARGETING:

Types: little hard target capability; "moderately hard"[12] military bases and industry; all targets in the USSR will be in range from submarines operating in the Atlantic, almost all targets from the Pacific;[13] like POSEIDON, TRIDENT RVs will be committed in support of NATO[14]

Selection Capability: unknown

Retargeting: rapid on-board retargeting to another pre-planned target set, more lengthy procedure when submarine is only given coordinates of new aim points[15]

Accuracy/CEP: 0.25 nm;[16] 0.2-0.3 nm[17]

COST: Each backfit of 16 TRIDENT I C4s into POSEIDON submarine cost $200 million.[18]

Program Cost: $17,148.4 m initial program (Dec 1982); $3712.3 m (TRIDENT I C4 backfit program)

Unit Cost: $6.934 m (FY 1980) (flyaway)

FY	Number Procured	Total Appropriation ($ million)
1977 & prior	48[19]	4404.6[20]
1978 & prior	144	5875.3[21]
1979 & prior	230	6959.3
1980	82	809.8
1981 & prior	384	8590.8[22]
1981	72	856.0
1982	72	954.7
1983	62	662.8
1984	43	597.7

COMMENTS: TRIDENT I missile carries a full payload to ranges comparable to maximum range of POSEIDON. This is principally due to more energetic propellants, the addition of a third stage motor, micro-electronics, and lighter materials.[23] Accuracy is on par with POSEIDON as development of TRIDENT I C4 was primarily oriented towards increasing range of SLBMs.

---

1 The stellar sensor will take a star sight during the post-boost phase of missile flight and will correct the post-boost vehicle flight path based on this star sight; SASC, FY 1980 DOD, Part 5, p. 2499; SASC, FY 1977 DOD, Part 12, p. 6549.
2 Paul H. Nitze, *op. cit.*
3 *Military Balance*, 1980-1981, p. 88; John Collins, *op. cit.*; Adelphi 140, p. 32.
4 ACDA, FY 1982 ACIS, p. 82; 4260 nm is given as range in ACDA, FY 1981 ACIS, p. 77.
5 SASC, FY 1981 DOD, Part 2, p. 509.
6 JCS, FY 1984, p. 16; JCS, FY 1981, p. 43, stated originally that TRIDENT "is capable of carrying a payload of seven RVs," but was followed the next year (JCS, FY 1982, p. 70) with the statement that TRIDENT "has independently targeted RVs" and not mentioning a number. Many sources give 8 as the RV loading (Paul H. Nitze, *op. cit.*, assumed "approximately 8 RVs"; *Military Balance*, 1980-1981, lists 8, and in FY 1984 the missile was listed with that number). Operational loadings of SLBMs are lower than maximum possible loadings; see ADCA, FY 1981 ACIS, p. 74.
7 Advanced Development work on the Mk-500 EVADER MaRV with acquisition readiness obtained in 1981 should a decision be made to deploy 45the EVADER. (See Reentry Vehicles.) ACDA, FY 1981 ACIS, p. 78; ACDA, FY 1982 ACIS, p. 82; ACDA, FY 1983 ACIS, p. 39.
8 The twelve POSEIDON submarines to be converted (in order) were: SSBN-657, -568, -655, -629, -630, -641, -627, -640, -632, -643, -634, -633.
9 *U.S. Missile Data Book*, 1980, 4th Ed., pp. 2-121; 288 missiles are being procured to support 12 converted POSEIDON submarines, 160 for launch tubes and 128 for testing and logistic support; ACDA, FY 1979 ACIS, p. 30. By the late 1980s, if SALT II limits on MIRVed launchers (1200 launchers) are extended and are met by reductions on other MIRVed launchers, the U.S. would have ten TRIDENT SSBNs with 240 launchers and 31 POSEIDON SSBNs with 496 launchers. The eventual number of TRIDENT SSBNs may be limited by compliance with the SALT II subceiling of 1200 MIRVed ICBM and SLBM launchers. Assuming 10 TRIDENTS and 12 POSEIDON SSBNs with C4 missiles: 736 missiles, 5888 to 7360 warheads. Operational loadings are lower, however, than maximum possible loadings.
10 DOD, Selected Acquisition Report, 30 June 1982.
11 DOD, FY 1984 Annual Report, p. 222.
12 HASC, FY 1982 DOD, Part 2, p. 759.
13 ACDA, FY 1980 ACIS, p. 44.
14 SASC, FY 1979 DOD, Part 1, p. 525.
15 HASC, FY 1982 DOD, Part 3, p. 130; SASC, FY 1980 DOD, Part 5, p. 2499.
16 Paul H. Nitze, *op. cit.*; UN Secretary General, *op. cit.*, assumes a CEP of 500 m (0.27 nm).
17 Colin S. Gray, *op. cit.*, p. 32.
18 HASC, FY 1982 DOD, Part 3, p. 129.
19 ACDA, FY 1979 ACIS, p. 32.
20 *Ibid.*
21 ACDA, FY 1980 ACIS, p. 49.
22 ACDA, FY 1983 ACIS, p. 47.
23 SASC, FY 1980 DOD, Part 5, p. 2499; USN, Strategic Systems Project Office, "FBM Facts: Polaris, Poseidon, Trident," 1978, p. 9.

5

TRIDENT II D5 Missile

TRIDENT II D5 Missile

An "improved accuracy program" for submarine-launched ballistic missiles began in Fiscal Year 1975; the program was prompted by the inquiries of the Secretary of Defense concerning the Navy's ability to maintain high accuracy in actual battle conditions. The research program was formulated to predict the type and magnitude of error contributors that limit accuracy, and to explore the conditions of submarine depth and speed within which missile accuracy and reliability could be maintained.

The resulting TRIDENT II missile, scheduled for deployment in TRIDENT submarines beginning in late 1988,[1] will be more accurate and have the capability of carrying more and larger warheads than the current TRIDENT I C4 missile. According to the DOD, "the TRIDENT II missile will nearly double the capability of each TRIDENT submarine."[2] The accuracy of the missile will give sea-based strategic forces the capability to attack any Soviet target; this represents a quantum jump in U.S. offensive nuclear capabilities. DOD plans to accelerate the initial deployment of the system to backfit the new missile into the ninth TRIDENT submarine, particularly as a hedge against late cancellation of the MX missile program.[3]

The purpose of the TRIDENT I missile development was essentially to increase the range of submarine-launched missiles to allow use of a larger patrol area. The purpose of the TRIDENT II is to increase the number of reentry vehicles to the POSEIDON level, so that even at extreme ranges, missiles can be fired with improved accuracy and increased warhead yield.[4]

The Reagan Strategic Program, announced 2 October 1981, stated that a new missile—the TRIDENT II D5—would be deployed in favor of alternative improvements to the present TRIDENT I (see Table 5.20).[5] A minimum of 480 operational missiles is planned for 20 submarines, each missile carrying 10 (or more) high yield warheads.

Table 5.20
Submarine-Launched Ballistic Missile Options[1]

TRIDENT I C4 with improved accuracy package (C4U):	Better CEP
TRIDENT I C4 long version with improved accuracy (C4L):	Better CEP, full length of TRIDENT launch tube, extended range
TRIDENT II D5:	Three stage scaled up TRIDENT I C4 with more warheads and greater accuracy
TRIDENT II D5 Clear Deck (CDD5):	New missile; Hard target kill across full spectrum, higher yield warheads.

1 HASC, FY 1982 DOD, Part 3, pp. 135-136; SASC, FY 1981 DOD, Part 6, pp. 3517-3518; HAC, FY 1982 DOD, Part 9, p. 202.

---

1 The TRIDENT II program is being accelerated under Reagan Administration plans from an IOC of 1989.
2 DOD, FY 1983 RDA, p. VII-7.
3 *Ibid.*
4 SASC, FY 1980 DOD, Part 1, p. 343.
5 HAC, FY 1982 DOD, Part 9, p. 202.

TRIDENT II D5 Missile

DESCRIPTION:	Large submarine-launched ballistic missile with greater range/payload capability and improved accuracy over the present SLBMs.
CONTRACTORS:	Lockheed Missiles and Space Co. Sunnyvale, CA (prime)
SPECIFICATIONS:	
Length:	45.8 ft
Diameter:	74.4 in
Stages:	3
Weight at Launch:	circa 126,000 lb
Fuel:	solid
Guidance:	stellar-aided inertial; NAVSTAR reception in missile;[1] digital computer; options include terminally-guided MaRV.
Throwweight/Payload:	6000 lb (max)[2]
Range:	4000 nm at full payload,[3] 6000 nm with reduced RVs[4]
DUAL CAPABLE:	no
NUCLEAR WARHEADS:	high yield (475 Kt) version of W87/Mk-21 ABRV designated Mk-5 by the Navy;[5] W78/Mk-12A[6] and MaRV (designated Mk-600)[7] is also under consideration; capability will be to carry more (and larger) warheads than the current TRIDENT I;[8] most probably 9-10 large warheads/missile;[9] reportedly capable of carrying 10-15;[10] RV of missile is designed to accept different warheads "tailored to the target assignment,"[11] testing of several warheads, of which one might be selected,[12] testing has already been completed;[13] yield in the 150-600 Kt range.[14]
DEPLOYMENT:	
Launch Platform:	OHIO class SSBNs starting with SSBN-734, the ninth TRIDENT submarine
Number Planned:	914 total missiles for 20 TRIDENT submarines;[15] 857 missiles[16]
Location:	Bangor, Washington; Kings Bay, Georgia
HISTORY:	
IOC:	1988;[17] Dec 1989[18]
1975	improved accuracy technology program initiated
end 1980	advanced development started
Oct 1981	TRIDENT II D5 missile chosen for development
Mar 1982	UK decides to acquire the TRIDENT II rather than the TRIDENT I missile[19]

5

TRIDENT II D5 Missile

Jun 1982	plans to install TRIDENT I C4 missile in TRIDENT class submarines pending TRIDENT II backfit modified so that TRIDENT II D5 deployment is the initial equipping[20]
1988	first TRIDENT submarine backfitted with TRIDENT II

TARGETING:

Types:	all hardened targets across the full spectrum (hard silos, command and control facilities)[21]
Selection Capability:	unknown
Retargeting:	instant retargeting

Accuracy/CEP: 400 ft[22]

COST:
Program Cost: $37,645.1 m (Dec 1982);[23] $12,900 m (FY 1983)[24]

FY	Number Procured	Total Appropriation ($ million)
1979 & prior	—	20.0[25]
1980	—	25.6
1981	—	96.5[26]
1981 & prior	—	143.2[27]
1982	—	239.5
1983	—	369.7

---

1 SASC, FY 1980 DOD, Part 5, p. 2506; AW&ST, 9 March 1981, p. 33.
2 ACDA, FY 1983 ACIS, p. 41; TRIDENT II will have 75 percent greater payload than TRIDENT I C4; HASC, FY 1982 DOD, Part 3, p. 129.
3 ACDA, FY 1982 ACIS, p. 84.
4 See for instance, *Military Balance*, 1980-81, p. 3.
5 ACDA, FY 1982 ACIS, p. 84; AW&ST, 17 January 1983, p. 26; see description of W87 under MX missile.
6 AW&ST, 22 March 1982, p. 18.
7 *Military Balance*, 1980-81, op. cit.
8 DOD is reportedly considering the option of dedicating one missile launcher on each TRIDENT submarine to a small communication satellite with booster to replace communication satellites destroyed in wartime; AW&ST, 13 April 1981, p. 15.
9 HASC, FY 1982 DOD, Part 3, p. 138; DOD, FY 1983 RDA, p. VII-7; AW&ST, 22 March 1982, p. 18; The SALT II limit would be 10. The option is being maintained for more than 10 warheads to be carried on the TRIDENT II D5 but actual RV loading is dependent on the size of the type warhead chosen. Fourteen is a common figure mentioned for maximum MIRVing although the pre-SALT figure was generally accepted as 17 (see for instance, *Projected Strategic Offensive Weapons Inventories of the U.S. and U.S.S.R., An Unclassified Estimate* (CRS, 77-59F, 24 March 1977).
10 Richard Halloran, *New York Times*, 6 February 1983, p. 17.
11 SASC, *Strategic Force Modernization Programs*, pp. 168, 172.
12 ACDA, FY 1983 ACIS, p. 38.
13 ACDA, FY 1979 ACIS, p. 31.
14 *Military Balance*, 1980-81, p. 3.
15 HAC, FY 1982 DOD, Part 9, p. 290.
16 HAC, FY 1983 DOD, Part 4, p. 599.
17 AW&ST, 22 March 1982, p. 18.
18 HAC, FY 1983 DOD, Part 4, p. 560.
19 SASC, FY 1983 DOD, Part 7, p. 4395.
20 GAO, "Information Regarding Trident II (D5) Missile Configured Trident Submarine Cost and Schedule" (MASAD-82-47), 3 September 1982.
21 SASC, *Strategic Force Modernization Programs*, p. 167.
22 AW&ST, 22 March 1982, p. 18; Richard Halloran, *New York Times*, 6 February 1983, p. with stellar inertial guidance system.
23 DOD, SAR Program Acquisition Cost Summary, as of December 31, 1982.
24 HAC, FY 1983 DOD, Part 4, p. 599; this figure is given for 857 missiles at $15 million apiece
25 ACDA, FY 1979 ACIS, p. 33.
26 SAC, FY 1982 DOD, Part 1, p. 426.
27 ACDA, FY 1983 ACIS, p. 47.

5
SUM; SSBN-X

Shallow Underwater Missile System (SUM)[1]

One alternative to the MX missile system is the idea of a Shallow Underwater Missile System (SUM). SUM is a small missile launching submarine that would be used as the vehicle for an externally mounted, encapsulated strategic missile. The SUM force would consist of small, non-nuclear powered diesel electric submarines operating in near-coastal waters off the east and west coasts of the continental U.S. Each submarine would carry two missiles horizontally mounted external to its pressure hull. In this way, 200 missiles could be deployed on 100 small submarines of 500-1000 ton displacement.

Proponents of SUM claim that the system would be less costly, less vulnerable, as accurate (using land-based guidance beacons), and as controllable (with short-range, reliable communications) when compared to the Multiple Protective Shelter version of the MX. Opponents argue that deployment could not occur before the early 1990s, that cost per surviving RV exceeds TRIDENT, that technical risks exist in submarine design, weight, and propulsion, and that manning costs are higher.[2] Furthermore, opponents contend that SUM submarines could not operate on the continental shelf because of a tidal wave phenomenon that would be caused by nuclear weapons, called the "Van Dorn effect," which would allow a few Soviet warheads to destroy all the submarines in a restricted patrol area.

SSBN-X

A SSBN-X program began in FY 1979 to investigate concepts and designs for future nuclear powered ballistic submarines (SSBNs). The program examined two concepts for cost-effective SSBNs in response to the excessive cost of the TRIDENT submarine: first, a new small submarine carrying encapsulated missiles, and second, a less expensive large SSBN, either a reengineered TRIDENT or a new 24-tube SSBN.[3]

In FY 1979-1981, approximately $25 million was appropriated in the SSBN-X program; design studies and preliminary work began in the following areas: alternative ship size and hull design, new propulsion plant, and new strategic weapon design. Much of the design and subsystems of a follow-on attack submarine are being used in the SSBN-X program.

The earliest possible start of SSBN-X work was projected as FY 1985.[4] During the Carter Administration, it was thought that it would not be until FY 1991 that such a ship would be available.[5] The Reagan Administration has not pursued SSBN-X development.

---

[1] Additional sources on SUM include the following: Sidney D. Drell and Richard L. Garwin, "SUM: The Better Approach to ICBM Basing," 25 April 1980; Office of Deputy Undersecretary of Defense for Research and Engineering, "An Evaluation of the Shallow-Underwater Missile (SUM) Concept," 9 April 1980 (reproduced in SASC, FY 1981 DOD, Part 6, pp. 3484-3504); Letters from Richard L. Garwin to Congressman J.F. Seiberling, 4 February 1980, and 7 April 1980.

[2] SASC, FY 1982 DOD, Part 7, p. 4067.
[3] ACDA, FY 1981 ACIS, pp. 80-81.
[4] SASC, FY 1981 DOD, Part 2, p. 565.
[5] Ibid., p. 608.

5
B-52 STRATOFORTRESS

Strategic Bomber Force
B-52 STRATOFORTRESS

Figure 5.21 B-52G bomber.

DESCRIPTION:	Long-range, heavy bomber used by the Strategic Air Command. Presently deployed and modified into three versions: B-52D, G, and H.
B-52D:	configured primarily for conventional bombing, being retired
B-52G:	planned as initial cruise missile carriers
B-52H:	planned as follow-on cruise missile carrier, most capable penetrator.

CONTRACTORS:	Boeing Aerospace Company Seattle, WA; Wichita, KS (prime)
	Pratt & Whitney (engines)
	Boeing Witchita (offensive avionics)
	IBM (navigation and weapons delivery computer)
	Teledyne Ryan (radar)
	Honeywell (navigation and radar)
	ITT Avionics

B-52 STRATOFORTRESS

(ECM)
Northrop
(ECM)
Westinghouse
(avionics)

SPECIFICATIONS:[1]

Dimensions:
Length: 156 ft (D/H); 158 ft (G)
Height: 48 ft (D); 40 ft 8 in (G/H)
Wingspan: 185 ft (37° fixed)

Takeoff Weight (max): 450,000+ lb (D); 488,000+ lb (G/H)

Takeoff Distance: 10,700 ft (G); 9900 ft (H)[2]

Powerplant: 8 PW J57-P43WB jet engines (G); 8 PW TF33-P-3 turbofans (H)

Ceiling: 50,000+ ft

Speed:
maximum: 0.95 Mach (at 50,000 ft);[3]
high cruise: 0.77 Mach (B-52G/H)[4]
low penetration: 0.53-0.55 Mach (B-52G/H)[5]
sea level: 0.59 Mach (B-52H)[6]
low withdrawal: 0.55 Mach (B-52G/H)[7]

Range: (depends on number of aerial refuelings)

B-52D: 5300 mi;[8] more than 6000 mi[9]

B-52G: 4500 nm (nuclear no refuel);[10] 6500 nm (high no refuel);[11] more than 7500 mi[12]

B-52H: 8600 mi;[13] more than 10,000 mi;[14] 7900 nm (high no refuel);[15] 5900 nm (nuclear no refuel)[16]

Aerial Refueling Capability: yes

Crew: 6 (pilot, copilot, navigator, radar navigator, electronic warfare officer, gunner)

Radar Cross Section: 90-100 sq m[17]

NUCLEAR WEAPONS: ALCM, SRAM missiles, B28, B43, B53, B57,[18] B61,[19] B83 bombs;[20] maximum load is 24 nuclear weapons. Typical load of B-52G/H would be 4 bombs and 6-8 SRAMs internal;[21] B-52G/Hs can carry up to 20 SRAM missiles, 6 under each wing and 8 in the bomb bay on rotary launcher; drag increase with external missile is approximately 15-25 percent.[22]

B-52D: 4 nuclear bombs, no SRAMs[23]

B-52G: being modified to carry up to 12 ALCMs, 6 under each wing, plans to deploy ALCM internally cancelled[24]

B-52H: to be modified to carry 20 ALCMs starting in 1985[25]

Figure 5.22 **B-52G** with SRAMs (AGM-69A) loaded under wing.

Nuclear Weapons Databook, Volume I 149

5

B-52 STRATOFORTRESS

Table 5.21
B-52 Bomber Force[1]

Model	Total Built	First Delivery	Last Delivery	Active Inventory (1983) Test	Active Inventory (1983) SAC
XB	1	1952	—	—	—
YB	1	1953	—	—	—
A	3	1954	1954	—	—
B	50	1955	1956	1	—
C	35	1956	1956	—	—
D	170	1956	1957	0	31
E	100	1957	1958	2	—
F	89	1958	1959	—	—
G	193	1958	1961	4	151
H	102	1961	1962	0	90
	744			7	272

1 SASC, FY 1980 DOD, Part 1, p. 332; JCS, FY 1984, p. 13.

DEPLOYMENT: 272 operational B-52s; 316 total in 20 squadrons, 31 in 3 training squadrons, backup and test; dispersal of alert force B-52s to more bases in peacetime is under consideration;[26] B-52 requires 150 foot wide runways to land, limiting the number of airfields capable of handling the plane.[27]

Number Deployed: 172 B-52G (169 operational, 151 of which are PAA, 4 test); 96 B-52H (90 PAA); 79 B-52D (75 operational (31 PAA));[28] 3 squadrons of B-52D were retired on 1 October 1982, the last two (31 PAA) will follow during 1983-1984.

HISTORY:
IOC: 1955

Sep 1947 — Boeing awarded contract for preliminary design of B-52

Nov 1951 — first B-52 prototype finished

Apr 1952 — first flight of YB-52 prototype

Aug 1954 — first flight of production B-52A

Jun 1955 — SAC receives first B-52

1956 — B-52D deployed

1958 — B-52G deployed

1961 — B-52H deployed

Oct 1962 — delivery of last B-52 (H mode

1974 — program to upgrade avioni weapons delivery, and d fenses of bomber force initiate

Sep 1980 — first flight of B-52 equippe with offensive avionics syste

Sep 1981 — first alert capability with o B-52G and 12 ALCMs at Griffi AFB, NY

Sep 1981 — Air Force directs cruise missi deployment on B-52H force

Oct 1981 — Reagan strategic program ca for the retirement of B-52 bombers in 1983

Oct 1982 — three squadrons of B-52Ds r tired[29]

end 1982 — first squadron of 14 B-52Gs ca rying 12 ALCMs under wings operational

FY 1986 — planned IOC of B-52H wi ALCM[30]

5
B-52 STRATOFORTRESS

Table 5.22
B-52 Modifications

Project	Model	Cost ($ million)
Already accomplished (through FY 1979)	All	3400
Ongoing (FY 1980-FY 1990)	(B-52D/G/H)	3300
Offensive Avionics System	B-52G/H	
Cruise Missile Carriage	B-52G	
ECM/Defensive Systems	B-52G/H	
(ALQ-117 improved, ALQ-122, ALQ-155, ALR-46)		
Functionally Related Observable Differences	B-52G	
Fuel Savings	B-52G/H	
Tail Warning System	B-52G/H	
Reliability & Maintainability	B-52D/G/H	
Future (FY 1983-FY 1990)	(B-52G/H)	4400
EMP Hardening and Thermal/Blast Protection	B-52G/H	
B-52H Cruise Missile	B-52H	
ECM	B-52G/H	
Reliability & Maintainablity	B-52G/H	
Proposed		
Reengining (with PW2037 turbofan)	B-52G	4200

Sources: AW&ST, 30 November 1981, p. 54; HASC, FY 1981 DOD, Part 4, Book 2, pp. 1886-1887; SAC, FY 1981 DOD, Part 5, pp. 1657-1658.

FY 1987 avionics modification program planned for completion

COST:

B-52 OAS Program: $1777.9 m (Dec 1982)

B-52 CMI Program: $611.0 m (Dec 1982) (see Table 5.8, Bomber Forces Funding)

Annual Operations: $948 m (FY 1980)[31]
$1891 m (FY 1982)[32]

FY	Number Procured	Total Appropriation ($ million)
1980	-	479.5[33]
1981	-	597.8[34]
1982	-	615.6

COMMENTS: At least 187 B-52 aircraft are in inactive storage at Davis-Monthan AFB, AZ. The FY 1982 budget request included $12.7 m to install a new monitor and control system for nuclear weapons in B-52 aircraft.

Boeing Fact Sheet, "Background Information, Boeing B-52 Stratofortress," November 1981; SASC, Military Implications of the SALT II Treaty, Part 4, p. 1608.
SASC, FY 1982 DOD, Part 7, p. 4329.
Military Balance, 1980-81, p. 90.
HAC, FY 1982 DOD, Part 2, p. 269; SASC, FY 1982 DOD, Part 7, p. 4329.
Ibid.
Ibid.
Ibid.
Military Balance, 1980-81, p. 90.
Air Force Fact Sheet, "B-52," 1 April 1980.
SASC, FY 1982 DOD, Part 7, p. 4329.
Military Balance, 1980-81, p. 90.
Air Force Fact Sheet, op. cit.
Military Balance, 1980-81, p. 90.
Air Force Fact Sheet, op. cit.
SASC, FY 1982 DOD, Part 7, p. 4329.
Ibid.
CRS, "Bomber Options for Replacing B-52s" (IB 81107), p. 18.

18 AFR 0-2, p. 45.
19 Military Applications of Nuclear Technology, Part 1, p. 7.
20 SASC, FY 1983 DOD, Part 7, p. 4242.
21 SASC, FY 1982 DOD, Part 7, p. 4284.
22 SASC, Strategic Force Modernization Programs, p. 162.
23 SASC, FY 1982 DOD, Part 7, p. 4284.
24 HAC, FY 1983 DOD, Part 4, p. 588.
25 DOD, FY 1984 Annual Report, p. 223.
26 DOD, FY 1983 RDA, p. VII-7.
27 SASC, Strategic Force Modernization Programs, p. 36.
28 SASC, FY 1983 DOD, Part 7, p. 4556.
29 Ibid., p. 4589.
30 ACDA, FY 1983 ACIS, p. 65.
31 Including military personnel; SASC, FY 1982 DOD, Part 7, p. 4002.
32 SASC, FY 1982 DOD, Part 7, p. 4337.
33 B-52 avionics modification for cruise missile.
34 SASC, FY 1982 DOD, Part 7, p. 4289.

5
FB-111

FB-111

Figure 5.23 FB-111 with SRAMs loaded under wing.

DESCRIPTION:	Variation of the F-111 tactical fighter used by SAC as a medium bomber. It is designed for low altitude, high speed penetration.	Takeoff Weight (max):	110,600 lb
		Takeoff Distance:	6200 ft
		Powerplant:	2 PW TF 30-P-7 turbofan jet engines
MODIFICATIONS:	None		
CONTRACTORS:	General Dynamics (prime) Pratt & Whitney (engine)	Ceiling:	60,000+ ft
		Speed: maximum: high cruise: low penetration: low withdrawal:	2.5 Mach (36,000 ft)[2] 0.77 Mach[3] 0.85 Mach[4] 0.55 Mach[5]
SPECIFICATIONS:[1] Dimensions: Length: Height: Wingspan:	75 ft 6.5 in 17 ft 70 ft at 16° sweep 34 ft at 72.5° sweep		

5
FB-111

Range:[6]	(depends upon aerial refuelings); 2900 nm (nuclear loaded, no refuel); 3200 nm (high, no refuel); 4300 nm (high, 1 refuel); 5200 nm (nuclear loaded, 1 refuel); 4700 mi[7]	**COMMENTS**:	FB-111 is reportedly used in attacking heavily defended and large-area targets.[14] Unlike other bombers, low-level missions at night, or even adverse weather, can be flown without crew interface. A 30 percent alert rate with 8 FB-111s and 5 KC-135 tanker aircraft is maintained at both bases.[15] The FY 1982 budget request included $2.7 million to install a new nuclear weapons monitoring and control device in FB-111 aircraft. Due to its high speed, small size, and low level terrain following capability, the FB-111 will remain a better penetrator than the B-52 throughout the 1980s.[16]
Aerial Refueling Capability:	yes		
Crew:	2 (pilot, navigator-bombadier)		
NUCLEAR WEAPONS:[8]	SRAM missiles, B43, B61 bombs; B83 (future); maximum load: 6 bombs or 6 SRAM;[9] 4 SRAMs carried on external pylons, capacity for 2 in bomb bay; 6 bombs in bomb bay in lieu of SRAMs; three external stations on each wing, two in the weapons bay; two outboard fixed pylons can carry tanks, but not weapons.[10]		
DEPLOYMENT:	Pease AFB, NH; Plattsburgh AFB, NY		
Number Deployed:	60+ FB-111A total, 56 in 4 operational squadrons (1983)		
HISTORY:			
IOC:	1969[11]		
Oct 1969	first FB-111A delivered to SAC		
1968-1971	76 FB-111As produced[12]		
1990s	FB-111As transferred to tactical inventory as ATB is deployed[13]		

---

1 SASC, FY 1982 DOD, Part 7, p. 4329; SASC, *Military Implications of the SALT II Treaty*, p. 1608.
2 External SRAMs limit performance.
3 HAC, FY 1982 DOD, Part 2, p. 269; SASC, FY 1982 DOD, Part 7, p. 4329.
4 *Ibid*.
5 *Ibid*.
6 SASC, FY 1982 DOD, Part 7, p. 4329.
7 *Military Balance, 1980-1981*, p. 90.
8 AFR 0-2, p. 45; FB-111 does not carry the B28.
9 SASC, FY 1982, DOD, Part 7, p. 4329.
10 David R. Griffiths, "FB-111 Bombers Playing Crucial Role," AW&ST, 16 June 80.
11 JCS, FY 1981, p. 42.
12 Of 76 FB-111s built, 11 had crashed as of March 1981.
13 DOD, FY 1984 Annual Report, p. 224.
14 David R. Griffiths, *op. cit*.
15 *Ibid*.
16 SASC, FY 1980 DOE, Part 2, p. 465.

5
Short-Range Attack Missile

Short-Range Attack Missile (SRAM) (AGM-69A)[1]

Figure 5.24 Short-Range Attack Missile (SRAM) (AGM-69A).

DESCRIPTION: Defense suppression, supersonic, ballistic trajectory air-to-surface missile deployed on B-52 and FB-111 bombers. It can reverse directions in flight up to 180 degrees.

CONTRACTORS:
Boeing Aerospace Co.
Seattle, WA (prime)
General Precision (guidance)
Lockheed (propulsion)
Thiokol Corp
Brigham City, UT (propulsion)
Singer-Kearfott Div (guidance)
Universal Match Corp, Unidynamics Div (fuse system)
Rockwell International, Autonetics Div (aircraft computer)
Litton Industries (inertial measurement unit)
Stewart-Warner Electronics Div (terrain sensor)
Delco Electronics (missile computer)

SPECIFICATIONS:
Length: 14 ft (4.27 m)

Diameter: 17.7 in

Stages: 1

Weight at Launch: 2240 lb[2] (at launch)

Propulsion: LPC-415 solid propellant, 2 pulse rocket engines

Speed: Mach 3.5+

Guidance: inertial with terrain clearance sensor

Range: 160-220 km at high altitude; 56-80 km at low altitude

DUAL CAPABLE: no

NUCLEAR WARHEADS: one W69 (similar to the W68, the warhead on MINUTEMAN III), 170-200 Kt

DEPLOYMENT: B-52G/H: up to 20 SRAMs, 12 in 3 round clusters under the wing and 8 on a rotary dispenser in the aft bomb bay, typical load is 6-8 SRAMs. FB-111A: up to 6 SRAMs, 4 under the wing and 2 internally; typical load is 2 SRAMs. (See Table 5.7.)

Number Deployed: 1140 operational;[3] some 1500 missiles delivered,[4] with some 1300 remaining in service.

Location: 1020 at B-52G/H bases; 120 at FB-111 bases.[5] (see Chapter Four)

HISTORY:
IOC: August 1972

1964 — Air Force develops requirement for SRAM

Oct 1966 — Boeing selected as prime contractor for SRAM

Jul 1969 — first powered flight

Jan 1971 — production authorized

Jul 1971 — flight test program completed

5

Short-Range Attack Missile

Mar 1972	first production missile delivered to Air Force	**COST**:	$290,000 (FY 1975) (flyaway)

FY	Number Procured	Total Appropriation ($ million)
1980 & prior	1500	1196.7

Jul 1975 — 1500th and final SRAM delivered to the Air Force

Jun 1977 — with cancellation of B-1, development of an upgraded B model SRAM was cancelled

1980 — 1152 SRAMs in 19 B-52 and FB-111 squadrons[6]

Late 1980s — SRAM replaced by Advanced Strategic Air-Launched Missile

TARGETING:

Types: heavily defended targets; air defense missile sites, radar, airfields, defensive installations

Selection Capability: air-burst and contact fuze;[7] missile can be launched at subsonic or supersonic speed, from high or low altitude

Retargeting: can be retargeted aboard the aircraft prior to launch[8]

Accuracy/CEP: "very good CEP"[9]

COMMENTS: Boeing has proposed to add a second solid motor on the end of the missile and upgrade the guidance to include an air-to-air mission to compete with ASALM, to be designated SRAM-L.[10]

---

1 See Boeing Fact Sheet, "Background Information, SRAM," February 1982.
2 GAO, Draft Study for B-1.
3 "Primary Airvehicle Authorized" as of January 1980; HASC, FY 1981 Mil Con, p. 431.
4 U.S. Missile Data Book, 1980, 4th Ed., p. 2-92.
5 HASC, FY 1981 Mil Con, p. 431; HAC, FY 1982 DOD, Part 2, p. 101.
6 HAC, FY 1981 DOD, Part 2, p. 288; Les Aspin, "Judge Not by the Numbers Alone," *The Bulletin of the Atomic Scientists*, June 1980, p. 31, lists 1250 SRAMs and *SIPRI Yearbook 1980*, p. 176, lists 1500 authorized through 1973.
7 *The World's Missile Systems*, 6th Ed., p. 116.
8 SAC, Fact Sheet, "Short Range Attack Missile," August 1981.
9 *Military Applications of Nuclear Technology*, Part 1, p. 9.
10 AW&ST, 10 March 1980, p. 15.

5

New Bombers

New Bombers

The search for a replacement for the B-52 began almost immediately after the bomber was deployed in the 1950s. Although the supersonic B-58 HUSTLER was developed, it proved unsatisfactory and no more than one hundred were procured. The B-58 was followed by the B-70, a long-range supersonic (Mach 3) bomber. The B-70 never got past the R&D stage because its cost, effectiveness, and vulnerability were not considered to offset any advantages of the emerging MINUTEMAN ICBM force. The B-70 was followed by the RS-70 project which was also cancelled due to excessive cost. This was followed by the Advanced Manned Strategic Aircraft (AMSA) program which continued studies through the 1960s and 1970s to develop a low flying supersonic bomber.

In June 1970, the DOD awarded contracts for the candidate AMSA bomber, the B-1. Although the design of the B-1 was completed by 1978, an uneven R&D program followed in which $6 billion were spent and 4 prototype planes were produced. On 30 June 1977, President Carter announced that production plans for the B-1 would be discontinued and that an upgraded B-52 force and other planes equipped with Air-Launched Cruise Missiles would supplant the need for a new penetrating airplane.

The FY 1981 Defense Authorization Act (P.L. 96-342) directed the Secretary of Defense to develop a strategic "multi-role bomber" for initial deployment by 1987. The program—called Long Range Combat Aircraft (LRCA)—was to consider a number of alternatives (see Table 5.23) both short and long term. On 2 October 1981, the Reagan Administration announced that a modified B-1 (designated the B-1B) would be the LRCA and that an Advanced Technology Bomber ("Stealth") would be developed for the 1990s. The plan is to procure 100 B-1Bs with the first squadron operational in FY 1985 and 135-150 ATBs starting in the early 1990s.

Although a variety of reasons, including the need for conventional bombing capabilities, were given to explain the need for the prospective LRCA. The primary justification for replacing the B-52 is the perceived military requirement for bombers to penetrate Soviet air defenses. But given that the deployment of long-range Air-Launched Cruise Missiles aboard the B-52 Bomber force greatly increases their ability to hit targets due to increased accuracy and defense evasion, the need for a bomber to penetrate the Soviet Union is hotly disputed. The age of the B-52 bomber, its capabilities at low altitudes, and improvements in Soviet defenses are used to justify a new airplane. Other operational requirements

Table 5.23
Candidate Systems for B-52 Bomber Replacement
Long-Range Combat Aircraft/Multi Role Bomber

System	Description	Status
Basic B-1	Supersonic, low altitude penetrating bomber	Upgraded to B-1B
B-1B	Improved wings, electronic equipment, longer range, heavier payload than B-1	Chosen as LRCA 2 October 1981, IOC in FY 1985
FB-111B/C	Stretched FB-111A and F-111D with longer fuselage, new engines, with SRAM	Originally favored by Congress and SAC in 1980, $6-8 billion program
Advanced Technology Bomber (Stealth)	Reduced radar cross section penetrating bomber	IOC in early 1990s
Cruise Missile Carrier Aircraft (CMCA)	Wide-bodied new ALCM transport[1]	Evolved into SAL/SWL, dropped in favor of B-1B
Strategic Weapons Launcher (SWL)[2]	Fixed-wing version of B-1 for standoff ALCM delivery as mid-term B-52 replacement	Advocated by Rockwell, dropped by AF in September 1977 in favor of SAL, unfunded by Congress
Strategic ALCM Launcher (SAL)	Fixed-wing version of B-1 for standoff and penetration as interim penetrator	Favored by Air Force and DOD as MRB, dropped for B-1B

1 Candidates included 707, DC-10, 747, L-1011, C-5, YC-14, YC-15 and C-141; ACDA, FY 1979 ACIS, p. 30.
2 AW&ST, 17 September 1979, pp. 14-15.

5
New Bombers

Table 5.24
New Bomber Funding ($ millions)[1]

	FY 1977	FY 1978	FY 1979	FY 1980	FY 1981	FY 1982	FY 1983	FY 1984
Cruise Missile Carrier Aircraft	—	—	—	24.0	—	—	—	—
Bomber Penetration Evaluation	—	—	—	54.9	—	—	—	—
B-1	767.0	443.4	55.0	55.0	—	2083.0	4787.0	6929.5
Long-Range Combat Aircraft	—	—	—	—	260.1	—	—	—

1 HAC, FY 1982 DOD, Part 2, p. 32; DOD, FY 1984 Annual Report, p. 225.

identified are better dispersal capabilities, base escape characteristics, and resistance to nuclear effects.[1]

The B-1B will use essentially the same "active defenses" (electronic countermeasures) as the present B-52, which has been continually updated with the most modern systems. It will incorporate many "passive defense" innovations not available when the B-52 was developed. These include smaller size, more efficient propulsion system, and materials advances which will decrease the aircraft's "radar cross section." This will reduce its susceptibility to detection and greatly aid penetration.

The B-1B, chosen as the near term penetrator, is of the same design as the basic B-1 bomber and is able to perform as either a penetration bomber, a cruise missile launch platform, or conventional bomber. The "core aircraft," which includes 85 percent of the design of the basic B-1, will have the following characteristics:[2]

- greater range, which allows intercontinental missions without aerial refueling,
- increased payload, including adding cruise missile capability, external stores, and enlarged forward weapons bay,
- reduction in supersonic maximum speed at high level (Mach 1.6 to Mach 1.2),
- reduction in maximum altitude (70,000 ft to 42,000 ft),
- abandonment of low level supersonic "dash" capability to high subsonic speeds at lower levels,
- offensive avionics system now being installed in B-52s, including upgraded radar and navigation system from F-15 and F-16,
- improved nuclear weapons effects hardening,
- new defensive avionics to include higher frequency jamming,
- reduced wing sweep (67.5° to 60°) and strengthened landing gear for heavier loadings,
- incorporated signature reduction design changes and ten-fold reduction in radar cross section, and
- increased takeoff gross weight limit (395,000 lb to 477,000 lb).

1 DOD, FY 1983 RDA, p. VII-2.
2 HAC, FY 1982 DOD, Part 9, pp. 82-83; SASC, *Strategic Force Modernization Programs*, p. 93.

5

B-1B Bomber

B-1B

Figure 5.25 **B-1** bomber.

DESCRIPTION:	Medium weight, intercontinental, penetrating, four seat, strategic bomber.	Takeoff Distance:	8300 ft (B-1B); 7500 ft (B-1)[1]
		Powerplant:	4 GE F101-100 turbofans
CONTRACTORS:	Rockwell International El Segundo, CA (prime/airframe) (See Table 5.26 for a list of major B-1B subcontractors)	Ceiling:	42,000 ft (B-1B); 70,000 ft (B-1)
		Speed:	
		low penetration:	0.85 Mach[2] (circa 646 mph)[3]
		high penetration:	Mach 2 (1320 mph); 1596 mph (B-1)[4]
SPECIFICATIONS:		high cruise:	0.72 Mach (B-1/B-1B)[5]
Dimensions:		low withdrawal:	0.42 Mach (B-1B); 0.55 Mach (B1)[6]
Length:	150 ft, 2.5 in		
Height:	33 ft, 7 in		
Wingspan:	136 ft, 8.5 in (15°), 78 ft, 2.5 in (67.5°)	Range:	6100 mi
		Aerial Refueling Capability:	yes
Takeoff Weight (max):	477,000 lb (B-1B); 395,000 lb (B-1)		

5

B-1B Bomber

Crew:	4 (pilot, copilot, offensive and defensive systems operators)
Radar Cross Section:	1 sq m[7]
NUCLEAR WEAPONS:	B28, B61, B83, SRAM, ALCM;[8] payload approximately twice that of B-52; drag increase with external missiles will be approximately 8 percent.[9] (See Table 5.25 for loading of bombers.)
DEPLOYMENT:	first base will be Dyess AFB, Texas where 26 B-1B will be deployed starting in late 1985[10]
Number Planned:	100 (under Reagan Administration plans) (1983)

HISTORY:

IOC:	1986; FY 1985[11] (B-1B)
Jun 1970:	development of B-1 begins
Dec 1974:	first flight (basic B-1)
Dec 1976:	production of B-1 started
Jun 1977:	basic B-1 cancelled by President Carter
Apr 1981:	flight testing of 4 B-1 R&D aircraft completed
Oct 1981:	decision taken by Reagan Administration to procure 100 B-1B bombers as near term penetrator
Jun 1985:	first B-1B production delivery
1986:	first B-1B squadron operational[12]
1987-1995:	B-1B serves as penetrator
1988:	FOC of 100 B-1B force[13]
1990s:	B-1B begins phase-in as cruise missile carrier as ATB is deployed

Figure 5.26 B-1 bomber.

COST:

Program Cost:	Original B-1: $21.5 billion (244 bombers)
	LRCA: $27.9 billion[14]
	B-1B (1981) (Administration): $19.7 billion[15]
	B-1B (1982) (Administration): $22 billion
	B-1B (1982) (CBO): $39.8 billion
	B-1B (Dec 1982) (SAR): $28.334 billion

FY	Number Procured	Total Appropriation ($ million)
1970-1980	4 (B-1A)	4758.7[16]
1982 & prior	1	2311.9[17]
1983	7	4787.0
1984	10	6935.4

5

B-1B Bomber

Table 5.25
Nuclear Weapons Loads for B-1B Bomber

		Capable Loadings			
		Internal Loadings		External	Total
Weapon	Weight	Mod-A[1]	Mod-B[2]		
B28	2540	12	—	8	20
B61	718	24	—	14	24-38
B83	2408	24	—	14	24-38
SRAM	2240	24	—	14	24-38
ALCM	3300	—	8-16[3]	14	22-30[4]

Typical Operational Loading

	ALCM Internal	SRAM External	Gravity Bombs	
Standoff Mission	8	14	—	—
Penetration Mission	—	—	8	16
Shoot and Penetrate	8	—	4	4

Source: GAO, Draft Study for B-1; HAC, FY 1982 DOD, Part 1, p. 321; SASC, *Strategic Force Modernization Programs*, pp. 91-92.
1 Configured with 3 180-inch internal weapons bays.
2 Configured with 1 265-inch and 1 180-inch weapons bay.
3 Enlarging the aft weapons bay for ALCM carriage allows for internal carriage of an additional 8 weapons.
4 Ibid.

Figure 5.27 B-1 bomber being refueled by KC-135 tanker aircraft.

COMMENTS: Targets for B-1B would cover the entire spectrum, from hard targets, and less than precisely located targets. Nuclear safety devices such as PAL and Command Disable were not part of the original B-1B, as they were not a SAC requirement,[18] but will be added with cruise missile carriage, and will include a new system called a "coded switch system."[19]

---

1 SASC, FY 1982 DOD, Part 7, p. 4329.
2 HAC, FY 1982 DOD, Part 2, p. 269; SASC, FY 1982, DOD, Part 7, p. 4329.
3 Radar cross section of new B-1B reduced by a factor of ten through the use of absorbtion materials and changes to engine inlets. AW&ST, 23 March 1981, pp.19-21; AW&ST, 11 May 1981, pp. 18-21.
4 *U.S. Military Aircraft Data Book*, 1981, p. 2-25.
5 SASC, FY 1982, DOD, Part 7, p. 4329.
6 Ibid.
7 CRS, "Bomber Options for Replacing B-52," (IB 81107), p. 18.
8 First 18 production B-1Bs will not have complete ALCM capability; HAC, FY 1982 DOD, Part 9, p. 257; SASC, FY 1983 DOD, Part 7, pp. 4153, 4242.
9 SASC, *Strategic Force Modernization Programs*, p. 162.
10 DOD, "Memorandum for Correspondents," 31 January 1983.
11 DOD, FY 1984 Annual Report, p. 281.
12 DOD, FY 1983 RDA, p. VII-6.
13 HAC, FY 1982 DOD, Part 9, p. 256.
14 LRCA baseline cost escalated for inflation; CRS, "Bomber Options for Replacing B-52" (IB 81107), p. 11.
15 HAC, FY 1982 DOD, Part 9, p. 188.
16 RDT&E, investment and operations cost in then year dollars; SASC, FY 1980 DOD, Part 1, p. 398.
17 B-1B funding including procurement of 1 aircraft in FY 1982; SASC, FY 1983 DOD, p. 4563.
18 HAC, FY 1982 DOD, Part 9, p. 220.
19 SASC, *Strategic Force Modernization Programs*, pp. 106, 151.

Table 5.26
Major B-1B Subcontractors

Company	Component
Aeronca, Inc. Middletown, OH	structural subassemblies
AVCO Corp. Nashville, TN	wings
B.F. Goodrich Co. Akron, OH	tires
Bendix Corp. Teterboro, NJ	avionics
Boeing Co.* Seattle, WA; Wichita, KS	avionics systems integration
Brunswick Corp. Marion, VA	radomes
Cleveland Pneumatic Co. Cleveland, OH	landing gear
Cutler Hammer, AIL Division* Deer Park, NY	defensive avionics
General Electric Co.* Binghampton, NY; Evandale, OH	engine components
Goodyear Aerospace Corp. Litchfield, AZ	windows
Goodyear Tire & Rubber Co. Akron, OH	wheel assembly
Hamilton Standard Div., UTC Windsor Locks, CT	air conditioning
Harris Corp. Melbourne, FL	avionics
Hercules, Inc. Taunton, MA	seals
Hughes-Treitler Mfg. Co. Garden City, NY	heat exchangers
Hydroaire Div. of Crane Co. Burbank, CA	anti-skid components
IBM Corp. Oswego, NY	on board computer
Instrument Systems Corp. Telephonics Huntington, NY	test system
Garrett Turbine Engine Co. Phoenix, AZ	power system
AiResearch Mfg. Co., Garrett Corp. Torrance, CA	computer
Kaman Aerospace Corp. Bloomfield, CT	rudders and fairings
Kearfott Div., Singer Co. Little Falls, NJ	avionics
Kelsey Hayes Co. Springfield, OH	launcher components
Martin Marietta Corp. Baltimore, MD	stabilizers
McDonnell Douglas Corp. Long Beach, CA	ejection seats
Menasco, Inc. Burbank, CA	nose gear
Parker Hannifin Irving, CA	avionics
Pittsburgh Plate & Glass Ind., Inc.	windshield
Simmonds Precision, Inc. Vergennes, VT	avionics
Sperry Corp. Phoenix, AZ	avionics
Sperry Vickers Co. Jackson, MS	pumps
Stainless Steel Products Co. Burbank, CA	air ducts
Sierracin Corp. Sylmar, CA	windshield
Sterrer Eng. and Mfg. Co. Los Angeles, CA	steering
Sundstrand Aviation Corp. Rockford, IL	rudder control
Sundstrand Data Control, Inc. Redmond, WA	test system components
TRW, Inc. Cleveland, OH	fuel pumps
United Aircraft Products, Inc. Dayton, OH	heat exchangers
Vickers Aerospace Co. Troy, MI	hydraulic pumps
Vought Corp. Dallas, TX	fuselage
Westinghouse Corp. Lima, OH	avionics

Sources: Council on Economic Priorities; *Aerospace Daily*, 26 October 1981, p. 301; ACDA, FY 1979 ACIS, p. 85; SAC Fact Sheet, "B-1B," December 1981.

* Associate Prime Contractors

5

Advanced Technology Bomber

Advanced Technology Bomber (ATB) ("Stealth")

On 22 August 1980, the Department of Defense formally announced that a technological advance involving aircraft design, absorbent materials, and electronics had resulted in reducing the detectability of future aircraft to radar, infrared (IR), and optical surveillance systems. The DOD announced that a "Stealth" bomber using such innovations would be developed. Reports of Stealth technology have appeared in *Aviation Week and Space Technology* since 1979 (29 January 1979, 16 June 1980), and a program of "strategic bomber enhancement" had been ongoing for many years.

Stealth was one of the original candidates for the B-52 replacement (LRCA). An "Advanced Technology Bomber," a new airplane design, rather than applying "Stealth" technology to a conventional bomber design, e.g., the B-1, was envisioned for an IOC of 1991.[1] However, the new technology was unable to meet a Congressionally mandated 1987 IOC.

The Air Force hopes to build 100-150 ATBs with an IOC in the early 1990s[2] to replace the B-1B (and remaining B-52s) as a penetrating bomber. The Congressional Budget Office has reported that a force of 132 ATBs will be deployed.[3] The Air Force contends that the ATB is necessary to ensure that the "strategic bomber force will continue to have the ability to penetrate Soviet air defenses into the next century."[4]

Stealth technology combines active and passive methods to reduce radar reflection and energy emissions. These techniques probably would include reductions in weight of aircraft and size of tail, addition of non-metallic and radar absorbing materials, modifying shapes and angles, advanced designs reducing engine exhaust temperatures, optical absorbers, active jammers, decoy transponders, and treating fuels to reduce infrared emissions.[5]

Northrop Corporation is the prime contractor to develop the ATB, with General Electric as a participant.[6] Also reportedly collaborating on Stealth research are Rockwell/Lockheed and Boeing/Northrop. Estimates for a 100-150 ATB program range from $22 billion to $40 billion. A recent DOD cost estimate for a 165 plane ATB force is $36 billion.[7]

COST:

FY	Number Procured	Total Appropriation ($ million)
1982	-	122[8]

---

1 HAC, FY 1982 DOD, Part 9, p. 217.
2 DOD, FY 1984 Annual Report, p. 224; First operational squadron has been stated as possible in 1990; SASC, FY 1982 DOD, Part 7, p. 3783; 1991 according to *Washington Post*, 13 March 1982, p. A8.
3 CBO, "Contribution of MX to the Strategic Force Modernization Program," n.d. (1982).
4 DOD, FY 1983 RDA, p. II-23.
5 *Ibid*.
6 SASC, FY 1983 DOD, Part 7, p. 4564.
7 HAC, FY 1982 DOD, Part 1, p. 322.
8 AW&ST, 9 March 1981, p. 23.

5
Strategic Defensive Systems

Strategic Defensive Systems

Only one nuclear system—the GENIE air-to-air missile—is presently used for the defense of the continental United States. GENIE, a dual-capable aircraft launched unguided rocket, is deployed at alert sites with interceptor aircraft throughout the country. Nuclear armed NIKE-HERCULES surface-to-air missiles, once widely deployed in the United States in the 1960s, have been dismantled and only remain as tactical air defense weapons in Europe. A limited anti-ballistic missile (ABM) system was briefly deployed from 1974-1976. Today ABM research is being greatly accelerated for future deployment in the United States.

Without an ABM system, the interception of bombers attacking the North American continent is the only U.S. nuclear defensive capability. Air defense is provided by U.S. and Canadian fighter interceptor aircraft that are maintained on alert at 23 sites in the continental United States, three in Canada, four in Alaska, one in Hawaii, and one in Iceland.[1] A variety of strategic interceptor aircraft models exist. Some models are designed solely for strategic defensive missions, and other models were selected for strategic air defense missions because of their air-to-air characteristics. Four aircraft are now used for strategic defense: F-106, F-4, F-15, and the Canadian CF-101. Eighteen of the new F-15s were given strategic interception missions in Fiscal Year 1982 and have been placed on peacetime alert at one location in the U.S.[2] The five remaining F-106 squadrons will be replaced with additional F-15s between FY 1983-1986,

Figure 5.29 **SPARTAN** missile test from Meck Island at Kwajalein Missile Range in the Marshall Islands.

and the Canadian CF-101 will be replaced by Canadian F-18s. Other Navy, Air Force, and Marine Corps aircraft would be given strategic defensive missions in crisis or wartime.

Ballistic Missile Defense

The U.S. Army spends several hundred million dollars a year on research and development to maintain a capability for deploying a strategic defensive system to destroy enemy reentry vehicles in flight. This research is presently being conducted within the constraints of the ABM treaty of 1972.[3]

The President's Strategic Program, presented in October 1981, accelerated ABM research and tied the development program closely to land-based MX deployment plans. According to one DOD official, "the more likely ballistic missile defense systems (chosen) to protect the

Figure 5.28 **SAFEGUARD** complex, where SPARTAN and SPRINT anti-ballistic missiles (ABMs) were deployed.

---

1 DOD, FY 1979 Annual Report, p. 121.
2 JCS, FY 1981, p. 44.

3 The treaty is one of two agreements signed at Moscow on 26 May 1972, known collectively as the SALT I agreements.

5

Strategic Defensive Systems

Table 5.27
BMD Funding (RDTE, $ million)

	FY 1981 & Prior[1]	FY 1982[2]	FY 1983[3]	FY 1984[4]	FY 1985[5]
Advanced Technology Development	1378.9	126.5	142.8	170.9	183.9
Systems Technology Development	1080.6	335.6	396.2	538.4	1380.0

1 ACDA, FY 1983 ACIS, p. 138.
2 Ibid.
3 DOD, *Program Acquisition Costs by Weapon System*, 31 January 1983.
4 Ibid.
5 DOD, RDT&E Programs (R-1), 31 January 1983.

land-based missiles would require a revision of the ABM treaty."[4] Deployment of an extensive ABM system to defend several fixed sites or a mobile ICBM system would require abrogation or modification of the 1972 ABM treaty, which limits the U.S. and the Soviet Union each to one ABM site.[5]

At the time of the signing of the ABM Treaty, the SAFEGUARD system was being deployed at Grand Forks, North Dakota, to protect the ICBM field there. The system was completed in 1974 at a cost of over $7 billion, but it was deactivated in 1976 because of its high cost and ineffectiveness. Even with the deactivation of the SAFEGUARD system's SPRINT and SPARTAN missiles, they remained Treaty accountable unless dismantled in accordance with the procedures in the Standing Consultative Commission. After dismantling the SAFEGUARD system, missiles and warheads were placed in storage in Army depots.[6] Both weapons will be retired in FY 1983-1985.

The deactivation of the SAFEGUARD system, the termination of interceptor flight tests, and a follow-on BMD system prototype in 1975 have led to a change of focus in the research program. The recent focus has been the definition and demonstration of options for ABM defense of MX and land-based strategic missiles. The Reagan Strategic Program, announced 2 October 1981, further focused research with the decision to deploy the MX in existing fixed silos. The pre-prototype demonstration program, begun in 1980 to provide options for enhancing ICBM survivability and for defending other strategic targets, was reoriented toward terminal defense of ICBM silos.[7] In FY 1985, BMD research will be doubled.[8]

Much of the BMD research program, which deals with radar, sensing, tracking, and guidance, is included in the Advanced Technology Program. The Systems Technology Program is involved in the prototyping and demonstration of potential BMD systems and is currently examining two systems: a nuclear armed Baseline Terminal Defense System (formerly Low-Altitude Defense System (LoADS)), with a missile designated SENTRY, and a non-nuclear "Exoatmospheric Overlay Defense."

---

4 SASC, *Strategic Force Modernization Programs*, p. 49.
5 The original treaty limited each side to two ABM deployment areas (one national capital area and one ICBM silo area) with restrictions to 100 launchers at each area. A protocol to the treaty signed in 1974 further restricted each side to only one ABM deployment area.
6 ACDA, FY 1982 ACIS, p. 441; HASC, FY 1982 DOE, p. 104.
7 HAC, FY 1982 DOD, Part 9, p. 347.
8 ACDA, FY 1983 ACIS, p. 129.

5
Strategic Defensive Systems

Table 5.28
Major Ballistic Missile Defense Contractors

Lincoln Laboratory MIT, Lexington, MA	TRW, Inc. Redondo Beach, CA	Rockwell International Corp. Anaheim, CA
Boeing Co. Seattle, WA	General Electric Co. Syracuse, NY	Hughes Aircraft Corp. Culver City, CA
Martin Marietta Corp. Orlando, FL	Lockheed Missiles and Space Co. Sunnyvale, CA	Electronic Space Systems Corp. Concord, MA
McDonnell Douglas Corp. Huntington Beach, CA	Teledyne Brown Engineering Huntsville, AL	Computer Development Corp. Minneapolis, MN
System Development Corp. Huntsville, AL	Raytheon Wyland, MA	IBM Gaithersburg, MD

Although SENTRY received the most attention, it was cancelled in February 1983 "as a result of shifting requirements within the BMD program leading to a change in focus on the technologies of interest."[9] Component development will be completed, but at a slower pace, and the SENTRY system will be kept available for possible deployment at a later date. Current interest is focused on:[10]

- developing operating rules for silo defense,
- developing command and control and operational procedures,
- beginning component preparation of subsystems, and
- selecting subcontractors for radar, vehicle, and support equipment.

The design of ABM warheads reportedly has always favored enhanced radiation designs to destroy incoming RVs with intense radiation. The SPRINT missiles of the SAFEGUARD system reportedly had enhanced radiation designs.[11] The nuclear warhead for the SENTRY missile is probably also an enhanced radiation design. DOD once considered taking the SPRINT missile warheads out of storage, refurbishing them, and using them in the SENTRY missiles.[12] Now, a newer generation warhead is planned. The warhead is described as a "small nuclear defensive warhead," with a "very small" yield.[13]

---

9 DOD, "Memorandum for Correspondents," 10 February 1983.
10 HAC, FY 1982 DOD, Part 9, p. 347.
11 George B. Kistiakowsky, "Enhanced Radiation Warheads, Alias the Neutron Bomb," Technology Review, May 1978.
12 AW&ST, 30 March 1981, p. 19.
13 HAC, FY 1983 DOD, Part 4, p. 572.

5
SENTRY

SENTRY

Figure 5.30 SPRINT missile, probably similar in size and characteristics to the newer **SENTRY** missile.

DESCRIPTION:	Very high acceleration, high velocity, nuclear armed, anti-ballistic missile.
CONTRACTORS:[1]	McDonnell Douglas (prime) Teledyne Brown (system engineering) Raytheon (system engagement controller) TRW (data processing) GTE/Sylvania (command, control, and communications) Martin Marietta Orlando, FL (missile)

SPECIFICATIONS:	
Length:	17 ft
Diameter:	unknown
Stages:	1
Weight at Launch:	unknown
Fuel:	solid
Guidance:	inertial with external guidance updates
Throwweight/Payload:	unknown
Range:	unknown
DUAL CAPABLE:	no
NUCLEAR WARHEADS:	small nuclear warhead;[2] 5 Kt range;[3] likely enhanced radiation warhead. Warhead is in Phase 2 Program Study. Phase 3 Development Engineering is being requested by DOD during FY 1983-84.[4]
DEPLOYMENT:	Layered defense initially would be provided with LoADS for intercepts within the atmosphere, with an overlay tier of interceptor missiles armed with non-nuclear warheads for target kills in space.[5]
Launch Platform:	fixed launcher
Number Planned:	some 500, one launcher per missile being protected
Location:	MINUTEMAN fields or MX bases[6]

5
SENTRY

HISTORY:
IOC: 1988[7]

1979 — LoADS warhead selection working group formed

Feb 1983 — SENTRY development terminated

TARGETING:
Types: reentry vehicles in flight

---

1 SASC, FY 1982 DOD, Part 7, p. 4131.
2 SASC, FY 1982 DOD, Part 7, p. 4128; HAC, FY 1983 DOD, Part 4, p. 572.
3 AW&ST, 8 March 1982, p. 28.
4 DOE, Budget Justification, FY 1983, p. 51.
5 AW&ST, 9 March 1981, pp. 24, 27.
6 Most likely option is for initial deployment to be around MINUTEMAN III fields near Grand Forks, ND, starting in the mid-1980s, followed by deployment within the MX basing areas.
7 Prior to termination; SASC, *Strategic Force Modernization Programs*, p. 95.

5

GENIE Rocket

Bomber Interception
GENIE (AIR-2A)

Figure 5.31 GENIE rocket, center, loaded into missile bay of F-106.

DESCRIPTION:	Short-range, unguided, nuclear capable air-to-air rocket designed for strategic interception of bombers and used by the Air Force.
CONTRACTORS:	McDonnell Douglas Astronautics Co. (prime) Thiokol Chemical Corp. (power plant) Hughes (fire control system) Aerojet General Corp.

SPECIFICATIONS:

Length:	9.6 ft
Diameter:	17.4 in
Stages:	1
Weight at Launch:	822 lb[1]
Propulsion:	solid propellant rocket motor
Speed:	Mach 3.0
Guidance:	no guidance system, fins and gyroscope stabilization
Range:	6 mi; 6.8 mi;[2] 6.2 mi[3]
DUAL CAPABLE:	yes
NUCLEAR WARHEADS:	one W25; 1.5 Kt range[4]

DEPLOYMENT:

Launch Platform:	CF-101, F-106A, F-4, F-15
Number Deployed:	thousands of missiles produced, 200 nuclear versions estimated presently operational (1983)
Location:	(see Table 4.6)

HISTORY:

IOC:	1957
Jul 1957	nuclear GENIE is tested in live firing at Indian Springs, Nevada by launching from F-89J airplane and detonated at 15,000 ft.[5]
1962	production of GENIE ended.

5
GENIE Rocket

TARGETING:

Types: bombers

Selection Capability: GENIE is designed to be fired automatically and detonated by the fire control system in the aircraft.[6]

Accuracy/CEP: not thought to be very accurate

COMMENTS: flight time varies between 4 and 12 seconds at ranges of 1.5 to approximately 6 miles at speed of 2100 mph (Mach 3).[7] Missile also known as "HIGH CARD," "DING DONG," and "MB-1."

---

1 The World's Missile Systems, 6th Ed., p. 46.
2 Nikolaus Krivinyi, World Military Aviation (New York: Arco, 1977), p. 222.
3 The World's Missile Systems, 6th Ed., p. 46.
4 The airborne test of the warhead was part of Operation Plumbob, 19 July 1957, see Michael J.H. Taylor and John W.R. Taylor, Missiles of the World, 6th Ed., p. 44.
5 Ibid.
6 Krivinyi, op. cit., p. 222.
7 Fact Sheet prepared by National Atomic Museum, Albuquerque, NM.

6
Cruise Missiles

Chapter Six
Cruise Missiles

Cruise missiles are unmanned, expendable flying vehicles programmed to carry explosives over a non-ballistic trajectory to their target. Using air as an oxidizer, the propulsion system is similar to that of a jet powered airplane. The missiles' engines thus propel cruise missiles in a similar way to aircraft and not over a ballistic path. Cruise missiles fly much slower than ballistic missiles, and thus can also utilize advanced guidance systems which make the present generation missiles extremely accurate.

Cruise missiles had their origin in World War II. The development of more accurate and autonomous ballistic missiles in the 1950s led to a significant reduction in cruise missile research for many years. The United States deployed nuclear armed cruise missiles in the 1950s (REGULUS and SNARK). Due to their large size, inaccuracy, and unreliable performance, they were abandoned in favor of ballistic missiles. Technological advances in the 1960s and 1970s in engine, warhead, and guidance miniaturization gave rise to the potential of a much smaller cruise missile airframe with increased range and higher accuracy. With the sinking of the Israeli destroyer *Elath* in 1967 by a Soviet SS-N-2 STYX cruise missile, the U.S. increased the pace of development of new cruise missile systems. Development of the first of the present generation of nuclear armed cruise missiles—the TOMAHAWK Sea-Launched Cruise Missile (SLCM)—was started by the Navy in 1972. The Air Force followed with the Air-Launched Cruise Missile (ALCM) in 1973.

Studies for the Navy's TOMAHAWK proceeded through the mid-1970s with the resulting design becoming the basic frame for both sea-launched and ground-launched applications. The Air Force's missile, which evolved from the Subsonic Cruise Armed Decoy (SCAD), resulted in a competition between the TOMAHAWK design by General Dynamics and a Boeing air-launched design (AGM-86B). The Boeing design won the competition and was chosen as the ALCM.

In January 1977, the cruise missile program received a new charter and greater emphasis with the establishment of the Joint Cruise Missile Project Office within the DOD. At the same time, the decision was made to begin full scale engineering development of long-range Air and Sea-Launched Cruise Missiles and to utilize the TOMAHAWK cruise missile for the Ground-Launched as well as Sea-Launched role.

The Air-Launched Cruise Missile, which had received more attention than the other missile programs, began deployment in late 1981. The Ground-Launched Cruise Missile for use in Europe is planned for deployment in late 1983, but it is unlikely that all 464 missiles earmarked for deployment in a December 1979 NATO agreement will be deployed. The Sea-Launched Cruise Missile will begin deployment in mid-1984 and will be fitted with both nuclear and conventional warheads aboard surface ships and submarines.

All long-range cruise missiles will be nuclear armed. The ALCM and SLCM utilize the same nuclear warhead design (W80) with an estimated yield of 200 Kt. These highly accurate missiles will be capable of destroying almost any target type in the Soviet Union. The GLCM warhead (W84) will have a lower (circa 50 Kt) yield primarily to make its deployment to Europe more palatable to Europeans by decreasing the potential for collateral damage with its use.

The TOMAHAWK GLCM will be carried on transporter-erector-launcher (TEL) vehicles where the ready missile (with the W84 nuclear warhead) will be stored in an aluminum canister in the four tube launcher. Both Ground-Launched and Sea-Launched missiles will be propelled from their launch tubes by a solid-fuel booster engine which is then jettisoned. Retracted wings and control fins will then extend and the air-breathing engine will ignite to provide propulsion to the target. The ALCM is designed for delivery from strategic aircraft, dropped from a bomb bay (internal) or from a pylon mounted on the wing (external).

Almost 9000 cruise missiles are now scheduled for deployment: at least 4348 ALCMs (including Advanced Cruise Missile replacements), 4068 SLCMs, and 560 GLCMs. Approximately 5000 will be armed with nuclear warheads. Only the SLCM will be dual capable. The total cost of the present cruise missile program is estimated at some $25 billion. Each missile will cost from $2-6 million. The nuclear armed ALCMs will go to the Strategic Air Command where they will be carried by B-52G bombers externally, B-52H bombers externally and internally, B-1B bombers, and the future Advanced Technology Bomber ("Stealth"). The SLCM, although conceived as a theater system, will be a strategic

6
Cruise Missiles

Table 6.1
Major TOMAHAWK Cruise Missile Contractors[1]

Company	Location	Product
Atlantic Research Corp.		GLCM, SLCM rocket motor
Boeing	Seattle, WA	GLCM, SLCM components
FMC/Northern Ordnance		SLCM armored box launcher
General Dynamics, Convair Div.	San Diego, CA	GLCM prime, SLCM prime/airframe
GTE Sylvania		GLCM LCC
Honeywell International	Minneapolis, MN	GLCM, SLCM radar altimeter
Kollsman Instrument Co.	Merrimach, NH	GLCM targeting, SLCM radar altimeter
Litton Guidance, Inc.	Woodland Hills, CA; Salt Lake City, UT	SLCM reference measurement
Litton Systems Limited	Toronto, Canada	SLCM reference measurement
Lockheed		GLCM, SLCM targeting
M.A.N.	West Germany	GLCM prime mover
Martin Marietta		SLCM VLS cannister
McDonnell Douglas Astronautics	St. Louis, MO	GLCM guidance/system hardware, SLCM guidance/theater mission planning system
Naval Surface Weapons Center	Dahlgren, VA	SLCM VLS software
Teledyne	Toledo, OH	GLCM, SLCM engine
Unidynamics	Phoenix, AZ	SLCM launcher
Vitro Laboratories		GLCM software, SLCM missile control system
Westinghouse		SLCM submarine launcher
Williams International	Walled Lake, MI	GLCM, SLCM engine

1 Information provided by General Dynamics and Joint Cruise Missile Program Office.

system.[1] One thousand nuclear armed SLCMs with ranges in excess of 1500 miles are planned for deployment on attack submarines and surface ships and would be part of the strategic reserve force and will be available for reconstitution and targeting" after a nuclear war.[2] GLCMs are planned for deployment in Europe for use as a theater system. All GLCMs will be nuclear armed.

The development and deployment of third generation cruise missiles falls under the "Advanced Cruise Missile Technology" (ACMT) program now underway. The program includes four separate elements: modifications to present cruise missiles with new components, development and deployment of a new and upgraded "Stealth" cruise missile, development of a new versatile air-to-air and air-to-ground supersonic cruise missile, and development of an intercontinental cruise missile (see Advanced Cruise Missiles section). Both the Air Force and the Defense Advanced Projects Research Agency (DARPA) have had formal programs for advanced cruise missiles since 1977. The short-range HARPOON cruise missile (described later in this chapter), while currently conventionally armed, is also under consideration to become a nuclear system.

The rapid progress of cruise missile technology and advances in Soviet defenses against low flying objects led to a late 1982 Defense Department decision to end ALCM (AGM-86B) procurement at 1499 missiles after FY 1983 rather than the 4348 planned and to pursue instead an Advanced Cruise Missile with a 1986 IOC to fulfill the remainder of the orders. The cost of the Advanced Cruise Missile program will probably not exceed the cost of the ALCM program. The number of ALCMs (both current design and advanced) to be procured will remain approximately the same.[3]

SASC, FY 1983 DOD, Part 7, p. 4354.
SASC, FY 1983 DOD, Part 7, p. 4517.

3 HAC, FY 1983 DOD, Part 4, p. 588.

6

Air-Launched Cruise Missile

Air-Launched Cruise Missile (ALCM) (AGM-86B)

Figure 6.1 Air-Launched Cruise Missile (AGM-86B).

DESCRIPTION:	Small subsonic, winged, long-range, turbofan powered, accurate, air-to-surface missile, for internal and external carriage on B-52 and B-1 strategic bombers.
CONTRACTORS:	Boeing Aerospace Co. Seattle, WA; Kent, WA (prime) (See Table 6.3 for list of major subcontractors for ALCM.)
SPECIFICATIONS:[1]	(AGM-86B)[2]
Length:	20 ft 9 in (249 in)
Diameter:	27.3 in
Stages:	1
Weight at Launch:	3300 lb,[3] 2900 lb[4]
Propulsion:	air breathing F-107-WR-10 turbofan engine, 600 lb thrust
Speed:	500 mph
Flight altitude:	100 ft above ground[5]
Guidance:	inertial navigation system, updated by terrain contour matching
Throwweight/ Payload:	240 lb[6]
Range:[7]	2500 km;[8] 1550 mi; 1350 nm 1600 nm[10]

6
Air-Launched Cruise Missile

Table 6.2
ALCM Chronology

Aug 1973	SCAD program converted with basic airframe and propulsion equipment taken over by the non-decoy ALCM
1976	Establishment of extended range ALCM requirement
Mar 1976	First test of powered flight ALCM (AGM-86A)
Jan 1977	DSARC II approves Boeing ALCM for full scale development
Jul 1977	Cancellation of B-1 increases importance of program; General Dynamics added for competitive full-scale engineering development[1]
Aug 1977	Advanced Cruise Missile Technology program begins
Feb 1978	Boeing AGM-86B and General Dynamics AGM-109 begin ALCM competition
Jun 1978	Limited Operational Capability of June 1980 cancelled by DOD
Jun 1979	First full scale development flight
Feb 1980	Final flight of ALCM competition
Mar 1980	Air Force selects Boeing AGM-86B as ALCM
Dec 1980	Boeing awarded first contract for production of 480 missiles
Jul 1981	First test launch of ALCM from OAS modified B-52G
Sep 1981	first cruise missiles deployed on B-52G at Griffiss AFB, NY (first alert capability)
Oct 1981	ALCM production increased from 3418 to 4348 missiles
Nov 1981	First full production missile completed by Boeing
FY 1982	Reagan Administration accelerates ALCM schedule and adds B-52H to program
Sep 1982	Advanced Cruise Missile proposals solicited by Air Force
Dec 1982	First squadron of 16 B-52Gs carrying 12 missiles fully operational (IOC)
Jan 1983	DOD reveals cancellation of ALCM after 1547 missiles and transfer to Advanced Cruise Missile Technology
Spring 1983	Selection of ACMT contractor expected
1984	Work at Boeing Plant on ALCM ceases
FY 1984	Planned retrofit of ECM package into ALCM to increase survivability[2]
FY 1986	IOC of Advanced Cruise Missile
FY 1989	B-52G/Hs attain a full ALCM capability
May 1989	Last delivery of ALCM planned under 3418 missile program
FY 1990	Final delivery of ALCM spares in 3418 program;[3] 3160-3300 ALCMs[4]

1 SASC, FY 1980 DOD, Part 5, p. 2491.
2 SASC, FY 1983 DOD, Part 7, p. 4589.
3 SASC, FY 1982 DOD, Part 7, p. 3802.
4 DOD, FY 1982 Annual Report, pp. 50, 114, v.

Figure 6.2 ALCM soon after drop from B-52 bomber.

DUAL CAPABLE: no

NUCLEAR WARHEADS: one W80-1, 200 Kt range (see W80 in Chapter Three)

DEPLOYMENT: B-52G: being modified to carry 12 ALCMs on external pylons; B-52H: To be modified to carry 12 ALCMs on external pylons and up to eight internally on rotary launcher;[11] B-1B: Capable of carrying up to 22 ALCMs.[12]

Number Deployed: approximately 350 (1983);[13] 1547 to be procured (Dec 1982); 4348 planned under Reagan Administration before adoption of advanced cruise missile in FY 1984;[14] 5369 planned under previous accelerated procurement program;[15] 3418 previously planned (circa FY 1981-1982) for procurement for B-52 force (FY 1978-FY 1987) including 24 developmental units;[16] 3020 planned before then (FY 1979).

Nuclear Weapons Databook, Volume I   175

6

Air-Launched Cruise Missile

Table 6.3
Major ALCM Subcontractors[1]

AiResearch Manufacturing Co. Torrance, CA	servo assembly	McDonnell Douglas Aeronautics East* St. Louis, MO	guidance
Aluminum Co. of America Corona, CA	airframe castings	Microcom Corp. Warminster, PA	telemetry transmitter
Anadyte-Kropp Chicago, IL	forgings	Northrop Corp. El Monte, CA	rate/acceleration sensor
Consolidated Control Corp. El Segundo, CA	arm/disarm device, fuzing	DEA, Inc. Denver, CO	fuel valves
Eagle Picher Industries Joplin, MO	batteries	Oklahoma Aerotronics Hartshorne, OK	C2 components
Explosive Technology Fairfield, CA	tube assembly	Pyronetics Devices Denver, CO	services
G&H Technology Santa Monica, CA	connector assembly components	Rosemont Minneapolis, MN	computers
Gulton Industries Albuquerque, NM	telemetry multiplexers	Sundstrand Aviation Rockford, IL	fuel pump
Hi Shear Corp. Torrance, CA	recovery system	Teledyne CAE* Toledo, OH	engine alternative
Honeywell	missile radar altimeter	Unidynamics/Goodyear Phoenix, AZ	actuator assemblies
Irvin Industries Gardena, CA	flight termination system	United Technologies Windsor Locks, CT	air cycle machines
Kollsman Instrument Company	missile radar altimeter	Wellman Dynamics Corp. Creston, ID	airframe castings
Lear Siegler Maple Heights, OH	generator	Williams International Research* Walled Lake, MI	engine
Litton* Woodland Hills, CA	guidance		
Litton Systems Canada Div.* Toronto, Canada	guidance		

* Associate contractors.
1 Under the ALCM program numerous contractors are "associate contractors" with whom the Air Force directly contracts; see AW&ST, 31 March 1980, p. 20.

Location: Griffiss AFB, NY (September 1981)
Wurtsmith AFB, MI (April 1983)[17]
Grand Forks AFB, ND (October 1983)
Ellsworth AFB, SD (January 1984)
Blytheville AFB, AR
Fairchild AFB, WA
Barksdale AFB, LA
Carswell AFB, TX
Castle AFB, CA[18]

HISTORY:
IOC: December 1982 (see Table 6.2 ALCM Chronology

TARGETING:
Types: Broad spectrum, including hard targets, ALCM may be used to deny an ICBM reload capability[19]

6
Air-Launched Cruise Missile

Table 6.4
ALCM Program Schedules[1]

	1979	1980	1981	1982	1983	1984	1985	1986	1987	1988	1989	1990	Total
ALCM Cumulative Deliveries													
Carter Program[2] (FY 1981-82)	—	10	22	209	689	1169	1649	2129	2609	3089	3394	—	3394
Reagan Program[3] (FY 1983)	—	10	22	206	680	1120	NA	NA	NA	NA	NA	NA	4348
Accelerated Program[4] (FY 1983)	—	—	21	319	799	1469	2189	2909	3629	4349	5069	5369	5369
B-52 Conversions													
B-52G:													
Carter[5] (FY 1981)	—	0	1	13	41	40	38	135	—	—	—	—	173
Reagan[6] (FY 1983) External	3	22	40	40	41	26	—	—	—	—	—	—	172
B-52H:													
Reagan[7] (FY 1983)	—	—	—	—	—	3	23	22	22	26	—	—	96

1 Does not take into account possible changes with conversion to Advanced Cruise Missile.
2 HASC, FY 1981 Mil Con, p. 431; HASC, FY 1982 Mil Con, p. 228.
3 SASC, FY 1983 DOD, Part 7, p. 4566.
4 SASC, FY 1982 DOD, Part 7, p. 3802; internal conversions of B-52G cancelled in FY 1983.
5 ACDA, FY 1981 ACIS, p. 122.
6 SASC, FY 1982 DOD, Part 7, p. 3802.
7 Ibid.

Selection Capability:	reportedly carries instructions for 10 different preselected targets;[20] ALCM can be armed from the bomber cockpit[21]
Accuracy/CEP:	reportedly 10-30 m;[22] 300 ft;[23] greater hard target kill capability than ICBMs, even MX[24]
COST:	
Total Program Cost:[25]	$3170.8 m (base year 1977 cost);[26] $5232.7 m (then year) (FY 1981) $9420.0 m (FY 1983)[27] $4327.6 m (Dec 1982)
Unit Cost:	$881,000 (FY 1981) (flyaway), $1.247 m (program)

FY	Number Procured[28]	Total Appropriation ($ million)
1977 & prior	—	268.3[29]
1978	24	381.5
1979	48	433.1
1980	225	477.1
1980 & prior	297	1470.3
1981 & prior	753	2119.7[30]
1982	440	799.3
1983	330	574.5
1984	—	152.5 (request)

Nuclear Weapons Databook, Volume I **177**

6

Air-Launched Cruise Missile

COMMENTS: ALCM-B (AGM-86B) is extended range alternative (20 inch fuel tank segment) of two originally considered concepts, with greater range and weight than ALCM-A (AGM-86A). ALCM has 1/1000th of radar return of B-52 bomber.[31]

---

1 Boeing Fact Sheet, "Background Information, AGM-86B Air Launched Cruise Missile," April 1982.
2 2500 km is "system operational range" where operational factors are taken into account; propulsion range is greater; HAC, FY 1980 DOD, Part 1, p. 759. Williams Research has designed a new engine that provides 30% thrust increase and possible 300 nm increase in range. Second generation CM is being developed with 800 nm increase in range over first generation AGM-86B.
3 GAO, Draft Study for B-1 (1982).
4 ACDA, FY 1979 ACIS, p. 60.
5 SASC, FY 1981 DOD, Part 2, p. 50.
6 *U.S. Missile Data Book*, 1980, 4th Ed., p. 2-5.
7 Range takes into account all operational limitations of the system to effectively engage the target (operational fuel, allowance for indirect routing, speed and altitude variations).
8 2500 km is "system operational range"; HAC, FY 1980 DOD, Part 1, p. 759.
9 *U.S. Missile Data Book, op. cit.*
10 HAC, FY 1982 DOD, Part 9, p. 246.
11 ACDA, FY 1983 ACIS, p. 67.
12 *Ibid.*
13 AW&ST, 17 January 1983, p. 101.
14 *Aerospace Daily*, 11 January 1982; DOD, Selected Acquisition Report, 19 March 1982; reports are that with conversion to ACMT, the number of all types of ALCMs to be deployed remains the same; HAC, FY 1983 DOD, Part 4, p. 588.
15 SASC, FY 1982 DOD, Part 7, p. 3802.
16 OSD (PA), "Memorandum for Correspondents," 2 May 1980; HASC, FY 1981 DOD, Part 4, Book 2, p. 1814.
17 First two ALCMs were received 12 June 1982; DOD, Selected Acquisition Report, as of 3 June 1982.
18 SASC, FY 1982 DOD, Part 7, p. 4291.
19 ACDA, FY 1980 ACIS, p. 27.
20 *Armed Forces Journal*, November 1976, p. 22.
21 HAC, FY 1983 DOD, Part 4, p. 588.
22 Senate Foreign Relations Committee/House International Relations Committee, *Analysis of ACIS Submitted in Connection with the FY 1978 Budget Request*, Joint Committee Print April 1977, p. 83.
23 Kosta Tsipis, "Cruise Missiles," *Scientific American*, February 1977, p. 29.
24 SASC, FY 1981 DOD, Part 2, p. 506.
25 The total program cost for the Boeing ALCM has been reduced with shift to the ACMT George Wilson, *Washington Post*, 16 February 1983, p. 1, suggests the cost could go to $4. billion.
26 SASC, FY 1980 DOD, Part 5, p. 2491.
27 SASC, FY 1983 DOD, Part 7, p. 4566.
28 Planned procurement rate under 3418 program was 480 per year after FY 1982; HASC, F 1981 DOD, Part 4, Book 2, p. 1822.
29 Includes funds for the SCAD, about half of which is considered directly applicable t ALCM.
30 SASC, FY 1983 DOD, Part 7, p. 4566.
31 SASC, FY 1981 DOD, Part 2, p. 50.

6 Ground-Launched Cruise Missile

Ground-Launched Cruise Missile (GLCM) (BGM-109)

Figure 6.3 **Ground-Launched Cruise Missile (BGM-109)** test firing.

DESCRIPTION: Long-range, all weather, accurate, surface-to-surface subsonic cruise missile for use in the European theater. GLCM is a version of the TOMAHAWK BGM-109 cruise missile (the Navy's SLCM).

CONTRACTORS: see Table 6.1, Major TOMAHAWK Cruise Missile Contractors

SPECIFICATIONS:

Length: 20.3 ft;[1] 219 in; (5.56 m)

Diameter: 20.4 in (52 cm);[2] designed to fit standard 54 cm torpedo tube; 2.5 m wingspan

Stages: 1

Weight at Launch: 1200 kg (2650 lb)

Propulsion: solid booster with air-breathing F107-WR-400 turbofan jet engine

Speed: Mach 0.7 (550 mph) (max)

Guidance: inertial navigation with Terrain Contour Matching (TERCOM) updates at periodic intervals, radar altimeter

Throwweight/Payload: 270 lb

Range: 1350 nm;[3] 2000-2500 km (3000 km achieved in tests);[4] 2500 km[5]

DUAL CAPABLE: no

NUCLEAR WARHEADS: one W84/missile, variable yield, low Kt, 10-50 Kt range (see W84)

DEPLOYMENT: GLCM firing unit ("flight") is composed of four transporter-erector-launchers (TELs), 16 missiles, two launch control vehicles (LCCs) (1 primary, 1 backup), 16 support vehicles, and 69 personnel. The ground mobile units will be air transportable (C-130 and C-141 aircraft).

Launch Platform: M.A.N. Tractor-semitrailer with launcher, erected to a 45-degree angle at launch

Number Planned: 565 missiles are planned for procurement; 137 TELs, 116 operational, 79 LCCs[6]

6

Ground-Launched Cruise Missile

Location: Six bases in Europe; two in United Kingdom: RAF Molesworth (24 launchers) and RAF Greenham Common (16 launchers); one in Italy: Comiso (Sicily) (28 launchers); one base in the Netherlands: Woensdrecht (12 launchers); one base in Belgium: Florennes (12 launchers); one base in Germany: Wueschein (24 launchers)

HISTORY:

IOC:	Dec 1983[7]
Jan 1977	Decision to develop ground-launched cruise missile made[8]
Oct 1977	Development begins
Dec 1979	First flight of prototype
12 Dec 1979	NATO agrees on deployment of 464 Air Force GLCMs to Europe
May 1980	First ground launch from transporter-erector-launcher
end 1980	Full scale engineering development
Feb 1982	Full testing of GLCM begins[9]
Dec 1983	IOC with initial deployment in UK
March 1984	Initial deployment in Italy[10]
end FY 1985	166 GLCM in Europe[11]
end FY 1988	464 GLCM in Europe[12]

TARGETING:

Types:	targets across the entire spretrum: missile sites, airfields, command and communications sites, nuclear storage sties, air defense centers in the Soviet Union and Eastern Europe[13]
Selection Capability:	Each missile sitting on quick reaction alert (QRA) will hold a series of targets.[14] Targets will be generated at three "mission planning" centers, one in U.K. and two on the continent. Each flight's launch-control center will maintain an additional series of programs for various targets.
Retargeting:	immediate for prepared programs of known target data; longer if target data must be prepared.[15] New program for new target and route is generated by mission planning system equipment.
Accuracy/CEP:	circa 30 m

COST:

Program Cost:	$3595.2 m (Dec 1982); $630 m (warheads) (DOE) (FY 1983)[16]
Unit Cost:	$814,000 (flyaway) (base year 1977)[17] $1.283 m (flyaway) (FY 1981) $2.341 m (program) (FY 1981)

6
Ground-Launched Cruise Missile

Figure 6.4 Part of **GLCM** convoy in highway test.

Figure 6.5 Field testing of **GLCM**, with missile launcher erect.

FY	Number Procured	Total Appropriation ($ million)[18]
1979 & prior	—	18.7[19]
1980 & prior	0	254.8[20]
1981	11	293.1[21]
1982	54	505.1
1983	84	562.1
1984	120	825.3
1985	120	637.0

COMMENTS: All-up round (missile, nuclear warhead, booster) is carried in cannister, 4 of which are mounted in a TEL, which weighs 77,900 lb, is 55 ft 8 in long, and has self-contained power. The LCC, which weighs 79,200 lb and is 56 ft 11 in long, contains communications and weapon control system. Peacetime QRA by one GLCM flight will be on Main Operating Base in hardened shelter. Wartime and crisis alert will be to dispersed sites in concealed positions.

---

1 AW&ST, 23 June 1980, pp. 24-25.
2 *U.S. Missile Data Book*, 1980, 4th Ed., p. 2-17.
3 Cited as nominal operational range, AW&ST, 21 June 1982, pp. 48-50.
4 DOD, FY 1982 Annual Report, p. 66, lists the GLCM range as 2000 km; DOD, FY 1981 RDA, p. VII-8 lists "operational range" at 2500 km.
5 SASC, FY 1980 DOD, Part 5, p. 2492.
6 Information provided by JCMPO; AW&ST, 26 June 1982, pp. 48-50; *Aerospace Daily*, 19 May 1980, p. 100; *U.S. Missile Data Book*, 1981, 4th Ed., p. 2-16; SASC, FY 1980 DOD, Part 5, p. 2493, refer to 696, the number planned prior to the NATO December 1979 decision; HAC, FY 1982 DOD, Part 3, p. 592.
7 SASC, FY 1982 DOD, Part 7, p. 3803; *Aerospace Daily*, 19 May 1980, p. 100; DOD, FY 1983 RDA, p. VII-13; IOC has slipped from May 1983; SASC, FY 1980 DOD, Part 5, p. 2492.
8 HASC, FY 1980 DOD, Part 3, Book 2, p. 2526; The Defense Systems Acquisition Review Council (DSARC) stipulated in January 1977 that an Air Force GLCM was to be adapted from TOMAHAWK and deployed on mobile launchers for the theater nuclear role, and a reprogramming of funds was requested to expedite operational status; see John Newbauer, "U.S. Cruise Missile Development," *Astronautics and Aeronautics*, September 1979, pp. 24-35.
9 SASC, FY 1983 DOD, Part 7, p. 4397.
10 DOD, FY 1983 RDA, p. VII-13.
11 *Aerospace Daily*, 19 May 1980, p. 100.
12 HASC, FY 1982 DOE, p. 45.
13 HAC, FY 1984 DOD, Part 4, p. 429.
14 HASC, FY 1982 DOD, Part 5, p. 480.
15 SASC, FY 1980 DOD, Part 6, p. 3469.
16 HAC, FY 1982 DOD, Part 7, p. 749; program cost has escalated from a $1109.0 million base year FY 1977 estimate.
17 SASC, FY 1980 DOD, Part 5, p. 2493.
18 Information provided by JCMPO unless otherwise noted, and current as of February 1982.
19 ACDA, FY 1980 ACIS, p. 139.
20 Information provided by JCMPO.
21 Includes increases of $109 million in FY 1981 and $47 million in FY 1982 to fund "cost overruns in the development and procurement of launch control segments" requested by the Reagan Administration.

6
W84

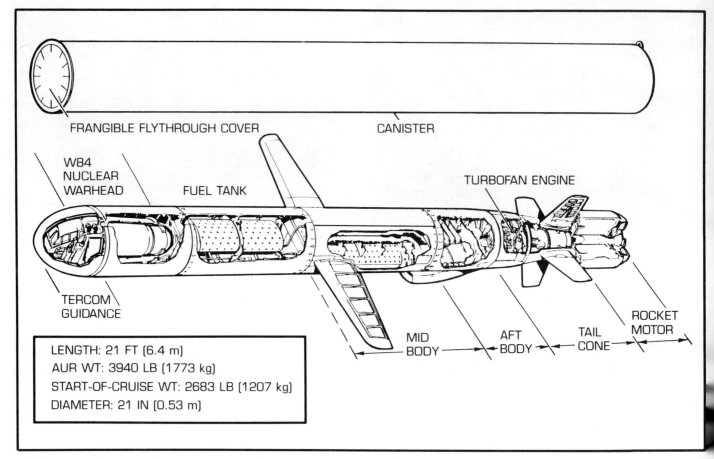

Figure 6.6 **BGM-109** cutaway diagram.

FUNCTION:	Warhead for the Ground-Launched Cruise Missile (GLCM).
WARHEAD MODIFICATIONS:	none known
SPECIFICATIONS:	
Yield:	variable,[1] low Kt, probably 10-50 Kt range
Weight:	light weight
Dimensions:	unknown
Materials:	oralloy as fissile material; IHE
SAFEGUARDS AND ARMING FEATURES:	CAT F PAL, command disable system, steel encased critical components, unique signal generator, final arming of warhead occurs only in target area.[2]
DEVELOPMENT:	
Laboratory:	LLNL
History:	
IOC:	Dec 1983
Sep 1978	Lab assignment[3]
Jan 1979	Phase 3 study initiated[4]
late 1983	initial deployment (Phase 5)[5]
Production Period:	1983-1987[6]

182 Nuclear Weapons Databook, Volume I

6
W84

DEPLOYMENT:

Number Planned: 464 operational missiles to be deployed; 565 missiles planned (1983)

Delivery System: TOMAHAWK GLCM (BGM-109) mounted on a four tube truck TEL

Service: Air Force

Allied User: none

Location: Deployment of 464 GLCM at six main bases in Europe is planned to begin in late 1983.

COMMENTS: W84 is presumed to be a modification of the B61 Mod 3/4 nuclear gravity bomb physics package and associated components.[7]

---

1 HASC, FY 1982 DOD, Part 2, p. 1009.
2 AF, "U.S. Air Force Ground Launch Cruise Missile," n.d. (1982).
3 Entered engineering development; HASC, FY 1980 DOE, p. 95; continued in Phase 3 during FY 1980; SAC, FY 1981 EWDA, p. 818.
4 DOE Budget Justification, FY 1983, p. 51.
5 Funds for production of W84 are included in the FY 1983 DOE Budget.
6 *Ibid.*
7 ACDA, FY 1979 ACIS, pp. 73, 75.

6

TOMAHAWK Sea-Launched Cruise Missile (SLCM) (BGM-109)

Figure 6.7 TOMAHAWK Sea-Launched Cruise Missile (BGM-109) in test over Mojave desert.

DESCRIPTION:	Long-range cruise missile capable of being deployed from a variety of air, surface ship, submarine, and land platforms.
MODIFICATIONS:	(see Table 6.5)
CONTRACTORS:	(see Table 6.1, Major TOMAHAWK Cruise Missile Contractors)
SPECIFICATIONS:	(BGM-109A)
Length:	219 in; 5.56 m
Diameter:	designed to fit standard 21 in torpedo tube
Wingspan:	104.4 in
Stages:	1
Weight at Launch:	1200 kg (2650 lb)[1]
Propulsion:	solid booster with air-breathing, F107-WR-400 turbofan jet engine, 600 lb thrust
Speed:	Mach 0.7 (550 mph) (max)
Guidance:	radar altimeter; inertial navigation with Terrain Contour Matching (TERCOM) which updates at periodic intervals

6
TOMAHAWK Sea-Launched Cruise Missile

Throwweight/Payload: 123 kg

Range: 123 mi (conventional land attack);[2] 2500 km (nuclear land attack)[3]

DUAL CAPABLE: yes

NUCLEAR WARHEADS: one W80-0/missile; 200-250 Kt[4] (see W80 in Chapter Three)

DEPLOYMENT:
Launch Platforms:[5] armored box launcher or Ex-41 VLS by December 1985;[6] SSN-594, SSN-637, SSN-688 class submarines; test platform is USS *Guitarro* (SSN-665); CALIFORNIA, VIRGINIA class cruisers; SPRUANCE class destroyers, reactivated battleships[7] (See Table 6.6)

Figure 6.8 First launch of **TOMAHAWK** missile from armored box launcher installed on the deck of U.S.S. *Merrill* (DD-976).

Table 6.5
TOMAHAWK SLCM Types[1]

Designation	Type	IOC	Front End[2]
BGM-109A	Land Attack Nuclear (TLAM/N)	Jun 1984	W80-0 nuclear warhead, INS/TERCOM
BGM-109B	Antiship Conventional	Aug 1984	BULLPUP warhead, active radar terminal seeker, midcourse guidance unit
BGM-109C	Land Attack Conventional (TLAM/C)	Aug 1984	BULLPUP warhead, INS/TERCOM, midcourse guidance terminal area optical scene matching, time delay fuze
BGM-109D	Combined Effects Bomblet		submunition dispenser, INS/TERCOM, midcourse guidance terminal area optical scene matching
BGM-109E	Reactive Case HE		active radar terminal seeker, midcourse guidance unit
BGM-109F	Airfield Attack Munition		INS/TERCOM, terminal area optical scene matching
AGM-109C	Air-Launched	Dec 1984	conventional warhead
AGM-109H	Air Force MRASM (airfield attack)		runway cratering submunitions, midcourse guidance, TERCOM, DSMAC II
AGM-109I	Air-Launched	Apr 1985	conventional warhead
AGM-109L	Navy MRASM dual mission (TOMAHAWK II)		WDU-188 (HARPOON) conventional warhead, IIR seeker, TERCOM, DSMAC II

1 Information provided by Joint Cruise Missile Project Office; SASC, FY 1982 DOD, Part 7, pp. 4088-4089; HAC, FY 1982 DOD, Part 9, p. 292.

2 All missiles use common aft end, same turbofan engine.

6

TOMAHAWK Sea-Launched Cruise Missile

Table 6.6
SLCM Deployments

Platform	No. to be Modified[1]	SLCMs
PERMIT (SSN-594) class submarines	unknown	8,[2] 12[3]
STURGEON (SSN-637) class submarines	22[4]	8,[5] 12[6]
LOS ANGELES (SSN-688) class submarines	56[7]	8,[8] 12[9], 31 with VLS
CALIFORNIA (CGN-36) class cruisers	7	16
VIRGINIA (CGN-38) class cruisers		16
USS Long Beach (CGN-9)	1	16
SPRUANCE (DD-963) class destroyers	24	16 on 2 Ex-41, VLS
Reactivated battleships (BB-61 class)	4	32 in 8 ABL,[10] VLS
TICONDEROGA (CG-47) class cruisers	all	24 on 3 Ex-41
BURKE (DDG-51) class destroyers	all	VLS

1 Programmed launch platforms; HASC, FY 1981 DOD, Part 4, Book 2, p. 1497; HASC, FY 1982 DOD, Part 2, p. 979; HAC, FY 1983 DOD, Part 2, p. 282.
2 Present torpedo tube launching allows for carriage of 8 SLCMs; Modified SSNs with VLS will be able to hold 20 SLCMs; information provided by Joint Cruise Missile Program Office.
3 VLS will allow 12 tubes for TOMAHAWK.
4 SASC, FY 1983 DOD, Part 6, p. 4043.
5 Present torpedo tube launching allows for carriage of 8 SLCMs; Modified SSNs with VLS will be able to hold 20 SLCMs; information provided by Joint Cruise Missile Program Office.
6 VLS will allow 12 tubes for TOMAHAWK.
7 SASC, FY 1983 DOD, Part 6, p. 4043.
8 Present torpedo tube launching allows for carriage of 8 SLCMs; Modified SSNs with VLS will be able to hold 20 SLCMs; information provided by Joint Cruise Missile Program Office.
9 VLS will allow 12 tubes for TOMAHAWK.
10 HASC, FY 1982 DOD, Part 3, p. 107.

Number Planned: 4068 SLCM planned in all versions, 1480 originally programmed; 196 nuclear missiles originally programmed,[8] 384 planned under early Reagan Administration;[9] increased to 1000 nuclear versions;[10] at least 190 planned for surface ships, 194 for submarines[11]

Location: worldwide deployment;[12] one-third of SLCM equipped attack submarines would be at sea on a day-to-day basis[13]

Figure 6.9 TOMAHAWK missile with inert warhead scores direct hit on a Navy ship target.

HISTORY:

IOC:	June 1984[14]
Jun 1972	development begins with direction for a long-range nuclear land attack missile[15]
Jan 1974	Navy selects General Dynamics and LTV to design a SLCM
FY 1976	TERCOM guidance first demonstrated
Jun 1976	first fully guided test flight
Jan 1977	advanced development completed, entered full-scale engineering development
Feb 1978	first successful submarine launch
Oct 1979	limited production begins
Mar 1980	first test launch from a surface ship
1981	60 flight tests through February 1981
Jun 1984	deployment of nuclear armed SLCM begins.[16]

6

TOMAHAWK Sea-Launched Cruise Missile

Table 6.7
SLCM Funding and Procurement[1]
($ millions)

	FY 1981 and Prior	1982	1983	1984	Total
Submarine-Launched					
Total Appropriation[2]	1082.8	331.5	434.5	433.7	4969.9
Quantity	46	62	70	165	1255
Surface Ship-Launched					
Total Appropriation	73.2	188.6	282.0	507.1	7859.6
Quantity	10	26	50	147	2739
Nuclear Peculiar Funding[3]	(—)	(8.0)	(15.0)	(32.0)	(?)
SLCM					
Total Appropriation	1156.0	520.1	605.4	940.8	12,829.5
Total Quantity	56	88	120	312	3994

1 Information provided by Joint Cruise Missile Program Office reflecting FY 1983 estimates.
2 Includes R&D, Procurement, and Operations and Maintenance.
3 HASC, FY 1981 DOD, Part 4, Book 2, p. 2313.

TARGETING:

Types: land targets, primarily naval related, ports, bases; also surface ships[17]

Selection Capability: Mission planner at theater level will consult interactive graphic display "theater planning package" to layout route for survivability and accuracy. Disc file present at each launching unit holds 1700-5000 land attack missions.[18]

Accuracy/CEP: circa 30 m

COST: (See Table 6.7)

Program Cost: $11,520.0 m (Dec 1982)
$12,829.5 m (FY 1983)

Unit Cost: $3.167 m (FY 1980) (flyaway);
$4.759 m (program)

---

1 ACDA, FY 1979 ACIS, p. 72.
2 HASC, FY 1982 DOD, Part 3, p. 327.
3 SAC, FY 1980 DOD, Part 4, p. 420; the nuclear warhead is "considerably smaller" than a conventional warhead, thus extending the range of SLCM; Sandia, *Lab News*, 18 September 1981.
4 AW&ST, 22 November 1976, p. 15.
5 For submarine launch, SLCM is loaded into a stainless steel capsule which protects it during handling and underwater launch. For surface ship applications, TOMAHAWK will initially be launched from a specially designed armored box launcher mounted on the deck.
6 SASC, FY 1983 DOD, Part 4, p. 4517.
7 DOD, FY 1981 RDA, p. VII-8.
8 HASC, FY 1981 DOD, Part 4, Book 2, p. 1497.
9 *Philadelphia Inquirer*, 4 December 1981, p. 1.
10 Michael Getler, *Washington Post*, 19 January 1983, p. A15.
11 *Philadelphia Inquirer*, 4 December 1981, p. 1.
12 ACDA, FY 1982 ACIS, p. 214.
13 ACDA, FY 1979 ACIS, p. 73.
14 Submarine-launched and ship-launched; SASC, FY 1982 DOD, Part 7, p. 4088.
15 SASC, FY 1980 DOD, Part 5, p. 2519.
16 DOD, FY 1983 RDA, p. VII-8; HASC, FY 1982 DOE, p. 144; SASC, FY 1983 DOD, Part 7, p. 4517.
17 A nuclear-armed antiship SLCM also could be deployed, but is not part of the current development program; ACDA, FY 1979 ACIS, p. 72.
18 Information provided by Joint Cruise Missile Planning Office.

6
HARPOON Missile

HARPOON Missile (AGM-84A/RGM-84A/UGM-84A)

Figure 6.10 Air-launched **HARPOON (AGM-84A)** conventional missile installed on wing of a P-3C ORION patrol aircraft.

DESCRIPTION: Medium range air/surface/sub-surface launched anti-ship cruise missile.

CONTRACTORS: McDonnell Douglas (prime)
Lear Seigler (cruise guidance)
Texas Instruments (terminal guidance)
Teledyne (turbojet)
Aerojet (booster)
Honeywell (radar altimeter)
IBM (on-board computer)

SPECIFICATIONS:

Length: air-launched: 151.2 in; ship/sub-launched: 182.2 in (with booster)

Diameter: 13.5 in

Stages: 1

Weight at Launch: air-launched: 1168 lb; ship/sub-launched: 1470 lb; 2200 lb

Propulsion: one J402-CA-400 turbojet sustainer engine augmented by a solid booster for ship/sub launch

Speed: subsonic (Mach 0.8) (max)

188 Nuclear Weapons Databook, Volume I

6
HARPOON Missile

Guidance:	inertial with radar altimeter and active radar mid-course and terminal guidance
Throwweight/ Payload:	510 lb, air and ship/sub launched
Range:	ship/sub-launched: 35 mi; air-launched: 120 mi
DUAL CAPABLE:	currently conventional only; nuclear option has been under consideration but has not been authorized; FY 1981 through FY 1983 budgets have not included any funds for the development or procurement of a nuclear warhead.[2]
NUCLEAR WARHEADS:	one/missile, not yet chosen.
DEPLOYMENT:	
Launch Platform:	armored box launchers containing a mix of TOMAHAWK and HARPOON missiles;[3] can be fired from STANDARD / TARTAR / TERRIER / ASROC ship launchers. BB, CG, CGN, DD-963, DDG, FF-1052, FFG-7, PHM class ships; P-3C, S-3, A-6E aircraft; SSN-594, -637 and -688 class nuclear attack submarines;[4] HARPOON will be deployed on B-52G bombers starting in 1984 for "sea control"[5]
Number Deployed:	2230 planned in program[6]

Figure 6.11 Ship-launched **HARPOON (RGM-84A)** missile.

HISTORY:

IOC:	1977
1968	development begins
Dec 1972	first flight
Jul 1973	Phase 1 Weapons Concept Study completed for Nuclear HARPOON[7]
Aug 1975	Phase 2 Weapons Feasibility Study completed[8]
Sep 1977	Phase 2A Advanced Engineering Study completed[9]
FY 1979	update conceptual and feasibility study for HARPOON nuclear warhead conducted[10]
FY 1980	nuclear HARPOON unfunded[11]

Nuclear Weapons Databook, Volume I **189**

6

HARPOON Missile

TARGETING:

Types: cruisers, destroyers, patrol craft, surfaced submarines, other shipping[12]

Selection Capability: unknown

COST:

Unit Cost: $397,000 (FY 1981) (flyaway); $803,000 (FY 1981) (program); $485,000 (FY 1978)

FY	Number Procured	Total Appropriation ($ million)
1979 & prior	699+	940.5
1980	240	151.1
1981	240	219.2
1982	240	230.4
1983	221	227.7
1984	330	305.2

COMMENTS: Nuclear warhead considered for the HARPOON has included a standard design and an "insertable nuclear component" concept.[13] This would be a warhead that could be converted from conventional high explosive to nuclear.[14]

---

1 When encapsulated for submarine launch; ACDA, FY 1979 ACIS, p. 169.
2 ACDA, FY 1981 ACIS, p. 367; HASC, FY 1980 DOE, p. 95.
3 AW&ST, 30 March 1980, p. 24.
4 ACDA, FY 1981 ACIS, p. 365.
5 AW&ST, 16 August 1982, p. 25.
6 U.S. Missile Data Book, 1980, 4th Ed., p. 2-24.
7 HAC, FY 1980 DOD, Part 2, p. 283.
8 Ibid.
9 Ibid.
10 ACDA, FY 1979 ACIS, p. 170.
11 SASC, FY 1980 DOD, Part 6, p. 2850.
12 ACDA, FY 1979 ACIS, p. 169.
13 ACDA, FY 1979 ACIS, p. 170; an insertable nuclear component would be useful, according to the Navy, for avoiding tradeoff between nuclear and conventional weapons when limited space aboard ships exists; HASC, FY 1980 DOD, p. 61.
14 SASC, FY 1978 ERDA, p. 31.

6
Advanced Technology Cruise Missiles

Advanced Technology Cruise Missiles

Four distinct programs are underway to upgrade the present generation of cruise missiles: modifications to deployed cruise missiles, development and deployment of a new "Advanced Cruise Missile," development of an intercontinental cruise missile, and development of a new bomber weapon to replace the SRAM. The formal "Advanced Cruise Missile Technology" (ACMT) program began in August 1977 to examine the next generation of cruise missiles. The program has the following broad goals:[1]

- increase in range up to 2300-2600 nautical miles, with options for further increases,
- increase in survivability through use of electronic countermeasures,
- use of "Stealth" technology to decrease missile detection ("reduced observables"),[2] and
- incorporation of new software and better "mission planning flexibility."

Modifications to the present cruise missile inventory to obtain these objectives have been under consideration since the beginning of the development program in 1977. In August 1980, the Air Force began an ALCM-L study. On 22 October 1980, DOD provided a program definition for the ACMT program. Boeing now suggests extending the useful life of the 1499 ALCMs already procured through FY 1983 by reducing the radar cross section of the engine inlet and body, upgrading the guidance software, adding an icing sensor, and improving the altimeter.[3]

Engine technology advancements using new fuels and design efficiencies are being studied by Williams International, Garrett Corporation, and Teledyne to obtain reduced fuel consumption, higher performance, and lower detection profiles. One plan is to replace the F107 engine with a new engine—the 14A6—which will provide 35 percent more thrust for 5 percent less fuel consumption and a 10 percent increase in range.[4] Boeing was awarded an engine improvement contract in 1980, but in 1981, DOD cancelled the engine improvement program because costs were too great. Emphasis was then shifted to further development of a new engine.[5]

Airframe design improvements using new materials for lower detection signatures and greater maneuverability are being investigated by General Dynamics and Boeing. The use of radar-absorbent materials and smoother, flatter designs in construction of the airframe would make cruise missiles more difficult to detect with current radar.[6] These so-called "Stealth" technologies could be partially applied to already deployed missiles, but would have the most significant applications in a new missile. The Air Force is also planning to retrofit electronic countermeasures packages aboard ALCMs and GLCMs during the 1985-1987 period.[7] The on-board active countermeasures would be designed to operate against interceptor aircraft and missiles.

Modifications to the present cruise missile force, particularly ALCMs, now seems to have lower priority than procurement of a new "Advanced Cruise Missile" incorporating all the new features. The FY 1984 Defense budget request to Congress ended Boeing ALCM procurement at 1499 of 4348 planned units[8] and shifted program focus to the new missile.

Accelerated development of the Advanced Cruise Missile may mean an IOC of as early as 1986.[9] The Air Force issued "requests for proposals" for an advanced cruise missile in September 1982[10] and expects to select a prime contractor in the spring of 1983.[11] The Air Force competition will be between Boeing, General Dynamics, and Lockheed.[12] Boeing won a competition with General Dynamics to become the ALCM contractor. General Dynamics is the contractor for the TOMAHAWK missile and has been a major participant in the Defense Advanced Research Projects Agency (DARPA) "TEAL DAWN" research program to develop a next-generation missile (see below). Lockheed, one of the major contractors in the secret stealth programs, has reportedly developed a stealth cruise missile.

At least 2000 advanced ALCMs will probably be procured starting in FY 1986 to augment and eventually replace the Boeing ALCM. Whether the new technologies will also be applied to Ground and Sea Launched missiles is still not clear, although it is known that the Navy is also developing a stealth cruise missile.[13] In FY 1981, an Advanced Cruise Missile Technology nuclear warhead Phase 1 conceptual study was underway within DOE to design a warhead to replace the W80 on the next generation of ALCM.

For many years, the DARPA has also been investigating cruise missile technology. Of particular interest is development of a new intercontinental cruise missile

---

1 HAC, FY 1983 DOD, Part 4, p. 588.
2 AW&ST, 10 March 1980, pp. 12-15.
3 Defense Week, 14 February 1983.
4 HAC, FY 1983 DOD, Part 4, p. 593.
5 Information provided the authors by Air Force Systems Command.
6 AW&ST, 8 November 1982.
7 Ibid., p. 18.
8 Richard Halloran, New York Times, 16 February 1983, p. 12; Defense Week, 1 February 1983, p. 1.
9 AW&ST, 23 August 1982; AW&ST, 1 November 1982, p. 13.
10 Defense Week, 14 February 1983.
11 AW&ST, 23 August 1982.
12 Richard Halloran, New York Times, 16 February 1983, p. 12.
13 AW&ST, 23 August 1982.

6

Advanced Technology Cruise Missiles

under the "TEAL DAWN" and the Advanced Cruise Missile Programs. In fact, a cruise missile with an intercontinental range of some 6000-8000 miles could compete quite strongly with the Air Force's plans for a quick follow-on. It is not clear whether the new missile will merely incorporate the advances into a new airframe or be completely new. Vought Corporation received a small Air Force contract in late 1982 to research guidance and other components for DARPA's intercontinental range cruise missile.[14]

The new missiles being developed by DARPA will be smaller, incorporate the latest stealth techniques, and have sensors to avoid detection and defensive systems. A new terminal homing unit and additional navigation aids will provide high accuracy. A "regenerative" engine which would channel some of the waste exhaust heat back into the engine cycle is being examined. High energy, jellied fuels could also add fuel savings and greater range. The most significant feature, however, would be the increase to supersonic speeds over 550 mph for the present ALCM. The Fiscal Year 1983-1984 DARPA Advanced Cruise Missile Program requested $63.6 million for the following:[15]

- Autonomous Terminal Homing: development of advanced sensors, day-night and adverse weather, precision guidance system, including an autonomous damage assessment capability,
- Advanced Delivery Concepts: development of techniques to counter threats to cruise missiles including "unconventional vehicle designs," increased range, and flight path optimization systems,
- Advanced Cruise Missile Engines: development of engines using new high energy fuels, increased thrust, and reduced fuel consumption,
- Cruise Missile Detection Technology: development of techniques (radar masking, clutter, propagation data, infrared background data) that limit the capability of defensive systems and enhance the design and countermeasures of cruise missiles, and
- Path Optimization Technology: development of new mission planning and onboard detection and routing systems to enhance the ability of cruise missiles and launching aircraft to evade defenses.

---

14 AW&ST, 31 January 1983, p. 13.
15 DARPA, "Fiscal Year 1983 Research and Development Program: A Summary Description," 30 March 1982.

6
Advanced Strategic Air-Launched Missile

Advanced Strategic Air-Launched Missile

The major program for the next generation of attack missiles for U.S. bomber forces is the Advanced Strategic Air-Launched Missile (ASALM), also known as the Lethal Neutralization System.[1] The objective of the ASALM program is to develop a supersonic cruise missile as an improved air-to-ground weapon with an anti-aircraft capability. While the ASALM program is primarily driven by developments in Soviet AWACs and future U.S. bomber forces, it is also influenced by the anticipated obsolescence of motors on the current SRAM missile. The missile technology could be used to provide the basis for a second generation, higher-speed, long-range ALCM.[2] The program was slowed by the Air Force in 1978-79 in order to accomplish a detailed mission analysis called Saber Mission A.[3] The analysis concluded that a multimode missile with air-to-air and air-to-surface capabilities was far superior to the present SRAM air-to-surface missile.

The ASALM program has its origin in the more than ten year old integral-rocket/ramjet propulsion system which can be used as a supersonic air breathing missile. Work on ASALM began in 1968 with competitive studies conducted by Boeing, Hughes/LTV, and Martin Marietta for the Bomber Defense Missile (BDM).[4] BDM evolved into the Multipurpose Missile (MPM) and later into ASALM, for which McDonnell Douglas and Martin Marietta competed for development. Much of the work has included studies and technical development in the areas of high-temperature structures, integral-rocket/ramjet propulsion, and inlet configuration. Prototype missiles have "flown" high velocity and high altitude trajectories in extensive wind tunnel testing and other simulations. Flight testing of the rocket/ramjet vehicle was accomplished from October 1979 to May 1980.

Unlike current generation cruise missiles, ASALM would be supersonic and capable of attacking ground targets as well as directly defending the bomber force. ASALM is seen as a penetration aid for U.S. bomber forces with improved air-to-ground capabilities. Its improved accuracy over the SRAM give it a significant capability to destroy enemy air defenses. ASALM would be designed to maneuver at sustained high speeds to evade enemy air defenses and be capable of flying a variety of trajectory profiles: all-high, all-low, and combination high-low. Finally, ASALM would be designed to maintain high speed in the terminal phase when high speed is essential for penetration of enemy point defenses.

The program has had technical problems and was scaled down for FY 1980-1982 with a refocus on basic technology. The ASALM program is looking at not only missile technology, missile flight testing, and subsystem evaluation, but also at electronic counter measures (ECM), decoys, and communications jamming. A large portion of ASALM funding is directed toward the difficult problem of developing an air-to-air guidance capable of attacking a Soviet AWACs once its radar has been shut down.[5] Martin Marietta is also testing ASALM as an "Outer Air Battle Missile" for the Navy to be used as a long-range anti-cruise missile system fired from the Vertical Launching System (VLS). A nuclear warhead for the ASALM, currently called the New Strategic Air-Launched Missile Warhead (formerly the Lethal Neutralization System), is in Phase 2, Program Study, at DOE laboratories. Another warhead program, the Bomber Defense Missile warhead, is in Phase 1 and thought to be for the ASALM.

The Air Force has also studied the feasibility of a cruise-ballistic missile, which after achieving altitude and speed converts over to a cruise mode.[6] The technology, however, is very difficult and the DOD states that it will be many years before a technology demonstration flight could be accomplished.[7]

---

1 Program has also been known as Counter SUAWACs.
2 ACDA, FY 1980 ACIS, p. 28.
3 HASC, FY 1981 DOD, Part 4, Book 2, pp. 1701-1702.
4 ASALM background provided by Martin Marietta.
5 SASC, FY 1982 DOD, Part 7,-p. 4303.
6 SASC, FY 1981 DOD, Part 5, p. 2709.
7 SASC, FY 1982 DOD, Part 7, p. 3998.

6
Advanced Strategic Air-Launched Missile

Advanced Strategic Air-Launched Missile (ASALM)

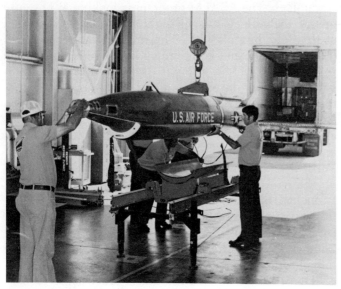

Figure 6.12 Advanced Strategic Air-Launched Missile (ASALM).

DESCRIPTION: Strategic supersonic medium-range cruise missile with air-to-air and air-to-ground capabilities, envisioned as the replacement for SRAM.

CONTRACTORS:
Martin Marietta Aerospace
Orlando, FL
(prime)
Raytheon
(missile/guidance)
McDonnell Douglas
(missile)
Martin Marietta
(airframe)
Hughes
(guidance)
Marquardt Co.
(ramjet propulsion)
United Technologies Corp.
(engine)
Thiokol
(fuel)
Rockwell
(guidance)
Litton Guidance & Control
(inertial navigation)
Delco
(subsystems)
Hercules, Inc.
(rocket propulsion)
Garrett AiResearch Mfg. Co.
(secondary power)

SPECIFICATIONS:

Length: 168 in[1]

Diameter: 25 in,[2] 21 in[3]

Stages: 1

Weight at Launch: 2700 lb;[4] 1800 lb[5]

Propulsion: integral rocket-ramjet engine

Speed: Mach 4

Guidance: passive updated inertial guidance, passive antiradiation homing capability, active radar terminal engagement in aerial intercept mode with frequency agility[6]

Throwweight/Payload: unknown

Range: over 200 mi; considerably less than ALCM but more than SRAM.[7]

DUAL CAPABLE: no

NUCLEAR WARHEADS: one/missile; two warheads possibly under development; W80 is a prospective candidate for use on the ASALM.[8]

DEPLOYMENT:
Launch Platform: B-52 (up to 7 internal/12 external),[9] FB-111, B-1B, ATB

Number Planned: 1200 (1983)

Location: bomber bases

6

Advanced Strategic Air-Launched Missile

HISTORY:

IOC: 1987[10]

Jun 1974 — McDonnell Douglas and Martin Marietta awarded contracts for concept formulation of ASALM[11]

Mar 1976 — Martin Marietta awarded contract for ASALM propulsion technology vehicle (PTV)[12]

Jul 1979 — Phase 2 feasibility study for ASALM warhead completed[13]

Oct 1979 — flight testing of supersonic propulsion technology vehicle begins[14]

Dec 1979 — program given go ahead

May 1980 — propulsion technology validation flight testing completed[15]

FY 1983 — captive flight testing

TARGETING:

Types: Soviet AWACs, interceptor airfields, air defense missile sites, radar

Accuracy/CEP: accuracy is not expected to be significantly degraded by the missile's high speed[16]

COST:

FY	Number Procured	Total Appropriation ($ million)
1977 & prior	-	38.8[17]
1978	-	37.2[18]
1979	-	39.0[19]

COMMENTS: FY 1982 defense budget changed the ASALM program to the Counter SUAWACS (Soviet Union AWACs) Technology Program (63318F). It was then changed to the Lethal Neutralization System Program in FY 1983.

---

1 *The World's Missile Systems*, 6th Ed., p. 96.
2 *Ibid.*
3 ACDA, FY 1980 ACIS, p. 29; ACDA, FY 1981 ACIS, p. 126.
4 ACDA, FY 1980 ACIS, p. 29; ACDA, FY 1981 ACIS, p. 126.
5 *The World's Missile Systems*, 6th Ed., p. 96.
6 AW&ST, 10 March 1980, p. 14.
7 ACDA, FY 1979 ACIS, p. 59; ACDA, FY 1980 ACIS, p. 29; ACDA, FY 1981 ACIS, p. 126; HAC, FY 1980 DOD, Part 6, p. 680; 1985-86.
8 *Ibid.*
9 *Ibid.*
10 Depending on the availability of warheads; ACDA, FY 1981 ACIS, p. 125; earliest IOC has also been referred to as 1989; HAC, FY 1980 DOD, Part 6, p. 680; 1985-86; ACDA, FY 1979 ACIS, p. 62.
11 Martin Marietta Release, 10 June 1974.
12 *Martin Marietta News*, 5 March 1976.
13 ACDA, FY 1980 ACIS, p. 30.
14 USAF, ASD, WPAFB Press Release (PAM #80-170).
15 *Ibid.*
16 ACDA, FY 1979 ACIS, p. 61.
17 *Ibid.*, p. 63.
18 *Ibid.*
19 *Ibid.*

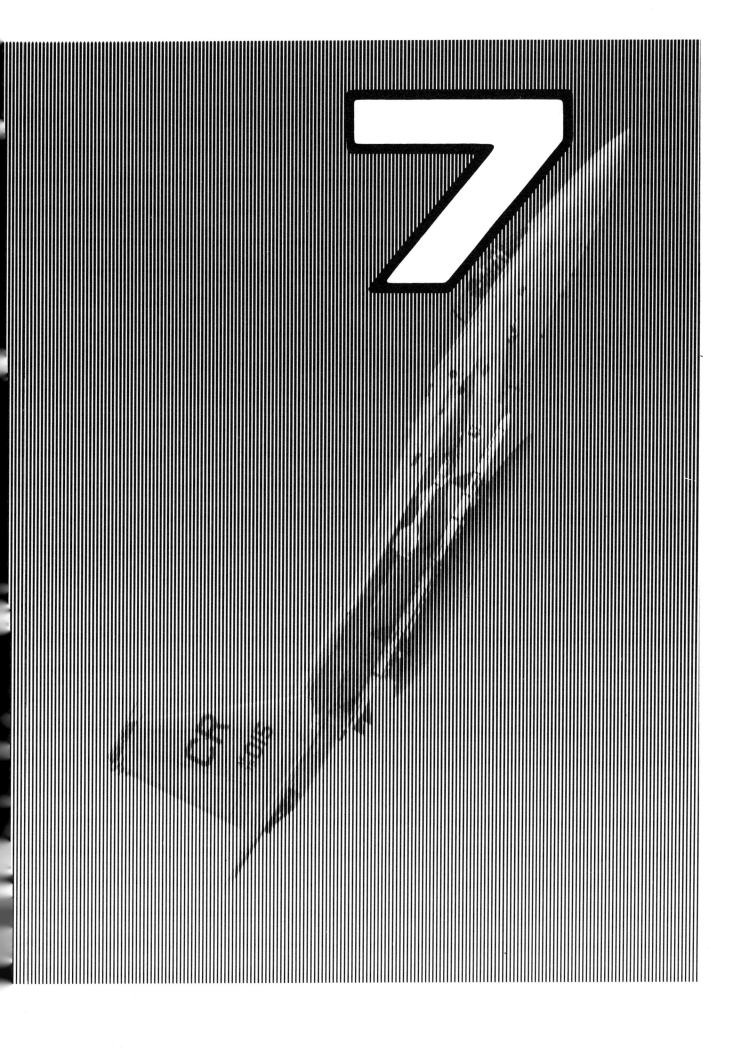

7

Nuclear Capable Aircraft and Bombs

Chapter Seven
Nuclear Capable Aircraft and Bombs

According to the Department of Defense, "any airplane that is designed to carry an ordinary bomb can, with the proper wiring and certification, also carry a nuclear bomb."[1] However, not every airplane or even tactical fighter is so certified. "Most dual capable aircraft have not been optimized for the nuclear strike mission and have deficiencies that limit their effectiveness in this role."[2]

Nuclear bombs are designed for delivery by aircraft either in a bomb bay (internal) or under the wing (external). Aircraft configured for nuclear weapons delivery have an Aircraft Monitoring and Control (AMAC) system installed to monitor and control fuzing, arming, and safing functions of the nuclear bombs. A permissive action link (PAL) or Nuclear Consent Switch is installed in the cockpit to release the weapon for detonation.

Fifteen tactical aircraft are currently modified to carry nuclear weapons (see Table 7.2). Four different kinds of bombs are used in the tactical air forces. These bombs, B28, B43, B57, and B61 (see Chapter Three), vary in yield from approximately 5 kilotons to over 1 megaton. The newest can be delivered at low altitudes at supersonic speeds. Currently there are no nuclear missiles deployed with tactical aircraft.

Nuclear Bombs

All deployed nuclear bombs can be delivered with a variety of options, including ground ("laydown") and airburst detonations. Four delivery and fuzing modes are most common: airburst/retarded, groundburst/retarded, air/full fuzing and ground/full fuzing (see Glossary). Table 7.1 describes the six nuclear bombs deployed or under development. Nuclear bombs must usually be dropped directly over their targets to assure accuracy. In order to achieve optimum heights of air burst with all nuclear bombs to avoid detonation too close to the ground, the delivery aircraft must fly at an altitude that is vulnerable to enemy air defenses. The newer bombs, the deployed B61 and the not yet deployed B83, allow the pilot to release the weapon at as low as 50 feet, activating a parachute-type (drogue) retard and a time-delay fuze.[3] When used at low altitudes, the laydown delivery method is extremely accurate.[4] The accuracy of the B61 and B83 bombs delivered in the laydown mode is reportedly averaging 600 ft CEP.[5] The older bombs, like the B28, B43, and B57, have a minimum delivery altitude of 300-600 feet.[6] They can be delivered "over the shoulder and at low or medium angle loft."[7]

The B83 "Modern Strategic Bomb" is the major new nuclear weapon under development for aircraft delivery. The bomb, will replace the older B28, B43, and B53 bombs. It is entering production in FY 1983 and is planned for deployment starting in 1984-1985 after a long and difficult development period. The B83's roots are in the B77, a very expensive strategic bomb under development in the 1970s. The B77 included improved safety features, but also included a capability for delivery at high speeds at extremely low altitudes.[8] The cost of the B77 grew so excessive that in FY 1979 the program was cancelled, and a modified B43 model took its place.[9] Congress, however, directed that FY 1978 and 1979 funds not be expended on a modified B43 and instead allocated funds for development of a cheaper new strategic bomb. The B83, initiated in FY 1980, is a modern strategic bomb which contains most of the essential features of the B77, but at reduced cost.[10]

The B83 is intended to "enhance the effectiveness of the strategic nuclear gravity bomb stockpile."[11] The primary reason for developing the B83 is to enable tactical and strategic aircraft to deliver their weapons while flying low level, supersonic evasion missions.[12] With a 150 foot low-level high speed delivery capability and yield in the megaton range, the B83 will be capable of destroying "hardened Soviet ICBM silo and launch complexes, command, control and communication installations, and nuclear storage sites."[13] The B83 is the first megaton yield bomb specifically designed for groundburst retarded ("laydown") delivery against hard targets.[14] The production schedule of the B83 is being increased to meet larger strategic bomber force requirements with deployment of the B-1B.[15]

---

1 SASC, FY 1982 DOD, Part 7, p. 389.
2 JCS, FY 1982, p. 78.
3 ACDA, FY 1979 ACIS, p. 92.
4 ACDA, FY 1980 ACIS, p. 169.
5 *Aerospace Daily*, 28 December 1978, p. 263.
6 ACDA, FY 1979 ACIS, p. 92; ACDA, FY 1980 ACIS, p. 169.
7 ACDA, FY 1979 ACIS, p. 92.
8 SASC, FY 1979 DOE, p. 41.

9 Cancellation was also tied to cancellation of the B-1.
10 HAC, FY 1980 DOD, Part 4, p. 667.
11 ACDA, FY 1983 ACIS, p. 65.
12 *Ibid.*
13 ACDA, FY 1981 ACIS, p. 114.
14 SANDIA, "Lab News," 12 June 1981.
15 SASC, FY 1983 DOD, Part 7, p. 4172.

7
Nuclear Capable Aircraft and Bombs

Table 7.1
Nuclear Bombs

Type	Weight (lb)	Yield (Kt)	Aircraft
B28	2027-2540	70-1450	A-7, F-4, F-100, F-104, B-52
B43	2060-2330	1000	A-4, A-6, A-7, B-52, F-4, F-100, F-104, F-111, FB-111
B53	8850	9000	B-52
B57	765	5-20	A-4, A-6, A-7, B-52, F-4, F-16, F-18, F-100, F-104, F-111, FB-111, P-3, S-3, SH-3
B61	less than 840	10-500	A-4, A-6, A-7, B-52, F-4, F-16, F-18, F-104, F-111, FB-111
B83	2408	1000+	A-4, A-6, A-7, B-1B, B-52, F-4, F-16, F-111, FB-111

Table 7.2
Nuclear Capable Tactical Aircraft

Type	Function	Service[1]	Nuclear Weapons
A-4	Short-range attack	MC	B43, B57, B61
A-6	Long-range attack	N	B43, B57, B61
A-7	Medium-range attack	ANG, N	B43, B57, B61
AV-8B	Medium-range fighter	MC	B57, B61
CF-101	Interceptor	Canada	GENIE
F-4	Medium-range fighter	AF, N	B28, B43, B57, B61
F-15	Interceptor/fighter	AF	GENIE, bombs
F-16	Medium-range fighter	AF, NATO[2]	B43, B61
F-18/A-18	Medium-range fighter/attack	MC, N	B57, B61
F-100	Medium-range fighter	NATO[3]	B28, B43, B57
F-104	Medium-range fighter	NATO[4]	B28, B43, B57, B61
F-106	Interceptor	AF	GENIE
F-111	Long-range fighter	AF	B43, B57, B61
P-3	Long-range Maritime Patrol	N	B57
S-3	Long-range Maritime Patrol	N	B57
SH-3	Short-range ASW Helicopter	N	B57
SH-60F	Short-range ASW Helicopter	N	B57
TORNADO	Medium-range fighter	NATO[5]	B57, B61

1 Nuclear capable versions.
2 Belgium, Netherlands.
3 Turkey.
4 Belgium, Greece, Italy, Netherlands, West Germany.
5 Italy, West Germany.

7
B83

B83

Figure 7.1 F-111 delivering **B83** bomb prototype.

FUNCTION: Modern high-yield strategic bomb, with improved low level delivery capability.[1]

WARHEAD MODIFICATIONS: none known

SPECIFICATIONS:

Yield: probably 1000+ Kt, "high yield,"[2] "megaton class"[3]

Weight: 2408 lb[4]

Dimensions:
  Length: 12 ft
  Diameter unknown

Materials: probably plutonium/oralloy mixed weapon; IHE (probably PBX-9502)[5]

SAFEGUARDS AND ARMING FEATURES: Category D PAL, nonviolent command disable;[6] weak link/strong link, one-point safe by the present criterion[7]

FUZING AND DELIVERY MODE: improved low-level, high speed delivery capabilities;[8] airburst, groundburst, full fuzing; new parachute design permits the B83 to be dropped at transonic and supersonic speeds (up to Mach 2), slowing down the bomb to 60 mph to withstand the shock of delivery at high speeds from altitudes as low as 150 feet and as high as 50,000 feet[9]

DEVELOPMENT:
Laboratory: LLNL

History:
  IOC: 1984
  Jan 1979 Lab assignment (Phase 3)[1] (through FY 1983)[11]
  1981 B83 enters Phase 4[12]
  1984 initial deployment (Phase 5)

Production Period: 1983-

DEPLOYMENT:
Number Planned: approximately 2500 (1983)

Delivery System: primarily carried by the B-1B, B-52, and FB-111 strategic bombers; F-4, F-111, A-4, A-6, A-7, and F-16 will be secondary carriers.[13] It will be the major gravity weapon for the B-1B.[14]

Service: Air Force, Navy

Allied User: none planned

COMMENTS: B83 is scheduled as a replacement for the older high-yield bombs, the B28, B53, and B43.[15] Because of the development of the B83, the production and development of the B77 was never executed.[16] The B77 was cancelled in 1978 and development was shifted to a variant of the B43Y1, then the B83. The B83 is still one of the more complicated and expensive bombs.[17]

---

1 HASC, FY 1982 DOE, p. 116.
2 SASC, FY 1981 DOE, p. 37.
3 Dennis Rockstroh, "A New Hydrogen Bomb Being Built," *San Jose Mercury*, 2 July 1981; Information also provided by Sandia Corporation.
4 Sandia, "Lab News," 12 June 1981; GAO, Draft Study for B-1.
5 ACDA, FY 1982 ACIS, p. 115; SASC, FY 1981 DOE, p. 37.
6 ACDA, FY 1982 ACIS, p. 115.
7 ACDA, FY 1981 ACIS, p. 115-116; ACDA, FY 1983 ACIS, p. 65: "One point safe means that the probability of achieving a nuclear yield greater than four pounds of TNT equivalent shall not exceed one in one million in the event of a detonation initiated at a single point in the high explosive system."
8 SAC, FY 1981 EWDA, Part 2, p. 726.
9 ACDA, FY 1981 ACIS, p. 115; ACDA, FY 1983 ACIS, p. 65.
10 Continued in Phase 3 in FY 1980; SAC, FY 1981 EWDA, p. 818.
11 DOE Justification, FY 1983, p. 51.
12 Sandia, "Lab News," 12 June 1981.
13 HAC, FY 1981 EWDA, Part 4, p. 2669; HASC, FY 1981 DOE, p. 119; SASC, FY 1981 DOE, p. 37.
14 GAO, Draft Study for B-1.
15 SASC, FY 1981 DOE, p. 37; SASC, FY 1983 DOD, Part 7, p. 4172.
16 HAC, FY 1980 EWDA, p. 2656.
17 SASC, FY 1981 DOE, p. 32.

7

Future Nuclear Capable Aircraft

Tactical Nuclear Air-Launched Missiles[1]

While the nuclear capability of tactical aircraft consists entirely of gravity bombs, missiles with standoff capabilities and improved accuracies are under development. The BULLPUP (W-45) and the WALLEYE (W-72), both retired in 1978-1979, were the last nuclear armed air-to-surface missiles to be deployed with tactical air forces. Although the WALLEYE missile resulted from air-delivered standoff weapons using terminal guidance developed during the Vietnam War, nuclear armed versions were never adopted in large numbers. Improved modifications of the B61 bomb (and development of the B83) were pursued instead.

In 1972, a research program—Tactical Air-to-Surface Munition (TASM)—began to investigate the possibility of an accurate standoff capability with nuclear bombs. In May 1974, the program was redirected toward the development of an Extended Range Bomb (ERB) which called for a single weapon with all-weather inertial guidance, terminal guidance, and return-to-target capability against mobile battlefield targets. The TASM/ERB program consists of two separate tracks: one to develop a new standoff weapon with new warhead and greatly increased accuracy, and the other to develop modification kits to provide presently stockpiled bombs with a standoff and return-to-target airburst delivery capability. This conversion would require the addition of canards and tail surfaces, a rocket propulsion system, inertial navigation system, flight computer, radar altimeter, and weapon control panel for preflight insertion of target data. The TASM/ERB would be compatible with the A-4, A-6, A-7, F-4, F-16, F-18, F-104, F-111, and TORNADO.

One candidate for the TASM is TIGER (Terminal Guided and Extended Range Missile), a guided nuclear bomb under development since 1972 at Sandia National Laboratories.[2] This weapon would allow for delivery of nuclear weapons at low altitudes, either outside of concentrated defense around fixed targets or against mobile targets, with one low level pass. TIGER would have extended range and a return-to-target capability by flying a circular trajectory, minimizing the delivery aircraft's exposure to air defenses. TIGER II is the present model being tested by Sandia as a candidate for the TASM/ERB. TIGER II will use field retrofit kits for the B61 bomb to demonstrate a standoff 90 meter CEP accuracy when delivered from low flying aircraft.

Three nuclear warheads are currently under development for TASM and other future air-delivered weapons. The TASM warhead was reported in Phase 1 during 1982 at DOE, with a yield of 10 Kt against both battlefield and fixed targets.[3] Also reported in Phase 1 during 1982 was the Advanced Tactical Air Delivered Weapon.[4] A feasibility study to design a nuclear warhead for the PHOENIX air-to-air missile was reported in FY 1983.[5]

Future Nuclear Capable Delivery Aircraft

A major expansion of tactical air forces with an increase in nuclear capability is scheduled for this decade. The Air Force tactical fighter force will have 40 full tactical fighter wings (26 active and 14 reserve) by 1985 and will build to full strength of 72 aircraft per wing, or 2880 aircraft. This is equivalent to an increase of some four wings between 1983-1988.[6] By FY 1990, 44 tactical fighter wings are planned, an addition of some 288 aircraft over FY 1982 levels.[7] A fourteenth aircraft carrier (USS *Roosevelt*) will join the Naval air fleet in 1988, which will add another air wing to the Navy's 13 wings. By the early 1990s, the Navy's 600 ship objective, built around 15 aircraft carrier battle groups, will add a 15th air wing.[8]

New aircraft will continue to enter the tactical inventory and replace older models. The tactical air forces will eventually stabilize with F-14, F-15, F-16, F/A-18, and AV-8B high performance aircraft. During the next 10 years, the A-4, A-7, F-4, F-100, F-104, and F-106 will be removed from the active inventory. Allied forces equipped with U.S. nuclear weapons will also undergo a major upgrade by the mid-1980s. The Canadian CF-101s will be replaced by CF-18s, and NATO nuclear armed F-104 aircraft will be replaced with F-16s and European-built TORNADOs.[9]

Rather than developing more high performance aircraft that have either air-to-air or air-to-surface roles, future tactical fighters will be dual role. The aircraft inventory was once composed solely of single role, highly specialized designs that were not capable of freely operating in other modes. In fact, some aircraft were specifically designed as nuclear weapons fighter bombers or interceptors with an internal weapons bay to carry only nuclear bombs or rockets. Only two aircraft of this type are still operational—the Marine Corps A-4M and the Air Force F-106. A third model (F-105) was retired in 1982.

---

1 Most of the information in this section is taken from Sandia, "TIGER: A Technology to Improve the Delivery Capability of Nuclear Bombs and the Survivability of the Delivery Aircraft," n.d.
2 AW&ST, 2 May 1981, p. 51.
3 Information provided by Sandia.
4 HAC, FY 1980 DOE, p. 95.
5 DOD, FY 1983 RDA, p. VII-14.
6 SAC, FY 1983 DOD, Part 4, p. 146.
7 HAC, FY 1983 DOD, Part 4, p. 146.
8 HAC, FY 1983 DOD, Part 5, p. 179.
9 SASC, FY 1983 DOD, Part 7, p. 4373.

7

Future Nuclear Capable Aircraft

As nuclear weapons became lighter and aircraft and air-to-air missile technology improved, allowing for greater versatility and payloads without sacrificing performance, air-to-surface ground attack aircraft were no longer designed only for bombing, but also for a variety of other roles. A portion of today's aircraft inventory, the so-called workhorses, is comprised of these versatile aircraft: F-111, A-6, A-7, and F-4. The new high performance aircraft first deployed in the 1970s—F-14, F-15, F-16, F/A-18, and AV-8A—largely concentrate on air-to-air or air-to-surface roles. The inclusion of nuclear weapons delivery neither influences design nor complicates other operations. Each of the new planes, with the exception of the F-14, is certified for nuclear weapons delivery.

As the older strike aircraft—specifically the F-4 and F-111—reach the end of their useful life because of attrition through accidents and old age, they will be replaced by a long-range dual role strike fighter which will augment the air-to-surface specialists. The Navy will introduce the F-18 for this role, and the Marine Corps plans the AV-8B. The Air Force, which has recently introduced the F-15 and F-16, has no plans to build another new aircraft until the early 1990s. Instead, the Air Force has established a new program—Tactical Fighter Derivative—to develop a modified F-16 designed to augment the aging F-111 and meet the F-4's requirements until the 1990s when the next generation of fighters is developed.

Tactical Fighter Derivative

The Tactical Fighter Derivative program, started in FY 1983, will examine upgraded dual role ground attack variants of two aircraft, designated the F-15E and the F-16E, which will incorporate improvements in range, payload, all-weather, and nighttime operations. According to the Air Force, the new aircraft will "double the target coverage of the [present] F-4" in Europe and make up for "critical deficiencies" in night/adverse weather operations.[10] Either the F-15 or F-16 candidate airplanes will be selected in FY 1983-1984, and 400 aircraft will be procured for the dual role.[11]

The F-15E derivative fighter will provide a full air-to-ground bombing capability with its greatly increased range and an upgraded and "missionized" rear cockpit for a weapons officer. An early prototype has been flying since 1980 under a McDonnell Douglas program, "STRIKE EAGLE". A new terrain following/terrain avoidance capability with greater ground target resolution and blind weapons delivery capability will be added to the F-15 radars. Nuclear capability would include control mechanisms added to five external weapons stations for nuclear weapons delivery.[12] Cost to develop the F-15E is estimated at $300-350 million. Procurement of 400 of these aircraft would cost $16 billion.[13]

The F-16E derivative fighter would use aerodynamic enhancements to improve the F-16's air-to-air characteristics and range. The new F-16E would employ a new "cranked arrow" (double delta) wing design in place of the current standard wing and horizontal tail. This would result in increased range and payload with more fuel capacity and greater lift. The new wing would also allow weapons to be carried "conformally" (close to the wing), which would reduce drag and give better fuel consumption. A section would be added to the fuselage for a second crew member and additional avionics. A computerized flight control system would be added. Finally, a new engine would be added, either a Pratt & Whitney F100 or General Electric F101 (the derivative fighter engine).[14] Cost to develop the F-16E is estimated at $776.1 million. Procurement of 400 aircraft would cost $12 billion.[15]

Advanced Tactical Fighter

A completely new airplane that is lightweight, reliable, easily maintained, and has increased combat radius and payload is now in development for the 1990s. The Air Force development program promises a "revolutionary change"[16] in capabilities through the incorporation of improved operating efficiencies and lower manufacturing costs, derived from new technology advances.[17] For example, composite structures would be used to achieve a very light weight. Very high speed integrated circuits (VHSIC) would also be used, as well as a new engine design and greater efficiency derived from the cruise missile program.[18] The advanced tactical fighter would incorporate three new major features:[19]

- Stealth technology: "significantly reduced radar and infrared detectability,"
- Supersonic Cruise: increased practical (sustained) operating speeds at both high and low altitudes, without penalties in maneuverability, and

---

10 HAC, FY 1983 DOD, Part 5, p. 629.
11 HAC, FY 1983 DOD, Part 5, p. 565.
12 SAC, FY 1983 DOD, Part 4, p. 249.
13 HAC, FY 1983 DOD, Part 5, p. 630.
14 HAC, FY 1983 DOD, Part 5, pp. 585-585, 631; SASC, FY 1983 DOD, Part 4, p. 150.
15 HAC, FY 1983 DOD, Part 5, p. 630.
16 HAC, FY 1983 DOD, Part 5, p. 586.
17 SAC, FY 1983 DOD, Part 4, p. 130.
18 HAC, FY 1983 DOD, Part 5, pp. 586-588, 633.
19 Ibid.

7
Future Nuclear Capable Aircraft

- Short Take Off and Landing: greatly increased flexibility with ability to operate from runways of less than 2000 ft.

The research and development program for the Advanced Tactical Fighter began in FY 1983. Full scale engineering development is planned for 1987.[20] The earliest possible IOC is 1993, and full scale operations are planned for the mid-1990s.[21]

Table 7.3

Future Tactical Fighter Aircraft Programs

Program	Description (Program Period)
Advanced Fighter Technology Integration (AFTI-16):	Future fighter aircraft testbed using modified F-16A (1978-present)
Advanced Tactical Fighter:	Next generation tactical fighter planned for 1990s with stealth technology, new engine, upgraded avionics (1980-present)
F/A-18L:	Northrop version of F-18 for land-based future lightweight fighter with increased payload (1980-present)
F-16/101:	Test version of F-16 powered by F100 (B-1) engine to determine its suitability as engine for advanced military aircraft (1980-1981)
F-16E/F-16 SCAMP/F-16XL:	Advanced versions of F-16 with new wing design, simplified flying controls, upgraded weapons load, additional fuel and storage space for future avionics and sensors, derivative fighter candidate (1978-present)
Forward Swept Wing:	DARPA sponsored Grumman tests of smaller, lighter weight, more efficient fighter designs (1980-present)
F-15E/STRIKE EAGLE:	Upgraded all-weather strike and interdiction model of F-15, designed for air-to-surface roles, derivative fighter candidate (1978-present)

---

20 SAC, FY 1983 DOD, Part, 4, p. 130.

21 HAC, FY 1983 DOD, Part 5, p. 566.

7
A-4 SKYHAWK

Nuclear Capable Aircraft
A-4 SKYHAWK

Figure 7.2 A-4M SKYHAWK.

DESCRIPTION:	Light, single-seat, single-engine, carrier-based, attack aircraft used by the Marine Corps.	IBM Corp. Federal Systems Oswego, NY (bombing computer)	
Nuclear Capable Versions:[1]	A-4D/E/M	**SPECIFICATIONS:**[2] Dimensions:	(A-4M)[3]
CONTRACTORS:	McDonnell-Douglas Corp. Long Beach, CA (prime/airframe) Pratt & Whitney Aircraft East Hartford, CT (engine) Hughes Aircraft Co. Canoga Park, CA (angle rate bombing system)	Length: Height: Wingspan: Takeoff Weight (max): Powerplant: Ceiling:	40 ft 3 in 15 ft 27 ft 6 in 25,500 lb 1 P&W J-52-P-408A turbojet[4] 57,570 ft; 40,800 ft[5]

Nuclear Weapons Databook, Volume I **205**

7

A-4 SKYHAWK

Speed:	(max) 650 mph (Mach 0.94) at 25,000 ft; 700 mph at sea level	1969	production began of A-4M
Range:	341 mi (550 km) (combat radius); 403 mi (648 km) (combat radius)[6]	Apr 1970	A-4M first flight
		Nov 1970	first delivery of A-4M
Aerial Refueling Capability:	yes	1975	last Navy squadron disbande
		FY 1977	last A-4 procured
Crew:	1	Feb 1979	last A-4 delivered (2960th A- produced)
NUCLEAR WEAPONS:	one nuclear weapon; B28,[7] B43, B57,[8] B61; five weapons stations	COST:	$2.8 m (program) (TY)[9] $5.8 m (flyaway) (FY 1979)[10]

FY	Number Procured	Total Appropriation ($ million)
1982 & prior	158 (A-4M)	399.6[11]

DEPLOYMENT:

Number Deployed: 158 built (A-4M)

Locations: MCAS El Toro, CA; MCAS Cherry Point, NC; MCAS Iwakuni, Japan

Number per Squadron: 16 (UE)

HISTORY:

IOC: 1956 (A-4A); 1970 (A-4M)

1952 preliminary design of A-4 begins

Oct 1952 authority to start production

Jun 1954 first flight of A-4A

COMMENTS: Originally built as a dayligh only nuclear strike aircraft f use in large numbers from ai craft carriers, the A-4 has be updated for visual referen day or night attack. It has be retired from U.S. Navy us AV-8B will replace A-4M Marines starting in 1985.[12] 29 A-4 and TA-4 SKYHAWI were built between 1953 a 1979.

---

1 JCS, FY 1982, p. 76.
2 Information furnished by McDonnell Douglas, "Navy/McDonnell Douglas Skyhawk, Background Information," February 1979.
3 A-4M has an improved engine and weapons delivery capability.
4 Added power of new engine greatly enhanced short field (4000 ft runway) take-off capability.
5 U.S. Military Aircraft Data Book, 1981, pp. 2-3 - 2-6.
6 Ibid.
7 Ships and Aircraft of the U.S. Fleet, 11th Ed., p. 265.
8 Ibid.
9 U.S. Military Aircraft Data Book, op. cit.
10 Ibid.
11 Ibid.
12 HAC, FY 1983 DOD, Part 5, p. 252.

7
A-6 INTRUDER

A-6 INTRUDER

Figure 7.3 A-6 INTRUDER.

DESCRIPTION:	Long-range, two-seat, twin engine, carrier-based, all-weather attack aircraft used by the Navy and Marine Corps.
Nuclear capable versions:	A-6E[1]
CONTRACTORS:	Grumman Aerospace Bethpage, NY (prime) Pratt and Whitney Hartford, CT (engine)

SPECIFICATIONS:	(A-6E)
Dimensions:	
Length:	54 ft 7 in
Height:	16 ft 2 in
Wingspan:	53 ft (25 ft 4 in folded)
Takeoff Weight (max):	60,450 lb
Powerplant:	2 P&W J52-P-8A/B turbojets
Ceiling:	42,400 ft
Speed:	(max) 655 mph (Mach·0.86)

Nuclear Weapons Databook, Volume I **207**

7
A-6 INTRUDER

Range: 370-1125 mi (595-1810 km) (combat radius) (with carrier-based aerial refueling);[2] 1924 mi (combat range)[3]

Aerial Refueling Capability: yes

Crew: 2 (pilot, bombadier/navigator)

NUCLEAR WEAPONS: three nuclear weapons;[4] B28,[5] B43, B57,[6] B61; five weapons stations under wings with total capacity of 18,000 lb; HARPOON is also carried on A-6E TRAM

DEPLOYMENT:
Number deployed: 332 A-6E;[7] 256 A-6;[8] 608 A-6 aircraft produced (1960-1980); 318 operational A-6Es planned

Locations: MCAS El Toro, CA; MCAS Iwakuni, Japan; MCAS Cherry Point, NC; NAS Oceana, VA; NAS Whidbey Island, WA

Number per Squadron: 10 (UE)[9]

HISTORY:
IOC: 1963 (A-6); 1972 (A-6E)

Apr 1960 first flight

1969 development begins on A-6E

Feb 1970 A-6E first flight

1975 final delivery of initial buy

1981 HARPOON capability added to A-6[10]

1984 production of A-6E completed

COST (A-6E):[11] $10.6 m (FY 1978) (flyaway) (TY);[12] $14.8 m (program);[13] $18.5 m (FY 1982) (flyaway); $24.6 m (FY 1983) (flyaway)[14]

FY	Number Procured	Total Appropriation ($ million)
1979 & prior	123	1351.9[15]
1980	6	159.8
1981	12	270.7
1982	12	293.1
1983	8	249.0
1984	6	239.0

COMMENTS: A-6 has low-level navigation and weapons delivery capability at night and in adverse weather.[16] Aircraft in Mediterranean are "dedicated" to SACEUR's Selective Strike Plan.[17] Aircraft also provide nuclear anti-surface ship capability with tactical bombs.[18]

---

1 Information provided by Grumman Aerospace Corporation, Bethpage, NY.
2 JCS, FY 1982, p. 77.
3 U.S. Military Aircraft Data Book, 1981, pp. 2-11, 2-14.
4 Adelphi No. 168, p. 32.
5 Ships and Aircraft of the U.S. Fleet, 11th Ed., p. 266.
6 Ibid.
7 As of 1 January 1982; HAC, FY 1983 DOD, Part 5, p. 249; inventory as of 1 January 1981 was 120; HASC, FY 1982 DOD, Part 2, p. 603.
8 As of January 1980; JCS, FY 1982, p. 78.
9 Number reduced from 12 to 10 due to depleted inventory of aircraft; HAC, FY 1980 DOD, Part 7, p. 455.
10 HAC, FY 1983 DOD, Part 5, p. 184.
11 Information provided by Grumman Aerospace Corporation, Bethpage, NY.
12 U.S. Military Aircraft Data Book, op. cit.
13 Ibid.
14 Ibid., p. 249.
15 U.S. Military Aircraft Data Book, op. cit.
16 Ibid., p. 78.
17 JCS, FY 1981, p. 46.
18 Ibid., p. 48.

7
A-7 CORSAIR II

A-7 CORSAIR II[1]

Figure 7.4 A-7 CORSAIR.

DESCRIPTION:	Lightweight, single-seat, single-engine, carrier and land based, visual attack aircraft with forward looking infrared, all-weather, and night capability used by the Navy (A-7E) and Air National Guard (A-7D/K).	**SPECIFICATIONS**:	(A-7E)
		Dimensions:	
		Length:	46 ft, 1.5 in
		Height:	16 ft, 2 in
		Wingspan:	38 ft 8.5 in
Nuclear capable versions:[2]	A-7A/B/D/E	Takeoff Weight (max):	42,000 lb
		Powerplant:	1 TA-41-A-2 turbofan
CONTRACTORS:	Vought Corporation (LTV, Inc.) Dallas, TX (prime/airframe) Detroit Diesel, Allison Division[3] Indianapolis, IN (engine)	Ceiling:	52,500 ft; 35,500 ft[4]
		Speed:	(max) 693 mph at sea level

Nuclear Weapons Databook, Volume I **209**

7

A-7 CORSAIR II

Range:	1123 mi (max) (combat radius clean); 236 mi (loaded with one hour loiter); 1000+ km (with carrier-based aerial refueling)[5]
Aerial Refueling Capability:	yes
Crew:	1; 2 (A-7K)
NUCLEAR WEAPONS:	B28,[6] B43, B57,[7] B61; reportedly capable of carrying 4 nuclear weapons;[8] eight weapons stations, six wing pylons and two missile stations; maximum capacity of wing pylons is 3500 lb.
DEPLOYMENT:	
Number deployed:	more than 1500 built; approximately 370 (A-7E) (Navy); 375 (A-7D) (Air National Guard); (A-7B) (Naval Reserve)
Locations:	NAS Cecil Field, FL; NAS Lemoore, CA; NAS Atsugi, Japan; other reserve bases.
Number per Squadron:	12 (UE) (Navy); 24 (UE) (Air Force)
HISTORY:	
IOC:	1967 (A-7A); 1969 (A-7E)
Sep 1965	first flight (A-7A)
Nov 1967	development of A-7E for Navy started
Feb 1968	first flight of A-7B
Apr 1968	first flight of A-7D
Nov 1968	A-7E first flight
Sep 1980	procurement of A-7E completed
Apr 1981	Air National Guard begins delivery of new two-seat A-7Ks
COST:	$5.3 m (flyaway) (FY 1977);[9] $4.4 m (program) (TY)[10]

FY	Number Procured	Total Appropriation ($ million)
1979 & prior	596 (A-7E)	2530.3[11]
1980	-	14.5
1981	-	31.2
1982	-	16.0

COMMENTS: Replaced the A-4 SKYHAWK and A-1 SKYRAIDER in the Navy. A-18 will replace A-7 starting in early 1983. Naval aircraft in Mediterranean are "dedicated" to SACEUR's Selective Strike Plan.[12] Naval aircraft also provide nuclear anti-surface ship capability with tactical bombs.[13] A-7 is also flown by Portugal (A-7P) and Greece (A-7H).

---

1 Information provided by LTV Corporation; background information is also available in "The Story of Sandy, SLUF and the Little Hummers," *Air International*, March 1982, pp. 121-125+, and April 1982, pp. 169-177+.
2 JCS, FY 1982, p. 76; ACDA, FY 1979 ACIS, p. 162.
3 Engine is Allison built Rolls-Royce designed turbofan.
4 *U.S. Military Aircraft Data Book*, 1981, pp. 2-15, 2-18.
5 JCS, FY 1982, p. 77.
6 *Ships and Aircraft of the U.S. Fleet*, 11th Ed., p. 266.
7 Ibid.
8 Adelphi No. 168, p. 32.
9 *U.S. Military Aircraft Data Book*, op. cit.
10 Ibid.
11 Ibid.
12 JCS, FY 1981, p. 46.
13 Ibid., p. 48.

AV-8B HARRIER II

Figure 7.5 AV-8B HARRIER.

DESCRIPTION:	Vertical or short take-off and landing (V/STOL) close air support attack aircraft planned for the Marine Corps.	Takeoff Weight (max):	29,750 lb
		Powerplant:	1 Rolls-Royce Pegasus II (F402-RR-404) turbofan
CONTRACTORS:	McDonnell-Douglas Corp. St Louis, MO (prime) Rolls Royce Ltd. Bristol, U.K. (engine)	Ceiling:	50,000 ft
		Speed:	(max) 684 mph
		Takeoff Distance:	0-1200 ft
SPECIFICATIONS:	(AV-8B)	Range:	163 mi (combat radius); 75-890 mi (combat radius)[1]
Dimensions: Length: Height: Wingspan:	46 ft 3 in 11 ft 6 in 30 ft 3 in	Aerial Refueling Capability:	yes
		Crew:	1

Nuclear Weapons Databook, Volume I **211**

7
AV-8B HARRIER II

NUCLEAR WEAPONS: one nuclear weapon; B61, seven weapons stations, three for heavy weapons, 9000 lb capacity

DEPLOYMENT:
Number Planned: 336 planned for U.S. Marine Corps

Locations: MCAS El Toro, CA; MCAS Cherry Point, NC

Number per Squadron: 20 (UE)

HISTORY:
IOC: September 1985[2]

Nov 1978 first flight of YAV-8B

COST:
Program Cost: $10,111.2 m (Dec 1982)

Unit Cost: $46 m (FY 1982) (flyaway)[3]

FY	Number Procured	Total Appropriation ($ million)
1981 & prior	-	785.64
1982	12	898.7
1983	21	1033.8
1984	32	1165.8

COMMENTS: AV-8B will replace 8 squadrons worth of A-4M and AV-8A aircraft. The AV-8B has twice the range and payload of the current AV-8A. The AV-8B upgrade includes modified engine and airframe and a new graphite wing. The AV-8A is not nuclear certified.

---

[1] Combat radius varies greatly depending upon weight of ordnance, mission profile, and use of vertical takeoff or 1200 ft short takeoff; HASC, FY 1982 DOD, Part 2, p. 610.
[2] HAC, FY 1983 DOD, Part 5, p. 186.
[3] HASC, FY 1982 DOD, Part 2, p. 609.
[4] U.S. Military Aircraft Data Book, 1981, pp. 2-111, 2-122.

7
CF-101B VOODOO

CF-101B VOODOO

Figure 7.6 F-101B VOODOO.

DESCRIPTION:	Long-range, nuclear armed, strategic interceptor used by Canada (F-101F).	Powerplant:	2 J57-PW-55 turbojets
		Ceiling:	52,000 feet
CONTRACTORS:	McDonnell-Douglas (prime/airframe)	Speed:	Mach 1.85 (max)
		Range:	1550 nm
SPECIFICATIONS:	(F-101B)	Aerial Refueling Capability:	no
Dimensions:			
Length:	67 ft 4 in		
Height:	18 ft	Crew:	2
Wingspan:	39 ft 7 in		
Takeoff Weight (max):	46,700 lb	**NUCLEAR WEAPONS**:	two GENIE (AIR-2A)[1]

Nuclear Weapons Databook, Volume I   **213**

7

CF-101B VOODOO

DEPLOYMENT:

Number Deployed: 66 (1982)

Locations: nuclear armed versions at Bagotville and Comox bases, Canada

HISTORY:

IOC: 1952 (U.S.)

1951   SAC develops requirement for long-range bomber escort

1952   TAC and ADCOM take delivery of F-101

1961   Canada receives F-101Bs capable of carrying GENIE

1961   F-101B interceptor deployed

1971   Canada receives new F-101Fs with new electronics

1981   Canada chooses F-18 HORNET to replace CF-101s

COST: $1.8 m (1961)

COMMENTS: remained in U.S. active service until 1974, and then with interceptor units in the Air National Guard until 1981

---

1 AFR 0-2, p. 46.

7
F-4 PHANTOM II

F-4 PHANTOM II

Figure 7.7 F-4D PHANTOM II.

DESCRIPTION:	Two-seat, twin-engine, all-weather, supersonic, multimission fighter, used by the Air Force, Marine Corps, and Navy.	**SPECIFICATIONS:**	(F-4E)
		Dimensions:	
		Length:	62 ft 11.75 in
		Height:	16 ft 3 in
		Wingspan:	38 ft 5 in
Nuclear capable versions:[1]	F-4C/D/E	Takeoff Weight (max):	60,630 lb
CONTRACTORS:	McDonnell Douglas St. Louis, MO (prime) General Electric (engines)	Powerplant:	2 GE J79-GE-15 (F-4 C/D) turbojets; 2 J79-GE-17 turbojets (F-4 E/G)
		Ceiling:	64,630 ft, 71,000 ft[2]
		Speed:	(max) 1500 mph (Mach 2.27) at 40,000 ft

Nuclear Weapons Databook, Volume I **215**

7

F-4 PHANTOM II

Range:	660 mi (1060 km) (combat radius); 1000 mi (1610 km) ground attack[3]
Aerial Refueling Capability:	yes
Crew:	2
NUCLEAR WEAPONS:	three pylons (centerline and wings) can carry nuclear weapons (one each) weighing up to 2170 lb;[4] B28RE[5], B43, B57,[6] B61, B83;[7] possibly GENIE in strategic interceptor force
DEPLOYMENT:	
Number deployed:	954 (Air Force);[8] 144 (Marine Corps, Navy)
Locations:	NAS Miramar, CA; NAS Oceana, VA; Clark AB, Philippines; Elmendorf AFB, AK; Spangdahlem AB, West Germany; Ramstein AB, West Germany; Osan AB, Korea; Taegu AB, Korea; Homestead AFB, FL; Seymour Johnson AFB, NC; Moody AFB, GA; Torrejon AB, Spain

HISTORY:	
IOC:	1961
May 1958	first flight
Jun 1967	F-4E first flight
1979	production completed of all F-4 versions
COMMENTS:	5057 F-4s produced, with last U.S. delivery in October 1979. It is probable that NATO nuclear capable F-4s are limited to Greek and Turkish forces.[9] F-4 is being replaced by F-18 and F-14 in the Navy and Marine Corps, and F-16 and F-15 in the Air Force. Two F-106 air defense National Guard units will receive F-4Ds in FY 1983-1984.[10]

---

1 Marine Corps F-4 versions (F-4J/S) are not nuclear capable; JCS, FY 1981, p. 49.
2 Norman Polmar, *World Combat Aircraft Directory*, p. 184.
3 *Ibid.*
4 *Air International*, November 1978, p. 215.
5 *Ships and Aircraft of the U.S. Fleet*, 11th Ed., p. 260; USAF, "Safety Rules for the F-4C/D/E B43/B57/B61 Weapon System," AFR 122-44, 11 July 1980.
6 *Ships and Aircraft of the U.S. Fleet*, 11th Ed., *op. cit.*
7 SASC, FY 1981 DOE, p. 37.
8 There are approximately 120 training F-4 aircraft assigned to NORAD for strategic defense in a contingency; SASC, FY 1980 DOD, Part 2, p. 440.
9 ACDA, FY 1979 ACIS, p. 144.
10 DOD, "Memorandum for Correspondents," 31 January 1983.

7
F-15 EAGLE

F-15 EAGLE

Figure 7.8 F-15 EAGLE.

DESCRIPTION:	Long-range, high performance, twin engine interceptor used by the Air Force.	Takeoff Weight (max):	56,000 lb (F-15A/B); 68,000 lb (F-15C/D)
Nuclear Capable Versions:[1]	F-15A/C, F-15E (derivative fighter)	Powerplant:	2 F100-PW-100 turbofan
		Ceiling:	65,000 ft
CONTRACTORS:	McDonnell Douglas (prime/airframe) Pratt & Whitney Aircraft (engines)	Speed:	(max) 1900 mph[2]
		Range:	1681 mi (combat range)[3]
		Aerial Refueling Capability:	yes
SPECIFICATIONS: Dimensions:			
Length:	63 ft 8 in	Crew:	1 (F-15A/C); 2 (F-15B/D trainers)
Height:	18 ft 6 in		
Wingspan:	42 ft 8 in		

Nuclear Weapons Databook, Volume I   **217**

7
F-15 EAGLE

Figure 7.9 F-15 EAGLE underside.

NUCLEAR WEAPONS: possibly GENIE (W25), five weapons stations capable of carrying more than 16,000 lb

DEPLOYMENT:
Number Deployed: 620;[4] 383 F-15A, 60 F-15B produced; 1400+ F-15 planned through 1990[5]

Locations: Elmendorf AFB, AK; Kadena AB, Japan; Langley AFB, VA; Bitburg AB, Germany; Eglin AFB, FL; Holloman AFB, NM; Soesterberg AB, Netherlands; First strategic interceptor unit at Langley AFB, VA; Langley and McChord AFB, WA earmarked for F-15 ASAT operations.[6]

Number per Squadron: 24 (UE)

HISTORY:
IOC: 1975

Dec 1969   development of F-15 started

Jul 1972   first flight of F-15A

Mar 1973   production started

Feb 1979   first flight of F-15C

Jun 1979   F-15C/D production began

COST:
Program Cost: $41,500.8 m (Dec 1982)

Unit Cost: $13.7 m (flyaway) (TY)[7]

FY	Number Procured	Total Appropriation ($ million)
1980 & prior	638	11,754.3[8]
1981	42	967.6[9]
1982	36	1187.7
1983	39	1553.9
1984	48	2266.8

7
F-15 EAGLE

COMMENTS: Although F-15 is not primarily for nuclear weapons use, it is nuclear certified and would be highly capable in the nuclear delivery mode. The F-15 is the only Air Force fighter able to carry and deliver air-to-surface weapons at supersonic speeds. It takes less than an hour to convert the air-to-air F-15 into air-to-surface role. Six squadrons of active force F-106s will be replaced by F-15s for strategic defensive forces.[10] In addition, in the event of a crisis, F-15s dedicated to peacetime training could be used for strategic interception.[11] F-15E STRIKE EAGLE, originally a company funded upgraded air-to-surface model, has been chosen as candidate in the derivative tactical fighter to augment and then replace F-111 and F-4 pending introduction of Advanced Tactical Fighter in the 1990s (see Introduction). The enhanced F-15E air-to-ground capability would be specifically to give the F-15 a nuclear weapons strike ission.[12]

AFM 50-5, Volume II, p. 3-85.
U.S. Military Aircraft Data Book, 1981, pp. 2-55, 2-58.
Ibid.
As of January 1982; SAC, FY 1983 DOD, Part 4, p. 548.
Total procurement upgraded from 729 F-15s in FY 1983; SAC, FY 1983 DOD, Part 4, p. 137.
HAC, FY 1983 DOD, Part 5, p. 548.

7 *U.S. Military Aircraft Data Book*, op. cit.
8 *Ibid.*
9 SASC, FY 1982 DOD, Part 3, p. 1540.
10 DOD, FY 1983 RDA, p. VII-9.
11 SASC, FY 1981 DOD, Part 2, p. 582.
12 *Ibid.*, p. 1617; SAC, FY 1983 DOD, Part 4, pp. 248-249.

7
F-16 FALCON

F-16 FIGHTING FALCON[1]

Figure 7.10 F-16 FALCON.

DESCRIPTION:	Lightweight, single-seat, single-engine, supersonic, multi-mission, air-to-air and air-to-ground fighter used by the Air Force and NATO Air Forces.		Delco (computers) Kaiser (radar and fire control) Singer-Kearfott (inertial system)
Nuclear capable versions:[2]	F-16A/B, F-16C/D (after 1984), F-16E (derivative fighter)	**SPECIFICATIONS:**	(F-16A)
CONTRACTORS:	General Dynamics[3] Fort Worth, TX (prime/airframe) Pratt & Whitney Aircraft East Hartford, CT (engine) Westinghouse Electric, Inc. (radar) Marconi-Elliott (flight control)	Dimensions: Length: Height: Wingspan: Takeoff Weight (max): Powerplant:	49 ft 6 in 16 ft 3 in 32 ft 33,000 lb[4] 1 F-100-PW-100 turbofan

7
F-16 FALCON

Figure 7.11 F-16 FALCON.

Ceiling: 50,000 ft

Speed: 1400 mph (max); 1520 mph[5] (max)

Range: 575 mi[6] (combat radius)

Aerial Refueling Capability: yes

Crew: 1

NUCLEAR WEAPONS: B43, B61;[7] reported capable of delivering up to five nuclear bombs on five of nine hardpoints;[8] later report indicates 3rd, 5th, and 7th stations used for nuclear bombs;[9] standard weapons configuration is one or two nuclear weapons;[10] general profile is one B61;[11] the B57 is prohibited from use on F-16A/B aircraft[12]

DEPLOYMENT:
Number Deployed: 365;[13] 1388 planned (1982); 1445+ planned (1983);[14] including 204 F-16B two-seaters; 348 produced initially for NATO; air-to-surface improvements incorporated into aircraft no. 786 and on.[15]

Locations: Kunsan AB, Korea; Shaw AFB, SC; Hill AFB, UT; Hahn AB, West Germany (first U.S. base in Europe).

Number per Squadron: 24 (UE)

HISTORY:
IOC: 1979

Apr 1972 — development of F-16 begins

Jan 1974 — first flight

Jun 1975 — four NATO counties announce joint program to procure F-16

Dec 1976 — full scale testing began

Sep 1977 — production started

Aug 1978 — first production aircraft accepted

Jan 1979 — first F-16 delivered to Hill AFB, UT

Jan 1979 — delivery of first European manufactured F-16

1982-1983 — F-16s arrive at Hahn AB, West Germany to take up nuclear roles in replacement of F-4s

COST:
Program Cost: $43,494.2 m (Dec 1982)

Unit Cost: $11.9 m (TY) (flyaway)

FY	Number Procured	Total Appropriation ($ million)
1977 & prior	-	751.3[16]
1978	105	1655.9
1979	145	1554.2
1980	175	1684.3
1981	180	2035.4
1981 & prior	605	7052.6[17]
1982	120	2294.5
1983	120	2334.1
1984	120	2279.5

7

F-16 FALCON

COMMENTS: F-16 will complement the F-15 in air superiority role and replace F-4 in air-to-surface mode. F-16 will replace 5 squadrons of F-106 in Air National Guard fighter interceptor units starting in FY 1986-1987.[18] Nuclear capable versions are also being produced for Belgium and the Netherlands. Although Denmark and Norway are receiving new F-16s, these will most likely not be nuclear certified. In the nuclear bombing role the weapon's delivery accuracy has been better than F-111.[19] F-16 could reach the western Soviet Union from bases in West Germany with a single nuclear weapon and aerial refueling.[20] Advanced versions of the F-16 (F-16E) are being considered for a derivative tactical fighter to augment and then replace F-111 and F-4 pending introduction of Advanced Tactical Fighter in 1990s (see introduction to this chapter).

---

1 Detailed background information on the F-16 is contained in Jay Miller, *General Dynamics F-16 Fighting Falcon* (Austin, TX: Aerofax, Inc., 1982).
2 JCS, FY 1982, p. 78; ACDA, FY 1979 ACIS, p. 141.
3 Consortium of 5 primary international companies assembling aircraft and producing components: Fokker (Netherlands), SABCA (Belgium), Fairay (Belgium), Per Udsen (Denmark), and General Dynamics. An additional 52 European subcontractors are involved in component production.
4 ACDA, FY 1979 ACIS, p. 141.
5 *U.S. Military Aircraft Data Book*, 1981, pp. 2-59, 2-62.
6 "The Texan Swing Fighter," *Air International*, November 1977, p. 223; ACDA, FY 1979 ACIS, p. 141.
7 HASC, FY 1981 DOD, Part 4, Book 2, p. 2318.
8 "The Texan Swing Fighter," *Air International*, November 1977, op. cit.
9 Jay Miller, *General Dynamics F-16 Fighting Falcon*, op. cit.
10 ACDA, FY 1979 ACIS, p. 141.
11 HASC, FY 1981 DOD, Part 4, Book 2, p. 2318.
12 USAF, "Safety Rules for the F-16A/B B57/B61 Weapon System," AFR 122-26, 30 January 1981.
13 As of January 1982; HAC, FY 1983 DOD, Part 5, p. 545.
14 HAC, FY 1983 DOD, Part 5, p. 545; the current DOD plan is 1445 F-16s through FY 1987, with more to follow on in later years.
15 HAC, FY 1983 DOD, Part 5, p. 648.
16 *U.S. Military Aircraft Data Book*, 1981, op. cit.
17 HASC, FY 1982 DOD, Part 2, p. 897.
18 "The Texan Swing Fighter," *Air International*, November 1977, op. cit.
19 HAC, FY 1981 DOD, Part 9, p. 145.
20 ACDA, FY 1979 ACIS, p. 142.

7
F/A-18 HORNET

F-18/A-18 HORNET

Figure 7.12 F-18 HORNET.

DESCRIPTION: Single-seat, twin-engine, supersonic carrier and land based all-weather fighter and attack aircraft used by the Marine Corps and Navy. Attack configuration (A-18) also capable of delivering nuclear weapons.

Nuclear capable versions:[1]   F-18A, A-18, CF-18

CONTRACTORS:
McDonnell-Douglas
St. Louis, MO
(prime/airframe)
Northrop Aircraft
Hawthorne, CA
(40 percent subcontracting)

General Electric
West Lynn, MA
(engine)
Hughes Aircraft Co.
Culver City, CA
(radar)

SPECIFICATIONS:

Dimensions:	
Length:	55 ft 7 in
Height:	15 ft 2 in
Wingspan:	40 ft 7 in
Takeoff Weight (max):	44,000+ lb
Powerplant:	2 F-404-GE-400 turbofans
Ceiling:	50,000 ft

Nuclear Weapons Databook, Volume I 223

7

F/A-18 HORNET

Figure 7.13 F-18 HORNET.

Speed: (max) 1368 mph[2]

Range: 400+ mi (645 km) (combat radius);[3] 840 mi (1350 km);[4] 550 nm (interdiction); 460 mi (combat radius)[5]

Aerial Refueling Capability: yes

Crew: 1 (2 in trainer version)

NUCLEAR WEAPONS:[6] two B57 or B61;[7] two of nine external weapons points on outboard wing stations capable of carrying nuclear bombs; BDU-11/12, BDU-20, BDU-36 nuclear practice bombs

DEPLOYMENT: (see Table 7.4)

Number Deployed: 27;[8] 1366 planned for U.S.; Canada is planning to buy 138 F-18s, partly to replace CF-101s, currently flying a nuclear armed air defense mission.

Locations: NAS Lemoore, CA (training); MCAS El Toro, CA (initial base)[9]

Table 7.4
F/A-18 Deployments[1]

Unit	Number of Squadrons	Squadron Aircraft	Total Procured
Marine Corps Fighter Squadron	12	144	258
Marine Corps Attack Squadron	8	160	278
Navy Fighter Squadron	6	72	161
Navy Attack Squadron	24	288	513
Navy Reconnaissance Squadron	1	36	74
Marine Corps Reconnaissance Squadron	1	21	38
Marine Corps TACA	1	30	44
			1366
Squadron Authorized Aircraft			751
R&D Aircraft			19
Training Aircraft			151
Total Operation			921
Total Pipeline			124
Attrition Aircraft			321
			1366

1 HASC, FY 1982 DOD, Part 2, p. 669; AW&ST, 19 January 1981, p. 25.

Number per Squadron: 12 (UE)

HISTORY:

IOC:	Dec 1982
Jul 1974	first flight (YF-17)
Nov 1975	development of F/A-18A begins
Nov 1978	first flight (F-18)
Nov 1980	first training squadron commissioned at NAS Lemoore, CA
Jun 1981	full production approved
1985	carrier deployment
1993	production completed

7
F/A-18 HORNET

COST:

Program Cost: $39,827.2 m (Dec 1982) (F-18 program, not counting YF-17 prototype costs)

Unit Cost: $22.5 m (FY 1982) (flyaway)
$25.1 m (FY 1983) (flyaway)

FY	Number Procured	Total Appropriation ($ million)
1977 & prior	-	491.8
1978	-	654.4
1979	9	1038.5
1980	25	1463.3
1981	60	2190.8
1982	63	2629.1
1983	84	2598.2
1984	84	2762.8

COMMENTS: F/A-18 will replace Navy's F-4, and A-7; Marine Corps' F-4. All-digital weapon control system provides greater bombing accuracy over the F-4.

---

1 JCS, FY 1982, p. 77; fighter and attack versions virtually identical in performance and characteristics, with common internal wiring and only external configuration adjustments; ACDA, FY 1979 ACIS, p. 159; A-18 will be equipped with five rather than three weapons store pylons. It will take 30 minutes to convert from attack to fighter versions and vice versa; information provided by McDonnell Douglas.
2 U.S. Military Aircraft Data Book, pp. 2-63, 2-66.
3 Norman Polmar, *World Combat Aircraft Directory, op. cit.*
4 Nikolaus Krivinyi, *World Military Aviation: Aircraft, Air Forces, Weaponry and Insignia* (New York: Arco, 1977), p. 175.
5 *U.S. Military Aircraft Data Book, op. cit.*
6 HAC, FY 1983 DOD, Part 5, p. 346.
7 General Accounting Office, "F/A-18 Naval Strike Fighter: Its Effectiveness is Uncertain" (PSAD-80-24), 14 February 1980.
8 As of 1 January 1982; HAC, FY 1983 DOD, Part 5, p. 257.
9 HAC, FY 1983 DOD, Part 5, p. 202.

7
F-100 SUPERSABRE

F-100 SUPERSABRE

Figure 7.14 F-100 SUPERSABRE.

DESCRIPTION:	Single-seat, single-engine, supersonic fighter bomber in use with the Turkish Air Force.	Takeoff Weight (max):	34,831 lb
Nuclear capable versions:	F-100D/F[1]	Powerplant:	1 P&W J57-P-21A turbojet
		Ceiling:	50,000 ft
CONTRACTORS:	North American (prime) Pratt & Whitney (engine)	Speed:	862 mph (max) (Mach 1.3) a 35,000 ft
		Range:	550 mi (885 km) (combat radius)
SPECIFICATIONS:	(F-100D)	Aerial Refueling Capability:	yes

Dimensions:
  Length:
  Height:
  Wingspan:
	54 ft 3 in[2]
	16 ft 2.5 in
	38 ft 9 in[3]
		Crew:	1

226 Nuclear Weapons Databook, Volume I

7
F-100 SUPERSABRE

NUCLEAR WEAPONS:[4]	one nuclear weapon; B28, B43, B57
HISTORY:	
IOC:	1954
May 1953	first flight
Jan 1956	F-100D first flight
COMMENTS:	1274 F-100D, 476 F-100C, built 1953-1959

F-100D (LW1), F-100D(1)/F(1)/F(CL) are designated as nuclear certified; USAF, "Safety Rules for the Non-U.S. NATO F-100D (LW1) and F-100D(1)/F(CL) B28/B43/B57 Weapon Systems"; AFR 122-71, 9 January 1979.

2 Norman Polmar, *World Combat Aircraft Directory*, p. 165.
3 *Ibid.*
4 *Ibid.*

7

F-104 STARFIGHTER

F-104 STARFIGHTER

Figure 7.15 F-104G STARFIGHTER in West German Air Force.

DESCRIPTION:	Single-seat, single-engine, daylight fighter-interceptor in wide use within Belgian, Dutch, Greek, Italian, and West German air forces for strike missions.	**SPECIFICATIONS:**	(F-104G)
		Dimensions:	
		Length:	54 ft 9 in
		Height:	13 ft 6 in
		Wingspan:	21 ft 11 in
Nuclear capable versions:[1]	F-104G/S[2] (F-104C, F-104D?)	Takeoff Weight (max):	28,770 lb
CONTRACTORS:	Lockheed (prime) General Electric (engines)	Powerplant:	1 GE J79-11A turbojet
		Ceiling:	58,000+ ft
		Speed:	913 mph (max) (Mach 1.2) at sea level; 1324 mph (Mach 2) at 39,375 ft

228 Nuclear Weapons Databook, Volume I

7
F-104 STARFIGHTER

Range:	808 mi (1300 km) (combat radius); 745 mi[3]	**HISTORY:**	
		IOC:	1958
Aerial Refueling Capability:	unknown	Feb 1954	first flight
		Oct 1960	F-104G first flight
Crew:	1		
NUCLEAR WEAPONS:[4]	one nuclear bomb[5]; B28, B43, B57, B61-2, -3, -4, and -5[6]	**COMMENTS:**	Being replaced by TORNADO in West German and Italian forces, and by F-16 in Belgian and Dutch forces.
DEPLOYMENT:			
Locations:[7]	Memmingen, West Germany; Buchel, West Germany; Norvenich, West Germany; Kleine Brogel, Belgium; Volkel, Netherlands; Rimini, Italy; Ghedi-Torre, Italy		

JCS, FY 1982, p. 76.
USAF, "Safety Rules for the Non-U.S. NATO F-104G/S B28/B43/B57/B61-2, -3, -4, and -5 Weapon System," AFR 122-71, 9 January 1980.
Norman Polmar, *World Combat Aircraft Directory*, p. 162.

4 USAF, AFR 122-71, *op. cit.*
5 Krivinyi, *op. cit.* p. 168.
6 USAF, AFR 122-71, *op. cit.*
7 Bases with USAF nuclear weapons.

7

F-106 DELTA DART

F-106 DELTA DART

Figure 7.16 F-106 DELTA DART after firing a training version of the GENIE rocket.

DESCRIPTION:	Single-seat, single-engine, all-weather, supersonic strategic interceptor used by the Air Force and Air National Guard.	**SPECIFICATIONS**:	(F-106A)
		Dimensions:	
		Length:	70 ft 8.75 in
		Height:	20 ft 3 in
Nuclear capable versions:[1]	F-106A[2]	Wingspan:	38 ft 3.5 in
		Takeoff Weight (max):	34,510 lb
CONTRACTORS:	General Dynamics/Convair (prime)		
	Pratt & Whitney (engine)	Powerplant:	1 J75-P-17 turbojet
	Hughes (fire control)	Ceiling:	57,000 ft
		Speed:	1525 mph (max) (Mach 2.3)
		Range:	365 mi (587 km) (combat radius)

7
F-106 DELTA DART

Aerial Refueling Capability:	no	**COMMENTS:**	337 aircraft produced, replacing the F-102. Under current DOD plans, five squadrons of active force F-106s will be replaced with F-15s assigned strategic defensive missions.[5] Air National Guard F-106s will also be modernized with F-4s and F-16s,[6] the first units receiving F-4Ds in late 1983.[7]
Crew:	1		

NUCLEAR WEAPONS: one GENIE (AIR-2A) (W25 warhead) air-to-air missile carried in an internal weapons bay[3]

DEPLOYMENT:
Number Deployed: 277 F-106A; 63 F-106B

Locations: See Table 4.6

HISTORY:
IOC: 1959

Dec 1956 — first flight

Jul 1959-Jul 1960 — production delivery

1988 — last F-106 unit deactivated[4]

---

1 JCS, FY 1979, p. 38.
2 F-106B is operational trainer. Norman Polmar, *World Combat Aircraft Directory*, p. 110.
3 HAC, FY 1983 DOD, Part 5, p. 549.

5 DOD, FY 1983 RDA, p. VII-9.
6 SASC, FY 1982 DOD, Part 7, p. 3838.
7 DOD, "Memorandum for Correspondents," 31 January 1983.

Nuclear Weapons Databook, Volume I **231**

7
F-111

F-111

Figure 7.17 F-111.

DESCRIPTION:	Long-range, two-seat, twin-engine, all-weather supersonic strike fighter used by the Air Force.	Takeoff Weight (max):	91,501 lb
		Powerplant:	2 TF-30-P/-3 turbofans
Nuclear capable versions:	F-111A/D/E/F[1]	Ceiling:	60,000+ ft
CONTRACTORS:	General Dynamics Fort Worth, TX (prime) Pratt & Whitney (engine)	Speed:	1650 mph (max) (Mach 2.5) at 49,000 ft; 915 mph (Mach 1.2) at sea level[2]
		Range:	1500 mi (2400 km) (combat radius)
SPECIFICATIONS: Dimensions: Length: Height: Wingspan:	73 ft 6 in 17 ft 1.5 in 63 ft (spread), 31 ft 11.5 in (swept)	Aerial Refueling Capability: Crew:	yes 2

7
F-111

NUCLEAR WEAPONS:[3] up to 3 nuclear bombs;[4] B43, B57, B61, B83;[5] 2 nuclear weapons on underwing pylons; also has internal bomb bay

DEPLOYMENT:
Number Deployed: 240; 455 built[6]

Locations: Cannon AFB, NM; Mountain Home AFB, ID; RAF Lakenheath, U.K.; RAF Upper Heyford, U.K.

HISTORY:
IOC: 1968

Dec 1964 first flight

1976 production completed

COMMENTS: F-111s are on nuclear armed quick reaction alert (QRA) at all times at two bases in the U.K.: Upper Heyford and Lakenheath.[7] F-111 has on-board radar for all-weather nuclear attack, including terrain following and ground mapping capabilities. Its low level navigation and weapons delivery capability allows bombing at night and in adverse weather.[8] It can conduct "direct" and "offset" bombing. Area targets for F-111 include: lines of communication, airfields, transportation terminals, bivouac areas, attack helicopter forward operating locations, supply depots, staging areas, choke points, and POL storage.[9]

---

1 All models of the F-111 are nuclear capable; SASC, FY 1982 DOD, Part 7, p. 3891.
2 Norman Polmar, *World Combat Aircraft Directory, op. cit.*
3 USAF, "Safety Rules for the F-111A/D/E/F B43/B57/B61 Weapon System," AFR 122-37, 11 July 1980.
4 All models of the F-111 are nuclear capable; SASC, FY 1982 DOD, Part 7, p. 3892.
5 SASC, FY 1981 DOE, p. 37.
6 SAC, FY 1983 DOD, Part 4, p. 245.
7 HASC, FY 1981 DOD, Part 4, Book 2, p. 2316.
8 JCS, FY 1982, p. 78.
9 HASC, FY 1981 DOD, Part 4, Book 2, p. 2316.

7
P-3 ORION

Figure 7.18 P-3 ORION.

DESCRIPTION: Long range, land-based four-engine, all-weather anti-submarine, and maritime patrol plane used by the U.S. and Dutch navies for nuclear weapons delivery.[1]

Nuclear capable versions: P-3A/B/C

CONTRACTORS: Lockheed Aircraft
Burbank, CA
(prime/airframe)
Detroit Diesel, Allison Division
Indianapolis, IN
(engines)
CTM
Buffalo, NY
(bomb racks)

SPECIFICATIONS: (P-3C)

Dimensions:
  Length: 116 ft 10 in
  Height: 33 ft 8.5 in
  Wingspan: 99 ft 8 in

Takeoff Weight (max): 142,000 lb

Powerplant: 4 T56-A-14 turboprop

Ceiling: 28,300 ft

7

P-3 ORION

Speed:	473 mph (max) at 15,000 ft; 237 mph at 1500 ft (patrol speed)	Sep 1968	first flight (P-3C)
Range:	1550 mi (2500 km) (radius), 3 hours on station at 1500 ft	1969	production of P-3 begins
		Nov 1981	Dutch Navy receives first of 13 P-3Cs
Aerial Refueling Capability:	no	FY 1983	procurement program for U.S. Navy continued after initial plans for its halting
Crew:	10 (normal complement)		
NUCLEAR WEAPONS:	two B57 nuclear depth charges;[2] also carries HARPOON; ten underwing stations, one station in bomb bay	COST:	$27.9 m (FY 1982) (flyaway)[5] $39.1 m (FY 1983) (flyaway)[6]

FY	Number Procured	Total Appropriation ($ million)
1979 & prior	195 (P-3C)[7]	2964.4[8]
1980	12	408.3
1981	12	459.8
1982	12	454.8
1983	6	317.7
1984	5	309.5

DEPLOYMENT:	24 active, 13 reserve squadrons stationed in the U.S., with overseas deployment sites
Number deployed:	200 P-3C;[3] 343 total[4] (18 of 24 active squadrons with P-3C)
Locations:	NAS Moffett Field, CA; NAS Barbers Point, HI; NAS Brunswick, ME; NAS Jacksonville, FL; numerous deployment sites worldwide
Number per Squadron:	9 (UE)
HISTORY:	
IOC:	1961 (P-3); 1969 (P-3C)
Aug 1958	first flight (prototype)
Apr 1961	first flight (P-3A)
1966	development of P-3C begins

COMMENTS: 17 of 24 active squadrons equipped with P-3C, remainder converted by FY 1991.[9] Present P-3 modification programs include improved sensor systems, HARPOON launch system, and navigation improvements. Patrol endurance of the ORION is up to 17 hours. Latest modification is P-3C Update III which will enter the Navy in May 1984.[10]

---

1 The P-3C began introduction in Dutch Navy in 1982, equipping one squadron. It assumed a nuclear role pending further governmental decision on future nuclear mission of Dutch forces.
2 Krivinyi, op. cit. p. 169.
3 As of 1 January 1982; HAC, FY 1983 DOD, Part 5, p. 264; as of 1 January 1981, there were 187; HASC, FY 1983 DOD, Part 2, p. 693.
4 JCS, FY 1982, p. 88.
5 As of 1 January 1982; HAC, FY 1983 DOD, Part 5, p. 264; as of 1 January 1981, there were 187; HASC, FY 1983 DOD, Part 2, p. 693.
6 HAC, FY 1983 DOD, Part 5, p. 191.
7 U.S. Military Aircraft Data Book, 1981, pp. 2-99 - 2-102.
8 Ibid.
9 JCS, FY 1983, p. 90.
10 HAC, FY 1983 DOD, Part 5, p. 191.

Nuclear Weapons Databook, Volume I **235**

7

S-3 VIKING

S-3 VIKING

Figure 7.19 S-3 VIKING.

DESCRIPTION:	Medium-range, twin-engine, carrier-based, maritime patrol and anti-submarine warfare aircraft used by the Navy.
Nuclear capable versions:	S-3A
CONTRACTORS:	Lockheed California Co. Burbank, CA (prime/airframe) General Electric West Lynn, MA (engines)

SPECIFICATIONS:	(S-3A)
Dimensions:	
Length:	53 ft 4 in
Height:	22 ft 9 in
Wingspan:	69 ft 8 in (29 ft 6 in folded)
Takeoff Weight (max):	52,530 lb
Powerplant:	2 TF-34-400B-GE-2 turbofans
Ceiling:	40,000 ft
Speed:	184 mph (296 kmh) (patrol loiter); 507 mph (816 kmh) at sea level
Range:	2300 mi (3700 km) (radius)

7
S-3 VIKING

Aerial Refueling Capability:	yes	1971	production begins
Crew:	4 (pilot, copilot, sensor operator, tactical coordinator)	May 1972	full production begins
		Jan 1972	first flight
NUCLEAR WEAPONS:	one B57 nuclear depth charge, three wing stations for weapons; future provisions for HARPOON	Mar 1978	production completed
		1983-1985	aircraft upgraded under weapon system improvement program and redesignated S-3B

DEPLOYMENT:

Number Deployed: 187 produced

Locations: NAS Cecil Field, FL; NAS North Island, CA

Number per Squadron: 10 (UE)

HISTORY:
IOC: 1974

Dec 1967   development of S-3A started

COST:

FY	Number Procured	Total Appropriation ($ million)
1981 & prior	179	3428.2[1]
1982	-	31.3[2]

COMMENTS: The VIKING's patrol endurance is over nine hours.

---

[1] *U.S. Military Aircraft Data Book*, 1981, p. 2-110.   [2] *Ibid.*

7

SH-3 SEA KING

SH-3 SEA KING

Figure 7.20 SH-3H SEA KING.

DESCRIPTION: Heavy helicopter, used for aircraft carrier-based anti-submarine warfare by the Navy.

Nuclear capable versions:[1] SH-3D/H

CONTRACTORS: Sikorsky (prime)

SPECIFICATIONS:
Dimensions:
  Length: 54 ft 9 in (fuselage)
  Height: 15 ft 6 in
  Wingspan: 62 ft (rotor diameter)

Takeoff Weight (max): 20,500 lb

Powerplant: 2 GE T58-GE-10 turboshafts

Ceiling: 14,700 ft

Speed: (max) 166 mph (267 kph) at sea level

Range: 625 nm

Aerial Refueling Capability: no

Crew: 4 (2 pilots, 2 systems operators)

NUCLEAR WEAPONS: one B57 nuclear depth bomb

DEPLOYMENT:
Number Deployed: 104 (SH-3H)[2]

Locations: NAS North Island, CA; NAS Jacksonville, FL

HISTORY:
IOC: 1961 (SH-3); 1966 (SH-3D)

Mar 1959 first flight of SH-3A

COMMENTS: SH-3 will be replaced by variant of SH-60 (SH-60F) in 1988.[3]

---

1 CANTRAC, p. G03.
2 HAC, FY 1983 DOD, Part 5, p. 314.
3 HAC, FY 1983 DOD, Part 5, pp. 312-314.

SH-60 SEAHAWK

Figure 7.21 SH-60 SEAHAWK.

DESCRIPTION:	Carrier-based, active sensor, inner zone anti-submarine helicopter to protect aircraft carriers; planned for the Navy.
Nuclear Capable Versions:	SH-60F
CONTRACTORS:	Sikorsky Aircraft Division, United Technologies Corp. Stratford, CT (prime) General Electric Lynn, MA (engine)

SPECIFICATIONS: (SH-60B)

Dimensions:	
Length:	64 ft 10 in
Height:	17 ft 2 in
Wingspan:	53 ft 8 in (rotor diameter)
Takeoff Weight (max):	21,844 lb
Powerplant:	2 GE T700-GE-401 turboshafts
Ceiling:	22,000 ft
Speed:	155 mph (max cruise)
Range:	50 nm (radius) with 3 hours on station

7
SH-60 SEAHAWK

Aerial Refueling Capability:	no	1986	first procurement of SH-60F planned
Crew:	4 (pilot, copilot, tactical officer, sensor operator)	COST:	Total Planned Cost: $3759.8 m[4] SH-60F variant R&D costs estimated at $87.7 m[5]
NUCLEAR WEAPONS:	B57 nuclear depth charge		

DEPLOYMENT:

Number Planned:	175[1] (1983)	FY	Number Procured	Total Appropriation ($ million)
Location:	NAS North Island, CA; NAS Jacksonville, FL	1982 & prior	23	2063.9[6]

HISTORY:		COMMENTS:	SH-60F is planned replacement for current SH-3H. SH-60F is modification of SH-60B Light Airborne Multipurpose System (LAMPS) Mk-III, planned for deployment aboard surface ships.
IOC:	1988[2]		
Dec 1979	first flight of prototype SH-60B		
1983	SH-60F program started to develop replacement for SH-3[3]		

---

1 Program cost for SH-60B; HAC, FY 1983 DOD, Part 5, p. 262.
2 *Ibid.*, p. 314.
3 *Ibid.*, p. 313.
4 *Ibid.*, p. 316.
5 *Ibid.*
6 *Ibid.*, p. 262.

7
TORNADO

TORNADO

Figure 7.22 **TORNADOs** being refueled by tanker aircraft.

DESCRIPTION: Multinationally developed (British, German, Italian) all-weather, low-level penetration fighter bomber.

CONTRACTORS: Panavia (British Aerospace, Messerschmnitt-Bolkow-Blohm, Aeritalia consortium) (prime)
Turbo-Union (Rolls Royce, Motoren, Fiat consortium) (engine)
Avionica (Elliott, Elektronik System Gesellschaft, SIA consortium) (components)

SPECIFICATIONS:[1]

Dimensions:
  Length: 16.7 m
  Height: 5.71 m
  Wingspan: 8.61 m (minimum); 13.92 m (maximum)

Takeoff Weight (max): 26,300 kg

Powerplant: Turbo-Union RB-199

Ceiling: unknown

Speed: Mach 1.1 (low flight); Mach 2.2 (high profile)

Range: 370-1250 km (combat radius)

Nuclear Weapons Databook, Volume I **241**

7
TORNADO

Aerial Refueling Capability:	yes	**COMMENTS:**	TORNADO uses improved attack sensors and has significantly greater nuclear strike radius than the present F-104.[2] Operating combat radius, however, appears similar to the F-104.[3]
Crew:	2		
NUCLEAR WEAPONS:	B28, B43, B57, B61		

DEPLOYMENT:
Number Planned: 647

HISTORY:
IOC: 1981

---

1 *Luftfahrt International*, May-June 1978, p. 4191.
2 JCS, FY 1982, p. 78.
3 See Alfred Mechtersheimer, *Rustung und Politik in der Bundesrepublik, MRCA Tornado* (Honnef: Osang Verlag, 1977), pp. 100-108.

Chapter Eight
Naval Nuclear Weapons

The current program to modernize and expand U.S. Naval forces includes a wide variety of nuclear weapons systems. The build-up, according to the Department of Defense, seeks "increased and more diversified offensive striking power ... increased attention to air defense ... [and] improvements in anti-submarine warfare."[1] The plan is to build-up to a "600-ship Navy" concentrating on "deployable battle forces." Numerous new ships will be built, centered around aircraft carrier battle groups, surface groups, and attack submarines. New, more capable anti-air warfare ships, such as the TICONDEROGA (CG-47) class cruiser and BURKE (DDG-51) class destroyers, will be deployed. New nuclear weapons and launching systems, as well as nuclear capable aircraft carrier based forces, form a major part of the program.

As of March 1983, the nuclear armed ships of the U.S. Navy consisted of all 13 aircraft carriers, one battleship, all 28 cruisers, all 71 destroyers, 73 of 96 nuclear attack submarines, and 61 of 87 frigates.[2] Nuclear weapons deployed within the Navy (see Table 8.1) include anti-submarine warfare rockets (both surface (ASROC with W44) and subsurface launched (SUBROC with W55)), anti-air missiles (TERRIER with W45), and bombs and depth charges (B43, B57, and B61) used by a variety of aircraft and helicopters, both carrier and land based (see Chapters Four and Seven).[3]

The various nuclear weapons systems that are under development or are being considered for tactical naval nuclear warfare include:

- A new surface-to-air missile nuclear warhead (W81) for the STANDARD-2 missile, soon to enter production,
- A long-range, land-attack nuclear armed Sea-Launched Cruise Missile (TOMAHAWK with W80-0), about to be deployed,
- A new ASW Standoff Weapon for submarines, under development since 1976,

Table 8.1
Nuclear Capable Ships and Submarines

Ship type	Nuclear Weapon*	Nuclear Weapons Supply and Transportation
Attack submarines	SUBROC, [HARPOON], [TOMAHAWK]	None
Aircraft carriers (A-6, A-7, S-3, SH-3)	B43, B57, B61, TERRIER	Limited supply for Battle Group
Battleships	ASROC, [STANDARD-2], [HARPOON], [TOMAHAWK]	None
Cruisers	ASROC, TERRIER/ [STANDARD-2ER], [HARPOON]	None
Destroyers (SH-3)	ASROC, TERRIER/ [STANDARD-2ER], B57, [TOMAHAWK]	None
Frigates	ASROC, [STANDARD-MR], [HARPOON]	None
Amphibious ships	None	Supply for Marine Corps
Supply ships	None	Resupply for Naval Ships

* Brackets indicate weapon is not yet nuclear armed.

---

[1] DOD, FY 1984 Annual Report, p. 139.
[2] Strategic submarines which carry ballistic missiles are part of the strategic force and are described in Chapter Five.
[3] HASC, FY 1980 DOE, p. 84.

8
Naval Nuclear Weapons

Table 8.2
U.S. Naval Forces (1983)

	SHIPS			AIRCRAFT		
	Active	Reserved	Nuclear Armed	Active	Reserved	Nuclear Armed
Strategic Ballistic Missile Forces	38	—	—	—	—	—
33 Strategic Submarines	33	—	33	—	—	—
5 Support Ships	5	—	—	—	—	—
Battle Group Forces	124	—	—	1350	152	?
13 Carriers with 14 Air Wings	13	—	13	1350	152	?
28 Cruisers, 74 Destroyers	99	3	102	—	—	—
Attack Submarines/Anti-Submarine Forces	96	—	73	—	—	—
5 Diesel, 91 Nuclear Subs	96	—	73	—	—	—
24 Patrol Squadrons (+13 Reserve)	—	—	—	320	115	435
Amphibious Warfare Forces Lift	59	5	—	—	—	—
3 Marine Divisions (+1 Reserve)	3	1	4	—	—	—
3 Marine Air Wing (+1 Reserve)	3	1	4	1120	215	?
Convoy Escort	87	4	61	—	—	—
87 Frigates with LAMPS (+4 Reserve)	87	4	61	—	—	—
6 Patrol Combatants	6	—	—	—	—	—
Mine Warfare Force						
3 Minesweepers	3	—	—	18	—	—
Logistics/Combat Support	70	—	—	53	—	—
Training/Utility	—	—	—	1370	134	—

Source: SASC, FY 1983 DOD, Part 6, p. 3601; U.S. Navy.

- A new surface and air delivered ASW nuclear warhead for use from surface ships or patrol aircraft, being developed,
- A nuclear Air-to-Air Missile warhead feasibility study by DOD and DOE (for the PHOENIX air-to-air missile), underway since FY 1982,[4]
- A long-range anti-air missile based on the Advanced Strategic Air-Launched Missile (ASALM), being designed for the vertical launching system,
- An Improved Nuclear Torpedo, which has been considered,[5]
- A nuclear armed HARPOON under continuing consideration, as the HARPOON continues widespread deployment,[6] and
- A Naval nuclear projectile for shipboard artillery, which has been considered.[7]

The TOMAHAWK cruise missile represents the most significant increase in nuclear capabilities within tactical naval forces. Although designated a part of the "strategic reserve force" (see Chapter Six), it is also planned as an anti-ship and land attack weapon (in both nuclear and conventional configurations) in support of naval and amphibious operations. The nuclear armed land attack version will be initially deployed in the summer

---

4 DOD, FY 1983 RDA, p. VII-14; DOD, FY 1984 RDA, p. V-12.
5 SASC, FY 1978 ERDA, p. 30.
6 Development work on a nuclear warhead for HARPOON began in 1975 (see Chapter Six).
7 SASC, FY 1980 DOD, Part 2, p. 833.

8

Naval Nuclear Weapons

of 1984 aboard attack submarines and on the battleship *New Jersey*.

TERRIER armed anti-aircraft cruisers and destroyers will be upgraded starting in 1984 with the nuclear armed STANDARD-2 missile as part of an increase in anti-air capabilities. The TICONDEROGA (CG-47) and BURKE (DDG-51) anti-air oriented ships will be armed with the STANDARD-2. The ASROC and SUBROC weapons will be replaced by two long-range ASW weapons starting in the mid and late 1980s—the Vertical Launch ASROC (VLA) for surface ship launch from the new Vertical Launching System (VLS) and the submarine ASW Standoff Weapon (ASWSOW) to replace the Submarine Rocket (SUBROC) deployed on attack submarines. According to a FY 1984 DOD report, "study continues on other aspects of current naval nuclear capabilites—strike, anti-surface, anti-submarine, and anti-air."[8]

The development of new naval tactical nuclear warheads has been controversial. A number of comprehensive studies of Navy tactical nuclear weapons and the utility of nuclear weapons in a war at sea have been conducted. These reviews originally followed a Presidentially-directed examination of the utility and arms control impact of the new naval nuclear warheads under development in FY 1979.[9] A considerable program developed during the mid-1970s for new naval nuclear weapons. In 1978, Secretary of Defense Brown precluded the Navy from spending any more money on these weapons pending outcome of the studies.[10] During 1978, 1979, and 1980, studies were conducted by the Under Secretary of Defense for Policy and the Center for Naval Analysis (for the Chief of Naval Operations) and came to contradictory conclusions.[11] The final Defense Department determination, completed on 2 January 1981,[12] concluded that developments in naval nuclear weapons should proceed.

The Reagan Naval Program

When the Reagan Administration took office, Naval forces numbered 479 ships, and plans were approved to increase shipbuilding to attain a level of over 614 ships by the 1990s.[13] Minimum force objectives of 15 battle groups, 100 attack submarines, four battleships, 100-110 frigates, and "adequate numbers of AEGIS-capable and AEGIS-interactive battlegroup escorts" were established.[14] By FY 1990, 13 aircraft carriers, 218 surface combatants, and 91 nuclear attack submarines will be in the force. This is an increase of one carrier, 31 surface combatants and four submarines over the FY 1981 inventory.[15]

The current five year shipbuilding plans call for three new aircraft carriers, reactivating four World War II era battleships, continued building of the LOS ANGELES class (SSN-688) attack submarines and conversion to attack configuration of eight old POLARIS submarines, continued construction of the TICONDEROGA (CG-47) class cruisers, and initiation of building a new BURKE (DDG-51) class of destroyers. Offensive striking power will continue to be concentrated in large aircraft carriers with aircraft capable of delivering nuclear bombs and depth charges, and their accompanying ships (carrier battle groups) which carry the entire range of anti-air and anti-submarine weapons. Two aircraft carriers were funded in the FY 1983 budget and the "service life extension program" will continue to upgrade older carriers. The Navy's goal is to increase the number of carrier battle groups to 15, from the current 13.

According to the Navy, "considerable effort to spread our (offensive) capability among a variety of warship platforms" is being made.[16] The reactivation of four IOWA class battleships during the 1980s will be a major addition to offensive naval forces, making it "the most effective offensive surface combatant in any navy today excluding our carriers."[17] Armed with anti-air and anti-submarine nuclear weapons, and both the HARPOON and TOMAHAWK cruise missiles, these ships will form the nucleus of naval task forces, operating independently of aircraft carriers. Installing vertical launch TOMAHAWK in the LOS ANGELES (SSN-688) class attack submarine as well as giving "virtually every surface combatant its own cruise missile capability" will further increase offensive capabilities.[18]

---

8 DOD, FY 1984 RDA, p. V-12.
9 SASC, FY 1980 DOD, Part 6, p. 2849.
10 SASC, FY 1982 DOD, Part 7, p. 3897.
11 HAC, FY 1980 DOD, Part 2, p. 283; HASC, FY 1981 DOD, Part 4, Book 2, p. 2281.
12 SASC, FY 1982 DOD, Part 7, p. 3897.
13 HASC, FY 1983 DOD, Part 4, p. 33.
14 SASC, FY 1983 DOD, Part 6, p. 3631.
15 *Ibid.*, p. 3632.
16 SASC, FY 1983 DOD, Part 6, p. 3601.
17 HASC, FY 1983 DOD, Part 4, p. 29.
18 *Ibid.*

8

Attack Submarines

Attack Submarines

The Navy attack submarine force (SSNs) numbers 96 submarines (March 1983), and comprises 14 classes (11 nuclear and 3 diesel) built between 1952 and the present. A force total of 100 nuclear attack submarines is planned for the 1980s.[1] There are no additional non-nuclear submarine classes presently planned. Virtually all of the operational (non-training) submarines, except five, are nuclear powered and more than half carry nuclear weapons. Five of the nuclear attack submarine classes are SUBROC capable.[2] Seventy-three are armed with the nuclear SUBROC anti-submarine missile. Even though primarily an ASW weapon, all submarines have 21-inch torpedo tubes capable of accommodating the HARPOON and nuclear armed TOMAHAWK cruise missiles.

The latest model attack submarine class, the LOS ANGELES (SSN-688), continues under construction, with 21 deployed as of March 1983, 39 authorized through FY 1983, and 56 planned in the current shipbuilding (FY 1984-1988) program.[3] Sixteen new submarines of this class are planned under current shipbuilding plans: three in 1984, four in FY 1985 and 1986, and five in FY 1987 and 1988.

The LOS ANGELES class incorporates a larger nuclear propulsion plant than earlier submarines and contains improved sonar, electronics, and quieting equipment. It also includes a more extensive weapons capability than any other submarine, with HARPOON being added and TOMAHAWK planned.[4] Initially, TOMAHAWK will be fired from torpedo tubes, but new construction submarines starting with SSN-719 will incorporate 12 bow mounted tubes for the vertical launching system in the front part of the submarine.[5] The older STURGEON (SSN-637) class submarines will also receive TOMAHAWK, although not the vertical launch system.

The high cost of the LOS ANGELES class attack submarines and numerous contract and production problems has led to various proposals to examine alter-

Table 8.3
Nuclear Capable Attack Submarines[1]

	Class[2]				
	LOS ANGELES (SSN-688)	**LIPSCOMB (SSN-685)**	**NARWHAL (SSN-671)**	**STURGEON (SSN-637)**	**PERMIT[5] (SSN-594)**
NUMBER DEPLOYED:[3]	21[4]	1	1	37	13
NUCLEAR WEAPONS:	SUBROC [HARPOON] [TOMAHAWK]	SUBROC [HARPOON]	SUBROC [HARPOON]	SUBROC [TOMAHAWK]	SUBROC [HARPOON]
TORPEDO TUBES:	4 21-inch	4 21-inch	4 21-inch	4 21-inch	4 21-inch
SPECIFICATIONS:					
Length:	360 ft	365 ft	314 ft	—	—
Displacement:	6000 t (surface) 6900 t (submerged)	5800 t (surface) 6480 t (submerged)	4450 t (surface) 5350 t (submerged)	3640 t (surface) 4650 t (submerged)	3740-3800 t (surface) 4300-4600 t (submerged)
Draught:	32 ft 4 in	—	26 ft	29 ft 6 in	25 ft
Speed:	30+ knots (submerged)	circa 18 knots (surface) circa 25 knots (submerged)	circa 25 knots (surface) circa 30 knots (submerged)	circa 20 knots (surface) circa 30 knots (submerged)	circa 20 knots (surface) circa 30 knots (submerged)
Crew:	127	120	120	121-134	120
IOC:	1976	1974	1969	1967	1962

1 All submarines have 21-inch torpedo tubes capable of firing HARPOON and TOMAHAWK cruise missiles. TOMAHAWK will be deployed with nuclear warheads in mid-1984. HARPOON is still under consideration as a nuclear system (see Chapter Six).
2 TULIBEE, SKIPJACK, and SKATE classes, and converted POLARIS, do not carry nuclear weapons.
3 As of March 1983.
4 Only SSN-699 and before (12 submarines) are SUBROC armed; HAC, FY 1983 DOD, Part 2, p. 166.
5 Lead ship of the class, the *Thresher*, was lost at sea on 10 April 1963.

---

1 HASC, FY 1983 DOD, Part 4, p. 228.
2 LOS ANGELES, LIPSCOMB, NARWHAL, STURGEON, and PERMIT classes.
3 HASC, FY 1983 DOD, Part 4, p. 229; DOD, FY 1984 Annual Report, p. 149.
4 By March 1981, 20 submarines are converted to carry the HARPOON; HAC, FY 1982 EWDA, Part 7, p. 303.
5 DOD, FY 1984 Annual Report, p. 144.

8
Attack Submarines

natives. A smaller and slightly slower, more specialized submarine—designated FA-SSN (fleet attack SSN)—was examined during the Carter Administration, but has received little attention in the Reagan Administration.

Instead, a new submarine "to capture the latest advances in technology" is in the preliminary stages of research, with initial construction envisioned for the late 1980s.[6]

---

6 *Ibid.*, p. 150.

8

LOS ANGELES Class Submarine

LOS ANGELES Class Submarines (SSN-688)

Figure 8.1 U.S.S. Los Angeles (SSN-688) attack submarine.

DESCRIPTION: High-speed, nuclear powered, attack submarine.

CONTRACTORS: Newport News Shipbuilding
Newport News, VA
Electric Boat Division,
General Dynamics
Groton, CT

SPECIFICATIONS:
Length: 360 ft

Displacement: 6000 t (surface); 6900 t (submerged)

Draught: 32 ft 4 in

Propulsion: water-cooled pressurized (S6G) nuclear reactor; core life of 10-13 years

Speed: 30+ knots (submerged)

Crew: 127 personnel

Armament: 4 21-inch torpedo tubes

NUCLEAR WEAPONS:[1] SUBROC (W55);[2] HARPOON capable; TOMAHAWK cruise missile from torpedo tubes; will incorporate 12 vertical launching system tubes starting with SSN-719[3]

Nuclear Weapons Databook, Volume I **249**

8
LOS ANGELES Class Submarine

Fire Control: Mk-113-10, Mk-117; new sonar and fire control system reduces time required to compute passive ranges

DEPLOYMENT:
Number Deployed: 21 (March 1983); at least 42 anticipated

HISTORY:
IOC: 1976

FY 1970   first SSN-688 authorized

Nov 1976   SSN-688 delivered to Navy

end 1981   15 SSN-688 class commissioned[4]

COST:
Program Cost: $29,375.6 m (Dec 1982)

Unit Cost: $500-$600 m

FY	Number Procured	Total Appropriation ($ million)
1978 & prior	32	6960.3[5]
1979	1	
1980	2	863.3
1981	2	1139.5
1982	2	1612.7
1983	2	1688.3
1984	3	2174.1
1985	4	3057.2

COMMENTS: SSN-688 class was a follow on to the SSN-637 class. New submarines (SSN-721 hull and later) will incorporate 12 vertical launch system tubes in the existing forward main ballast tank area for TOMAHAWK and other future weapons.[6]

---

1 The HARPOON designed for use in the SSN-688 has no potential nuclear capability; ACDA, FY 1979 ACIS, p. 165.
2 ACDA, FY 1979 ACIS, p. 164.
3 SSN-719 and thereafter will incorporate VLS; SASC, FY 1983 DOD, Part 2, p. 1063; DOD, FY 1984 Annual Report, p. 144.
4 JCS, FY 1982, pp. 89-90.
5 ACDA, FY 1979 ACIS, p. 165.
6 HASC, FY 1982 DOD, Part 3, pp. 174-176.

8
Aircraft Carriers

Aircraft Carriers

U.S. aircraft carriers are the world's largest warships. They serve as floating airfields and contain the most complete array of offensive and defensive weapons. The Navy operates 13 aircraft carriers, four of which are nuclear powered (one carrier is undergoing an extensive modernization and is out of commission). The ships are multi-mission attack platforms from which 80-100 aircraft and helicopters operate. Three also have nuclear capable TERRIER surface-to-air missile systems. At any one time, four carriers are forward deployed: in the Mediterranean Sea, the Far East, and the Indian Ocean. The Reagan Administration's naval program has led to revived support for the nuclear-powered large deck carrier. The nuclear powered carriers have an estimated steaming life of some 13 years and one million miles before refueling. The original nuclear powered carrier, the *Enterprise* (CVN-65), was commissioned in 1961 and has eight nuclear reactors. The three later carriers, *Nimitz* (CVN-68), *Eisenhower* (CVN-69), and *Carl Vinson* (CVN-70), each have two reactors.

A number of the older carriers are slated to undergo a "service life extension program" (SLEP) to add 15 additional years of operation. The *Saratoga* (CV-60) is the first ship to undergo SLEP and will be completed in FY 1983. The second carrier, the *Forrestal* (CV-59), will begin SLEP in FY 1985, and one carrier will enter the SLEP every two years thereafter. Three additional nuclear powered aircraft carriers of the NIMITZ class (CVN-71, CVN-72, and CVN-73) are under construction. The last two were included in the FY 1983 Defense Budget and will be deployed in 1990 and 1992, respectively. Another NIMITZ class ship is planned for the FY 1988 budget. Both the *Midway* (CV-41) and *Coral Sea* (CV-43), carriers which saw service in the Second World War, will probably be retired in 1990. The Navy's goal, with new construction and SLEP and taking retirements into account, is to have 15 carriers by the 1990s.

The typical carrier air wing is composed of 90 aircraft, including A-6 and A-7 attack aircraft, F-4 or F-14 fighter aircraft and interceptors, E-2 early warning aircraft, S-3 anti-submarine patrol aircraft, SH-3 anti-submarine patrol helicopters, and a variety of transport, refueling, and intelligence collection planes (see Chapter Seven for descriptions of nuclear capable aircraft). Naval aircraft—A-6, A-7, and F-4—are currently certified to carry B43, B57, and B61 nuclear bombs. These range in yield from five kilotons to over one megaton. The B57 is also a nuclear depth bomb carried aboard the S-3 and SH-3. No nuclear armed air-to-air or air-to-surface missiles are now deployed, although a nuclear warhead for the

Table 8.4
Nuclear Capable Aircraft Carriers

	Class				
	NIMITZ	**ENTERPRISE**	**KITTY HAWK**	**FORRESTAL**	**MIDWAY**
NUMBER DEPLOYED:	3[1]	1	4	4[2]	2
NUCLEAR WEAPONS:	B43, B57, B61	B43, B57, B61	B43, B57, B61 (TERRIER)	B43, B57, B61	B43, B57, B61
SPECIFICATIONS:					
Length:	1092 ft	1123 ft	1047-1062 ft	1039-1046 ft	979 ft
Displacement (t):					
(standard)	81,600	75,700	61,000	60,000	52,500
(combat load)	93,400	89,600	80,800-82,000	78,000	64,000
Nuclear Reactors:	2 A4W	8 A2W	—	—	—
Speed (knots):	30+	30+	30+	33	33
Crew:	3073-3131	3157	2879-2924	2865-2911	2523-2613
Air Wing Personnel:	2627	2628	2500	2400	1945
IOC:	1975	1961	1961	1955	1945
COST ($ million):	—	451.3	263.8 avg.	200.6 avg.	?

1 Three ships of the NIMITZ class (CVN-71, CVN-72, and CVN-73) are also under construction.

2 The *Saratoga* is undergoing service life extension and is out of commission.

8

Aircraft Carriers

Table 8.5
Nuclear Weapons and Systems on Aircraft Carriers

	TERRIER	A-6*	A-7*	S-3**	SH-3**
Carl Vinson		x	x	x	x
Nimitz		x	x	x	x
Dwight D. Eisenhower		x	x	x	x
Enterprise		x	x	x	x
Kitty Hawk	x	x	x	x	x
Constellation	x	x	x	x	x
America	x	x	x	x	x
John F. Kennedy		x	x	x	x
Forrestal[1]		x	x	x	x
Saratoga[2]		x	x	x	x
Ranger		x	x	x	x
Independance		x	x	x	x
Midway		x	x		
Coral Sea		x	x		

* Attack planes carry B43, B57, and B61 nuclear bombs.
** ASW planes (and SH-3 helicopter) carry B57 bomb. New version of S-3 (S-3B) will carry HARPOON.

1 The *Forrestal* will enter service life extension in FY 1985.
2 The *Saratoga* is undergoing service life extension and out of commission until FY 1983.

PHOENIX air-to-air missile, carried aboard F-14 long-range interceptors, is being developed.

Future developments in the weaponry aboard aircraft carriers include the introduction of F-18 aircraft to replace A-7 and F-4 aircraft, adoption of the SH-60B LAMPS helicopter for carrier use (designated SH-60F) to replace the SH-3, and modernization of the S-3 with an S-3B starting in the late 1980s.

NIMITZ Class Aircraft Carriers (CVN-68)

Figure 8.2 U.S.S. *Nimitz* (CVN-68) with F-4 PHANTOM, A-7 CORSAIR, and A-6 INTRUDER nuclear capable aircraft aboard.

DESCRIPTION:	Large deck, nuclear powered, multi-mission Attack Aircraft Carrier.
CONTRACTORS:	Newport News Shipbuilding Company
SPECIFICATIONS:	
Length:	1092 ft
Displacement:	81,600 t (standard); 93,400 t (combat load)
Draught:	37 ft
Propulsion:	2 water-cooled pressurized (A4W) nuclear reactors
Speed:	30+ knots
Crew:	3073-3131 (ship); 2627 (air wing)
NUCLEAR WEAPONS:	B28, B43, B61 nuclear bombs; B57 nuclear depth charges/bombs
NUCLEAR CAPABLE AIRCRAFT:	A-6, A-7, S-3, SH-3; Marine Corps A-4 and A-6 also periodically operate off aircraft carriers
DEPLOYMENT:	
Number deployed:	3 (1983)
HISTORY:	
IOC:	1975

Nuclear Weapons Databook, Volume I 253

8
NIMITZ Class Aircraft Carriers

COST:
CVN-71 Program: $2564.4 m (Dec 1982);
CVN-72/73 Programs: $7267.2 m (Dec 1982)

FY	Number Procured	Total Acquisition ($ million)
1982	-	554.5
1983	2	6554.1
1984	1	98.2

8
Surface Combatant Ships

Surface Combatant Ships

The wide variety of naval warfare responsibilities requires a mix of surface combatants, designed either for single or multiple tasks such as strike, anti-air, anti-ship, or anti-submarine warfare. Surface combatants are used as part of battle groups supporting aircraft carriers, surface action groups centered around the newly reactivated battleships, amphibious forces, convoys, and underway replenishment groups. There are currently four different types of surface combatants: battleships, cruisers, destroyers, and frigates. Battleships are the largest surface warships except for aircraft carriers and provide offensive strike with a wide variety of missile systems and guns. They can also operate in offensive support of amphibious operations. Cruisers and destroyers take part in all naval operations and can team together to form surface action groups or operate independently. These ships are designed to operate in a "high threat" environment with a wide array of both offensive and self-defense weapons, command and control facilites, engineering redundancy, and endurance. Frigates, the smallest of the surface combatants and less capable and more costly than cruisers or destroyers, are designed as open ocean escorts for anti-submarine warfare.

Two nuclear weapon systems are currently deployed with surface combatant ships: the anti-submarine warfare ASROC missile and the TERRIER surface-to-air missile. Both weapons are being upgraded with new nuclear armed replacements. The ASROC will be replaced with the vertical launch ASROC (VLA) for deployment on those ships which will incorporate the vertical launching system. The TERRIER will be replaced by the STANDARD-2 nuclear air defense missile, which will supply the current TERRIER armed ships and will arm other destroyers and cruisers.

Table 8.6
Nuclear Capable Surface Ships (1983)

Type/Class	Number of Ships	Nuclear Weapons*
Battleships	1	[HARPOON], [TOMAHAWK]
Cruisers	(28)	
TICONDEROGA	1	ASROC, [HARPOON], [TOMAHAWK], [STANDARD]
VIRGINIA	4	ASROC
CALIFORNIA	2	ASROC
TRUXTON	1	ASROC, TERRIER/[STANDARD-ER]
BELKNAP	1	ASROC, [HARPOON] TERRIER/[STANDARD-ER]
JOSEPHUS DANIELS	8	ASROC, TERRIER/[STANDARD-2ER], [HARPOON]
BAINBRIDGE	1	ASROC, [HARPOON] TERRIER/[STANDARD-ER]
LEAHY	9	ASROC, TERRIER/[STANDARD-ER]
LONG BEACH	1	ASROC, TERRIER/[STANDARD-ER]
Destroyers	(71)	
BURKE (DDG-51)	0	[VL ASROC], [STANDARD-ER], [HARPOON], [TOMAHAWK]
KIDD	4	ASROC, [HARPOON]
FARRAGUT	10	ASROC, TERRIER/[STANDARD-ER]
DECATUR	1	ASROC
CHARLES F. ADAMS	23	ASROC
SPRUANCE	31	ASROC, SH-3 Helicopter (B57)
FORREST SHERMAN	2	ASROC
Frigates	(87)	
OLIVER HAZZARD PERRY	26	[HARPOON]
BROOKE	6	ASROC
KNOX	42	ASROC, [HARPOON]
GARCIA	10	ASROC
GLOVER	1	ASROC
BRONSTEIN	2	ASROC
Patrol Combatants	(6)	
PEGASUS	6	[HARPOON]

* Potential and future nuclear weapons are indicated in brackets.

8

Battleship Reactivation

Two anti-ship and anti-land cruise missiles—HARPOON and TOMAHAWK—are also planned for widespread deployment. Although the HARPOON will be the most widely deployed, plans to arm this short-range cruise missile with a nuclear warhead seem to have been deferred. The dual capable TOMAHAWK cruise missile, which will begin deployment in 1984, is planned for a large number of surface ships.

The present shipbuilding plan (FY 1984-1988) includes 24 new surface combatants and three battleship reactivations. The largest program is for the TICONDEROGA (CG-47) class cruiser, three of which are planned for every year except for 1988 when two are planned. One SPRUANCE (DD-963) class destroyer will be built in FY 1988. A new class of ship, the BURKE (DDG-51) class destroyer, will begin construction in FY 1985. One ship is planned for that year, three in FY 1987, and five in 1988. Nuclear powered cruisers will stabilize at six ships including those already under construction. The requirement exists for 27 TICONDEROGA (CG-47) class cruisers. Other newer destroyers will include four KIDD (DDG-993) class ships originally ordered by Iran, and 37 SPRUANCE (DD-963) class ships. Frigate requirements total 101 ships, although the present five year shipbuilding program does not include any new construction.[1]

Battleship Reactivation

The Reagan Administration has instituted a plan to reactivate four World War II era IOWA class battleships and outfit them with nuclear armed TOMAHAWK and potentially nuclear armed HARPOON cruise missiles, as well as with anti-air and anti-submarine weapons. The four 59,000-ton battleships are the *New Jersey* (BB-62), the *Iowa* (BB-61), *Missouri* (BB-63), and *Wisconsin* (BB-64). All of the ships were commissioned in 1943-1944, and were in combat during World War II and Korea. The *New Jersey* was reactivated in 1968-1969 during the Vietnam War. The reactivation of these ships will mean a significant increase in surface ship strike capabilities. They will operate as the lead ships of surface groups or as part of an amphibious assault operation.

Eventually the nuclear capbility of the heavily armed ships will be improved when three of the nine 16-inch guns and armored boxed launchers on the battleships are removed and the vertical launching system (VLS) is installed (see Vertical Launching System later in this chapter). The VLS will have the capability of firing a variety of nuclear armed missiles, including STANDARD-2, vertical launch ASROC, HARPOON, and TOMAHAWK. Initially, 32-60 missiles will be carried for the VLS, with long term plans to equip firing modules for 320-400 missiles.[2]

The first of the reactivated ships, the *New Jersey*, was commissioned in December 1982. Its first deployment, scheduled for early 1983, will test the HARPOON and TOMAHAWK missiles which are deployed in armored boxed launchers. The *Iowa*, the next ship to be reactivated, was funded in FY 1983, and plans are to fund the third ship, the *Missouri*, in FY 1985, and the fourth ship, the *Wisconsin*, in FY 1986.[3] Battleship reactivation will cost $1876.6 million as of 31 December 1982.

Figure 8.3 U.S.S. *New Jersey* (BB-62) battleship.

---

1 HAC, FY 1983 DOD, Part 4, p. 122.
2 AW&ST, 2 March 1981, p. 15.

3 DOD, FY 1984 Annual Report, p. 143.

8
TICONDEROGA Class Cruisers

Figure 8.4 U.S.S. *Ticonderoga* (CG-47) on high speed maneuvers during sea trials.

TICONDEROGA Class Cruisers (CG-47)

The TICONDEROGA is a new class of cruiser employing the hull and gas turbine propulsion system of the SPRUANCE (DD-963) class destroyer and the new AEGIS fire control and anti-air defense system. The TICONDEROGA ships will be the most heavily armed cruisers in the Navy. Although their primary mission will be to provide air-defense for carrier task groups, they will also be highly capable of mounting attacks against land, surface ships, and submarines. The TICONDEROGA class will operate with aircraft carrier task/battle groups, amphibious task groups, underway replenishment groups, or convoys.

The primary weapon for the AEGIS system is the nuclear capable STANDARD-2 (SM2) missile system. The ship also carries the nuclear armed ASROC and the potentially nuclear capable HARPOON anti-ship missile. Conventional weapon systems include the PHALANX anti-aircraft gun, two 5-inch deck guns, and depth charges. Two SEAHAWK surveillance and ASW helicopters are also carried.

8

TICONDEROGA Class Cruisers

TICONDEROGA Class Cruisers (CG-47)

Figure 8.5 U.S.S. *Ticonderoga* (CG-47) cruiser.

DESCRIPTION:	Specialized anti-aircraft, but also anti-ship or land attack battle group, guided missile escorts.	Speed:	30+ knots
		Crew:	360
CONTRACTORS:	Litton/Ingalls Pascagoula, MS General Electric (propulsion)	**NUCLEAR WEAPONS**:	ASROC;[1] HARPOON, TOMAHAWK capable, STANDARD-2MR;[2] 122 missiles in vertical launching system, excluding first five ships of class
SPECIFICATIONS:		Fire Control:	Mk-7 AEGIS; Mk-99 missile directors, Mk-116 ASW
Length:	563 ft 4 in		
Displacement:	9100 tons	**DEPLOYMENT**:	
Draught:	31 ft	Number Deployed:	1 (1983); 25 by the early 1990s
Propulsion:	gas turbines	**HISTORY**: IOC:	Jan 1983

8
TICONDEROGA Class Cruisers

Jan 1980   keel of *Ticonderoga* (CG-47) laid

COST:
Program Cost: $23,033.1 m (Dec 1982)

FY	Number Procured	Total Appropriation ($ million)
1980	1	834.4
1981	2	1941.5
1982	3	2927.7
1983	3	2926.8
1984	3	3707.3

COMMENTS: Later TICONDEROGA class ships (CG-52 on) will have two Ex-41 vertical launch sytems in place of Mk-26 launchers.

---

1 CG-47/48 will fire ASROC from deck Mk-26 launcher; CG-49 and later ships will fire vertical launch ASROC (VLA) from Ex-41 vertical launcher.
2 HARPOON is fired from Mk-141 launchers in CG-47/48; STANDARD-MR is fired from Mk-26 Mod 1; all will be converted for firing from Ex-41 vertical launcher in later ships.
3 VLS will be included in all but the first five CG-47 class cruisers; DOD, FY 1984 Annual Report, p. 144.
4 HASC, FY 1983 DOD, Part 4, p. 59.

8
Nuclear Capable Cruisers

Table 8.7
Nuclear Capable Cruisers

	Class				
	TICONDEROGA (CG-47)	**VIRGINIA (CGN-38)**	**CALIFORNIA (CGN-36)**	**TRUXTON (CGN-35)**	**LONG BEACH (CGN-9)**
NUMBER DEPLOYED:	1	4	2	1	1
NUCLEAR WEAPONS:	ASROC [STANDARD-MR] [HARPOON] [TOMAHAWK]	ASROC [STANDARD-MR] [HARPOON] [TOMAHAWK]	ASROC [STANDARD-MR] [HARPOON] [TOMAHAWK]	ASROC TERRIER [STANDARD-2ER] [HARPOON]	ASROC TERRIER [STANDARD-2ER] [HARPOON]
SPECIFICATIONS:					
Length:	563 ft 4 in	585 ft	596 ft	564 ft	721 ft
Displacement: (combat)	9100 t	11,000 t	10,150 t	9200 t	17,350 t
Speed:	30+ knots	30+ knots	30+ knots	30+ knots	30+ knots
Crew:	360	519	533	538	983
IOC:	1983	1976	1974	1967	1961

	Class			
	BELKNAP (CG-26)	**JOSEPHUS DANIELS (CG-27)**	**BAINBRIDGE (CGN-25)**	**LEAHY (CG-16)**
NUMBER DEPLOYED:	1	8	1	9
NUCLEAR WEAPONS:	ASROC TERRIER [STANDARD-2ER] [HARPOON]	ASROC TERRIER [STANDARD-2ER] [HARPOON]	ASROC TERRIER [STANDARD-2ER] [HARPOON]	ASROC TERRIER [STANDARD-2ER]
SPECIFICATIONS:				
Length:	547 ft	547 ft	565 ft	533 ft
Displacement: (combat)	7930 t	7930 t	8580 t	7800 t
Speed:	33 knots	33 knots	30+ knots	32 knots
Crew:	418	418	circa 480	circa 415
IOC:	1964	1965	1962	1962

8
BURKE Class Destroyers

BURKE Class Destroyers (DDG-51)

DESCRIPTION: Guided missile area anti-air specialized destroyer with anti-submarine, anti-air, and anti-ship missions.

CONTRACTORS: Not yet chosen

SPECIFICATIONS:

Length: circa 500 ft

Displacement: 8500 t[1]

Draught: unknown

Propulsion: gas turbine

Speed: circa 30 knots

Crew: 304

NUCLEAR WEAPONS: two vertical launcher systems; VL ASROC, STANDARD-2 (MR), TOMAHAWK,[2] cannister launch HARPOON (eight)

Fire Control: Mk-99 missile directors

DEPLOYMENT:
Number Planned: 49,[3] 60,[4] 63[5]

HISTORY:

IOC: 1995

FY 1985: construction slated to begin on first DDG-51 ship

COST:
Program Cost: $10,953.5 m (Dec 1982)

Unit Cost: $729 m[6]

FY	Number Procured	Total Appropriaton ($ million)
1982	-	52.9
1983	-	138.3
1984	-	210.5
1985	1	1493.5

COMMENTS: BURKE class is planned replacement for a number of destroyers and cruisers scheduled for retirement by 2000; CG-16, CG-26, DDG-2 and DDG-37 classes. Ship will be deployed as part of battle groups, battleship surface action groups, amphibious groups, and underway replenishment groups.[7]

---

1 HASC, FY 1983 DOD, Part 4, p. 73.
2 SASC, FY 1983 DOD, Part 6, p. 3815.
3 Norman Polmar, *Ships and Aircraft of the U.S. Fleet*, 12th Ed., op. cit., p. 93.
4 DOD, FY 1984 Annual Report, p. 146.
5 *Ibid.*, p. 122.
6 HASC, FY 1983 DOD, Part 4, p. 151.
7 HASC, FY 1983 DOD, Part 4, p. 122.

8

Nuclear Capable Destroyers

Figure 8.6 U.S.S. *Spruance* (DD-963) destroyer.

Table 8.8
Nuclear Capable Destroyers

	Class						
	BURKE (DDG-51)	**KIDD (DDG-993)**	**FARRAGUT (DDG-37)**	**DECATUR (DDG-31)**	**ADAMS (DDG-2)**	**SPRUANCE (DD-963)**	**FORREST SHERMAN (DD-931)**
NUMBER DEPLOYED:	—	4	10	1	23	31	2
NUCLEAR WEAPONS:	(VL ASROC) [HARPOON] [STANDARD] [TOMAHAWK]	ASROC [HARPOON]	ASROC [HARPOON] [STANDARD-2ER]	ASROC	ASROC [HARPOON]	ASROC [HARPOON]	ASROC
SPECIFICATIONS:							
Length:	circa 500 ft	563 ft	512 ft	407 ft	437 ft	563 ft	418-425 ft
Displacement:	8500 t	8140 t	5800 t	4150 t	3370 t	7800 t	4050 t
Speed:	circa 30 knots	30+ knots	33 knots	32.5 knots	31.5 knots	30+ knots	32.5 knots
Crew:	circa 325	338	399	333-344	333-350	288-302	319-332
IOC:	1995	1981	1960	1967	1960	1975	1955

OLIVER HAZARD PERRY Class Frigates (FFG-7)

DESCRIPTION: Open ocean escort ships for convoys, naval ships, and amphibious groups.

CONTRACTORS: Bath Iron Works
Todd Shipyards
San Pedro, CA; Seattle, WA

SPECIFICATIONS:
Length: 445 ft

Displacement: 3710 t

Draught: 24 ft 6 in

Propulsion: gas turbines

Speed: 28 knots (sustained)

Crew: 179

NUCLEAR WEAPONS: HARPOON, STANDARD-MR

Fire Control: Mk-13 weapon direction system, Mk-92 weapon fire control system

DEPLOYMENT:
Number Deployed: 26 (1983); 53 planned through FY 1984-1988 shipbuilding plan

HISTORY:
IOC: 1977

Jun 1976 — keel of *Oliver Hazard Perry* laid

FY 1984 — no new FFG-7 class frigates requested, ending procurement program

COST:
Program Cost: $9822.3 m (Dec 1982)

Table 8.9
Nuclear Capable Frigates

	\multicolumn{6}{c}{Class}					
	OLIVER HAZARD PERRY (FFG-7)	**BROOKE (FFG-1)**	**GLOVER (FFO1098)**	**KNOX (FF-1952)**	**GARCIA (FF-1040)**	**BRONSTEIN (FF-1037)**
NUMBER DEPLOYED:	26	6	1	42	10	2
NUCLEAR WEAPONS:	[STANDARD-MR] [HARPOON]	ASROC [STANDARD-MR]	ASROC	ASROC [HARPOON]	ASROC	ASROC
SPECIFICATIONS:						
Length:	445 ft	414 ft	414 ft	438 ft	414 ft	371 ft
Displacement:	3710 t	3245 t	3426 t	4100 t	3400 t	2650 t
Speed:	28 knots	27 knots	27 knots	27+ knots	27 knots	24 knots
Crew:	179	253-256	316	257-266	258-270	209
IOC:	1977	1966	1965	1969	1964	1963

8

Naval Nuclear Missile Systems

Naval Nuclear Missile Systems

The nuclear armed ASROC and TERRIER naval missiles, and potential and future nuclear armed missiles (HARPOON, STANDARD-2, and TOMAHAWK), are fired from a variety of shipboard launchers. These launchers weigh between 9000-500,000 pounds, and carry from four to 60 missiles (see Table 8.10). They are designed as either launcher "boxes" in which a small number of missiles are stored and launched from within the tubes of a self-contained unit, or from either single or twin "rail" launchers in which below deck "magazines" of missiles are fed up to the launcher rails and then fired. Another launcher, called an armored box launcher, is being specially configured for TOMAHAWK Sea-Launched Cruise Missiles.

Vertical Launching System:[1]

A new modular magazine and launch system designed to carry and launch nuclear capable STANDARD-2, vertical launch ASROC, ASW Standoff Weapon, HARPOON, and TOMAHAWK missiles below deck is being developed by the Navy and Martin Marietta for installation in a wide variety of ships (see Table 8.11). The Vertical Launching System (VLS) Ex-41 design allows for a higher number of missiles to be available for fire than with rail and box launchers; a simultaneous anti-ship, anti-air and anti-submarine capability; increased magazine size; reduced manning; increased rate of fire for surface-to-air missiles; reduced reaction time; improved reliability; and enhanced survivability over the current shipboard launchers.[2]

The VLS can be arranged in basic eight-cell modules with eight modules and 61 missiles comprising a ship's magazine. Seven modules would be for missiles and one would contain a crane module with only five cells, allowing for underway replenishment and reload. The VLS would be built into the ship and be flush with the deck. Each cell would carry and launch one missile. The missile would be shipped and launched from a canister which would fit into the cell. The canister would be reusable. The 61-missile VLS (Ex-41) magazine is designed to fit into the same space as the 44-missile Mk-26 Mod 1 guided missile launching system.

The surface ship VLS launcher required lengthening the launcher by three feet to accommodate TOMAHAWK. VLS would allow for naval missiles of 259 inches long, 22 inches in diameter, and weighing 2500 pounds. Martin Marietta has developed a derivative of the Advanced Strategic Air-Launched Missile (ASALM) with booster and folding fins as an "outer air battle missile" for firing from VLS against enemy jammers and cruise missile aircraft.[3]

Anti-Submarine Nuclear Weapon Systems

Although a nuclear armed torpedo (ASTOR) was once developed as a last ditch defensive weapon for submarines, the short range of torpedos fired in the water meant that the launching submarine might not escape

Figure 8.7 Destroyer U.S.S. *Agerholm* (DD-826) after firing a nuclear-armed **ASROC (RUR-5A)** missile during exercise Swordfish on 11 May 1962. The surface effects of the underwater nuclear blast are seen (left), and the peak is seen (right).

---

1 AW&ST, 30 March 1981, p. 24; HASC, FY 1981 DOD, Part 4, Book 2, pp. 1438-1446.
2 DOD, FY 1984 Annual Report, p. 146.
3 Information provided by Martin Marietta.

8
Naval Nuclear Missile Systems

Table 8.10
Nuclear Capable Shipboard Missile Launchers[1]

Type	Missiles*	Configuration	Ships
Mk-9 Mod 1	60 TERRIER	Twin	CG-6-7
Mk-10 Mod 0	40 TERRIER/[STANDARD-ER]	Twin	DDG-37-46
Mk-10 Mod 1	40 TERRIER/[STANDARD-ER]	Twin	CGN-9
Mk-10 Mod 2	80 TERRIER/[STANDARD-ER]	Twin	CGN-9
Mk-10 Mods 3-4	40 TERRIER/[STANDARD-ER]	Twin	CV-63, -64, -66
Mk-10 Mods 5-6	40 TERRIER/[STANDARD-ER]	Twin	CG-16-24, CGN-25
Mk-10 Mod 7	60 TERRIER/[STANDARD-ER]/ASROC	Twin	CG-26-34, CGN-25
Mk-10 Mod 8	60 TERRIER/[STANDARD-ER]/ASROC	Twin	CGN-35
Mk-11 Mod 0	[42 HARPOON]	Twin	DDG-2-14
Mk-13 Mod 4	[40 HARPOON]	Single	FFG-7
Mk-16 Mods 1-6	8 ASROC[2]	8-Tube	CG, DD, FF
Mk-26 Mod 0	[24] ASROC	Twin	CGN-38-41
Mk-26 Mod 1	[44] ASROC	Twin	CGN-38-41, DDG-47
Mk-26 Mod 2	[64] ASROC	Twin	CGN-42 Class
Mk-112	[HARPOON]		
Mk-140 Mod 0	[4 HARPOON]	4-Tube	PHM-1, 3-6
Mk-141 Mod 0	[4 HARPOON]/[TOMAHAWK]	4-Tube	CG, DD
Ex-41	[TOMAHAWK] [ASW Standoff Weapon] [STANDARD-2]	Vertical Launching System	BB, SSN-719-, CG-52-, DD-963, DDG-51

* Brackets indicate weapon is not yet nuclear capable.
1 *Ships and Aircraft of the U.S. Fleet*, 11th Ed., p. 299.
2 Reloads available in some ships.

the effects of a nearby nuclear explosion. The destructive capability of nuclear weapons meant that a new type of nuclear weapon had to be developed for anti-submarine warfare (ASW). The new weapon, combining the launch characteristics of a torpedo with the free flight performance of a guided missile, resulted in longer range than that of a torpedo, whether fired from a surface ship or underwater from a submarine. SUBROC was the first tactical weapon system to use the new concept of underwater launch followed by rocket motor ignition, guided ballistic out-of-water airborne trajectory, and then water entry and underwater detonation.[1]

Today two nuclear armed ASW rocket are deployed: the SUBROC and the surface ship-launched ASROC. Both weapons are armed with low yield nuclear warheads. While the ASROC is dual capable, the SUBROC is only nuclear armed. A third ASW nuclear weapon, the low yield B57 nuclear depth bomb (see Chapter Three), is carried by the land-based P-3 aircraft and the aircraft carrier-based S-3 and SH-3.

Two new long-range ASW weapons to replace ASROC and SUBROC are being developed. They will enable naval forces to continue to destroy enemy submarines yet remain outside the range of enemy torpedos. The Vertical Launch ASROC (VLA) program was started in FY 1984 to convert the ASROC for launching from the new vertical launching system by 1987 or 1988.[2] VLA will provide an updated anti-submarine warfare rocket for vertical launch capable of faster reaction, greater range, and increased kill probability than the current ASROC. A new submarine ASW Standoff Weapon (ASWSOW) is also being developed to replace SUBROC deployed with attack submarines. The ASWSOW is projected for 1989-1990. Beginning in 1984, the number of SUBROC armed attack submarines will begin to decrease from the present strength of about 64.[3] SUBROC phaseout will take place over an extended period, due to rocket motor burnout and lack of nuclear warheads.[4]

---

1 Information provided by Goodyear Aerospace and contained in Goodyear *Profile*, Vol. VII, No. 1, 1969; Fall 1981.
2 DOD, FY 1984 Annual Report, p. 153; SASC, FY 1983 DOD, Part 6, p. 4074.
3 SASC, FY 1983 DOD, Part 6, p. 4075.
4 HASC, FY 1983 DOD, Part 4, p. 254.

8

Naval Nuclear Missile Systems

Table 8.11
Vertical Launching System Platforms

Ship Type	Number of Missiles	Type	Schedule
Battleships	36-60	TOMAHAWK	initial loading
(Reactivated)	320-400	TOMAHAWK	long term loading
		VL ASROC	
		STANDARD-2	
		HARPOON	
Cruisers			
CG-47 class	122	HARPOON	all but first five
	(61 + 61)	STANDARD-2	starting in FY 1988
		TOMAHAWK	
		VL ASROC	
CGN-36/38 class		VL ASROC	
Destroyers			
DDG-51 class	90	TOMAHAWK	FY 1990 IOC
	(61 + 29)	VL ASROC	
DDG-993 class		VL ASROC	
DD-963 class	61	TOMAHAWK	retrofitted in all,
		VL ASROC	starting in 1985
Frigates			
FF-1052 class		VL ASROC	
Submarines			
SSN-688 class	12	TOMAHAWK	retrofit by 1985
		ASW Weapon	
SSN-719 class	12	TOMAHAWK	fitted in SSN-719 and
		ASW Weapon	later, first delivery due in 1985

Two nuclear warhead development programs for antisubmarine warfare are in Phase 2, Feasibility Study, within DOE for FY 1984. The first is a subsurface delivered ASW Standoff Weapon warhead for the ASWSOW, and the second is a surface and air delivered anti-submarine warfare weapon, earmarked as a new common warhead for a B57 replacement and an ASROC replacement.

8
ASROC Missile

ASROC (RUR-5A)

Figure 8.8 **ASROC (RUR-5A)** missile being fired from U.S.S. Brooke (DDG-1).

Figure 8.9 **Mk-16** eight tube box launcher aboard Naval ship. This launcher is capable of firing the ASROC or HARPOON missiles.

DESCRIPTION:	Widely deployed ship-launched, unguided, range-controlled nuclear capable Anti-Submarine ROCket (ASROC) depth charge.
CONTRACTORS:	Honeywell, Inc. (prime) Naval Weapons Center (nuclear depth charge section)
SPECIFICATIONS:	
Length:	15 ft (180 in)
Diameter:	12 in[1]
Stages:	1
Weight at Launch:	under 1000 lb; 570 lb[2]
Propulsion:	solid propellant rocket motor
Guidance:	unguided in flight, depth charge descends to predetermined depth before detonating
Throwweight/ Payload:	unknown
Range:	1-6 nm[3]
DUAL CAPABLE:	yes
NUCLEAR WARHEADS:	one W44 on Mk-17 nuclear depth charge; 1 Kt range (see W44)
DEPLOYMENT:	
Launch Platform:	Mk-112 and other TERRIER launchers
Number Deployed:	850 nuclear warheads (1983)
Location:	aboard 65 frigates (FF, FFG), 78 destroyers (DD, DDG), and 27 cruisers (CG, CGN);[4] fitted to all Navy major surface ships until the OLIVER HAZARD PERRY class (FFG-7) frigates
HISTORY:	
IOC:	1961
1956	development began
Retirement Plans:	to be replaced with the Vertical Launch ASROC (VLA)

Nuclear Weapons Databook, Volume I **267**

8
ASROC Missile

TARGETING:

Types: primarily submarines, also capable against surface ships and land targets

COST:

FY	Number Procured	Total Appropriation ($ million)
1980	-	11.9
1981	-	5.3
1982	-	3.9[5]

COMMENTS: Enroute to target, the nuclear armed ASROC sheds its rocket motor at a predetermined signal and a steel band holding the airframe together is severed by a small explosive charge, allowing the depth charge to drop into the water.[6] Vertical Launch ASROC will be procured for CG-47, CGN-36/38, DD-963, DDG-993, DDG-51 and FF-1052 classes.

---

1 *The World's Missile Systems*, 6th Ed., p. 324.
2 *Ibid.*
3 *Ibid.*
4 Norman Polmar, 12th Ed. op. cit.; ASROC is carried on some 61 frigates: 2 BRONSTEIN, 1 GARCIA, 42 KNOX, 1 GLOVER, and 6 BROOKE class; 71 Destroyers: 4 KIDD, 10 FARRAGUT, 1 DECATUR, 23 ADAMS, 31 SPRUANCE, and 2 SHERMAN class; and all 2? cruisers.
5 HASC, FY 1982 DOD, Part 3, p. 321.
6 *Missiles of the World*, p. 15.

8
SUBROC Missile

SUBROC (UUM-44A)

Figure 8.10 **SUBROC (UUM-44A)** missile being loaded into attack submarine.

DESCRIPTION: Short-range, inertially guided, SUBmarine launched range-controlled nuclear depth charge anti-submarine ROCket; SUBROC breaks the surface of the water, travels through the air and reenters the water to attack submarines.

CONTRACTORS: Goodyear Aerospace Corporation
Akron, OH; Litchfield Park, AZ
(prime)
AiResearch
Los Angeles, CA
(auxiliary power)
Singer-Librascope
Glendale, CA
(fire control)
Singer Kearfott
Little Falls, NJ
(guidance)
Thiokol Corp.
Elkton, MD
(propulsion)
Naval Ordnance Laboratory
White Oak, MD
(concept feasibility/technical direction)

SPECIFICATIONS:

Length:	22 ft (264 in)[1]
Diameter:	21 in
Stages:	1
Weight at Launch:	4000 lb[2]
Propulsion:	solid fuel TE-260G boost motor
Speed:	supersonic
Guidance:	inertial guidance on depth charge; depth charge also controlled by small fins;[3] analog fire control system[4]
Range:	25-35 mi[5]

DUAL CAPABLE: no[6]

NUCLEAR WARHEAD: one W55; 1-5 Kt range (see W55)

DEPLOYMENT:

Launch Platform:	torpedo tube on 73 of 96 attack submarines (SSN-688, -685, -671, -637 and -594 classes)[7]
Number Deployed:	approximately 400 (1983); 4-6 SUBROC/submarines on patrol
Location:	submarines homeported in Pearl Harbor, HI; Groton, CT; San Diego, CA; Charleston, SC; and Norfolk, VA

Nuclear Weapons Databook, Volume I **269**

8

SUBROC Missile

HISTORY:

IOC:	1965
1958	Goodyear awarded first development contract for SUBROC
1964-1968	SUBROC produced
1972-1974	production line reopened[8]
1977-1981	224 SUBROC motors refurbished, adding 15 years service life[9]
FY 1982	service life extension of SUBROC proposed[10]
Retirement Plans:	plans are to phase out SUBROC with nuclear armed Subsurface Delivered ASW Standoff Weapon starting in 1989; "must be retired in the nineties,"[11] some earlier phase out will occur in certain submarines as the TOMAHAWK SLCM is deployed;[12] last submarines to phase out SUBROC will be SSN-594 and -637 classes[13]

TARGETING:

Types:	submarines primary targets; surface ships secondary; can also be used for subsurface-to-surface missions against land[14]
Selection Capability:	fire control system computes course, speed and range to target; missile velocity is determined by distance to target; at proper point, rocket motor separates from depth bomb, which continues toward target aided by aerodynamic fins, inertial guidance and depth fuze[15]
Retargeting:	unknown
Accuracy/CEP:	can destroy enemy submarines if explosion occurs within 3-4 miles

COMMENTS: SUBROC was first tactical weapon capable of underwater launch, rocket motor ignition guided airborne trajectory, and underwater detonation. SUBROC warhead, rocket motors and fire control system require significant rework and upgrade to continue system in service.[16] SUBROC, reported in short supply in the Navy,[17] is beginning phase out in FY 1983[18] and will be phased out over an extended period.[19] SSN-594 and -637 classes will carry SUBROC until retirement, other classes will give up SUBROC with deployment of TOMAHAWK.

---

1 *The World's Missile Systems*, 6th Ed., p. 330; much longer range than torpedos.
2 *Ibid.*
3 *Ibid.*
4 SUBROC uses an ANALOG fire control system which is incompatible with SSNs-700 and later, equipped with all-digital fire control; HAC, FY 1983 DOD, Part 2, p. 166.
5 *Ibid.*
6 JCS, FY 1981, p. 48.
7 Norman Polmar, 12th Edition, *op. cit.*, p. 16, states 66; SAC, FY 1983 DOD, Part 1, p. 100, states 63 subs are SUBROC capable; analog fire control system is not compatible with the all digital system in the late model 688 class SSNs; HAC, FY 1980 DOD, Part 6, p. 1111; HASC, FY 1982 DOD, Part 3, p. 176; SASC, FY 1983 DOD, Part 6, p. 3710.
8 SASC, FY 1982 DOD, Part 7, p. 3896.
9 HAC, FY 1983 DOD, Part 2, p. 166.
10 HASC, FY 1982 DOD, Part 2, p. 777.
11 SASC, FY 1983 DOD, Part 6, p. 4040.
12 SASC, FY 1982 DOD, Part 6, p. 3674.
13 HASC, FY 1983 DOD, Part 4, p. 254.
14 *Military Applications of Nuclear Technology*, Part 1, p. 23.
15 Goodyear Aerospace, "SUBROC, Flying Torpedo, Stalks Its Prey: From Sea to Shini Sea," *Profile*, Spring 1969.
16 HAC, FY 1980 DOD, Part 6, p. 1111.
17 HASC, FY 1982 DOD, Part 3, p. 176.
18 SAC, FY 1983 DOD, Part 1, p. 100.
19 SAC, FY 1983 DOD, Part 1, p. 639.

8

ASW Standoff Weapon

ASW Standoff Weapon

A new Anti-Submarine Warfare Standoff Weapon (ASWSOW) is being developed by the U.S. Navy to "counter the threat posed by hardened, deep-diving Soviet submarines using sophisticated countermeasures."[1] The program, an outgrowth of a FY 1980 Joint DOE/DOD/Navy Tactical Nuclear Anti-Submarine Warfare Phase 1 Study, will incorporate state-of-the-art weapons and guidance possibilities, new concepts of employment, and nuclear safety design. The nuclear warhead was initially designed to emphasize commonality to replace the various warheads (W55, B57, and W44) currently being used aboard ships, submarines, and aircraft for nuclear ASW.[2]

A feasibility study to determine the nuclear warhead for the Common ASW Standoff Weapon for deployment near the end of the decade was initiated by DOD and DOE in FY 1981.[3] The DOE FY 1984 Budget request, however, included a distinct new warhead development program for a new ASW weapon. Called the "subsurface delivered ASW Standoff Weapon" warhead, the nuclear warhead was evidently separated from the ASROC and B57 replacement programs for the ASW standoff weapon when two separate programs were once again established.

The ASW Standoff Weapon program was initiated in FY 1980.[4] In February 1980, Boeing Aerospace Company, Seattle, Washington, won a four-way competition and was awarded a $10 million contract to continue to develop the missile.[5] The amended FY 1982 Reagan defense budget restored full funding to the ASW Standoff Weapon program after it had been cut in half by the Carter Administration.[6] Development funding for FY 1982 was $35.4 million, $20.2 million was planned for FY 1983, and $28.0 was requested for FY 1984.[7] Development cost is estimated as $500-600 million (FY 1980).[8] Life cycle costs for deploying 1000 missiles aboard attack submarines were estimated, before program restructuring to include surface ships, at $2.6 billion. Inclusion of surface ships could increase program costs by $2 billion.[9]

Until October 1981, the standoff weapon was being developed primarily for STURGEON (SSN-637) and LOS ANGELES (SSN-688) class attack submarines. In October 1981, the program was restructured as the Common ASW Standoff Weapon to provide for deployment of a common weapon aboard surface ships as well as submarines. Boeing and Gould, Incorporated were contracted to conduct engineering development.[10]

The FY 1984 Defense budget request contained another change in the ASWSOW program. The weapon was again referred to as a SUBROC replacment, with the introduction of the Vertical Launch ASROC (VLA) as the current ASROC replacement, at least in the short term.

The ASW Standoff Weapon will be a dual capable all-digital, long-range, quick reaction missile.[11] It will have increased range over the SUBROC and will use state-of-the-art targeting capabilities.[12] The standoff weapon is being studied to permit the engagement of enemy submarines at or near maximum range of future sensors, including over-the-horizon.

---

1 JCS, FY 1981, p. 48.
2 HAC, FY 1980 DOD, Part 6, p. 1112.
3 DOD, FY 1983 RDA, p. VII-14; SASC, FY 1982 DOD, Part 7, p. 3882.
4 SASC, FY 1983 DOD, Part 6, p. 4043.
5 AW&ST, 4 May 1981, p. 19.
6 SAC, FY 1982 DOD, Part 1, p. 486.
7 DOD, FY 1984 Annual Report, p. 153.
8 SASC, FY 1980 DOD, Part 6, p. 2974.
9 GAO, "Improving the Effectiveness and Acquisition Management of Selected Weapon Systems," 14 May 1982, GAO/MASAD-82-34, pp. 52-55.
10 Six Navy laboratories are also involved in Common ASW Standoff Weapon development: Naval Underwater Systems Center, Newport RI; Naval Weapons Center, China Lake, CA; Naval Surface Weapons Center, Silver Spring, MD; Naval Ocean Surveillance Center, San Diego, CA; Naval Air Development Center, Warminster, PA; and Naval Coastal Systems Center, Panama City, FL; SASC, FY 1980 DOD, Part 6, p. 2964.
11 SAC, FY 1983 DOD, Part 1, p. 100.
12 SASC, FY 1980 DOD, Part 6, p. 2967.

8

Anti-Air Weapons Systems

Anti-Air Nuclear Weapon Systems

The Navy currently has one nuclear armed surface-to-air missile deployed: the TERRIER with the W45 warhead. Originally deployed in November 1955 aboard the cruiser *Boston* (CAG-1), TERRIER was developed in numerous modifications. Now obsolescent, it is retained in only the nuclear version. All conventional TERRIERs have been withdrawn.[1] TERRIER is deployed aboard three aircraft carriers and 31 cruisers and destroyers.

The TERRIER (and other naval surface-to-air missiles such as TARTAR and TALOS) has been almost completely replaced by the STANDARD missile, which underwent development in 1968 to standardize TERRIER and TARTAR. Today, four basic versions of the STANDARD missile are deployed: STANDARD-1 medium range (MR), STANDARD-1 extended range (ER), STANDARD-2 medium range (MR), and STANDARD-2 extended range (ER). The STANDARD-2ER, which began deployment in FY 1982, is compatible with the TERRIER fire control system and has been designated as the replacement missile for the remaining TERRIER missiles.

The W81 nuclear warhead for the SM-2 missile has been in development since 1978. According to the FY 1984 DOD Research Development and Acquisition Report, "progress continues on the DOD/DOE warhead development program to provide a nuclear option for the STANDARD missile (SM-2). The nuclear option provides a complementary capability to the conventional warhead to defend against the multi-threat array available to the Soviet Union to attack the U.S. fleet, particularly high-speed, air-to-surface missiles launched from the BACKFIRE bomber."[2] The warhead is planned for deployment in 1987. The nuclear version of the SM-2 missile—designated SM-2(N)—will be compatible primarily with the STANDARD/AEGIS and STANDARD/TARTAR systems.[3]

STANDARD missile improvements continue under the SM-2 Block II program completed in 1983. SM-2 Block II is a high-speed, mid-course correction guided missile which "will provide a quantum improvement in capability" over the SM-2 Block I for AEGIS class (CG-47) cruisers and TERRIER for other cruisers and destroyers.[4] A STANDARD-3 has also been examined with "considerable effort" expended during FY 1983 to define its characteristics and technology as the "next generation surface-to-air missile."[5] The cost of a new missile, however, was considered prohibitive, and an evolutionary upgrade of SM-2 Block II—designated SM-2 Block III—was instead proposed. Martin Marietta has also proposed converting the Advanced Strategic Air Launched Missile (ASALM) as a long-range "outer air battle missile" for use on ships equipped with the vertical launching system.

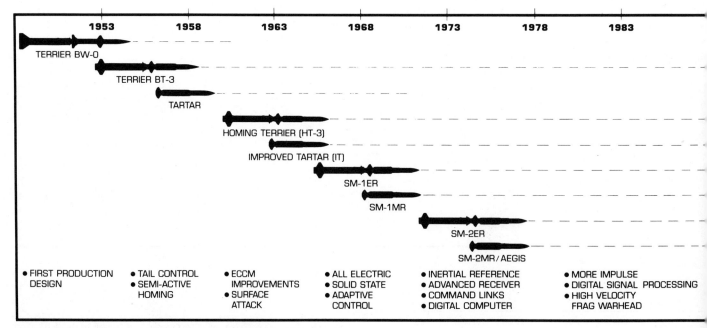

Figure 8.11 **Anti-Air Nuclear Weapon Systems.**

---

1 The W45 nuclear warhead, according to one source, was originally deployed to provide "the lethality for achieving the full effectiveness of a surface-to-surface mission;" General Dynamics, "Evolution of a Missile Family: 1949-1976," January 1967, p. 6.
2 DOD, FY 1984 RDA, p. V-12.
3 HASC, FY 1983 DOD, Part 3, p. 112.
4 SASC, FY 1983 DOD, Part 6, p. 3805.
5 SASC, FY 1983 DOD, Part 6, pp. 3655, 3708, 3804-3805.

TERRIER (RIM-2)

Figure 8.12 TERRIER (RIM-2) missiles on launcher.

DESCRIPTION: Short-range, surface-to-surface and surface-to-air nuclear missile aboard surface ships and some aircraft carriers in the Navy. The only remaining TERRIER missiles in the inventory are nuclear armed.[1]

CONTRACTORS: General Dynamics Pomona, CA (prime/guidance)
Allegany Ballistics Laboratory (motors)
Atlantic Research (booster)

SPECIFICATIONS:[2] (RIM-2F)
Length: 26.3 ft; (314 in) (with booster)
Diameter: 13.5 in
Stages: 2
Weight at Launch: 3093 lb; 3000 lb[3]
Propulsion: solid propellant fuel with booster
Speed: Mach 2.5

Guidance: target detected and tracked, missile readied and launcher aimed; beam-riding with semi-active terminal radar homing, proximity fuzing

Range: 20 nm+; 21.5 mi[4]

DUAL CAPABLE: yes[5]

NUCLEAR WARHEADS: one W45-0; 1 Kt range (see W45)

DEPLOYMENT:
Launch Platform: Mk-10 twin rail launchers loaded by automatic magazine; carried aboard 31 cruisers and destroyers, and 3 aircraft carriers.[6]

Number Deployed: approximately 310 nuclear warheads (1983)

HISTORY:
IOC: 1956[7]

1951 TERRIER program begins

1958 TERRIER BT-3A(N) version is fielded, first to carry a nuclear warhead[8]

1963 Advanced TERRIER with beam-riding homing guidance deployed

Retirement Plans: nuclear TERRIER will be replaced by the W81/STANDARD-2, which uses the same launcher.

TARGETING:
Types: aircraft, land targets,[9] limited anti cruise missile, and limited anti-ship capability[10]

8
TERRIER Missile

COMMENTS: Currently deployed version is BT-3A(N). TERRIER is the system name also associated with the shipboard fire control sytem which fires long-range STANDARD and TERRIER missiles.

---

1 HASC, FY 1982 DOD, Part 2, p. 768.
2 *The World's Missile Systems*, 6th Ed., p. 236.
3 *Missiles of the World*, p. 5.
4 *The World's Missile Systems*, op. cit.
5 Only RIM-2D (nuclear version) is retained; see also HASC, FY 1982 DOD, Part 2, p. 768.
6 SASC, FY 1981 DOD, Part 4, p. 2518.
7 The TERRIER version fielded in 1956 was the BT-3 (B-Beam rider, T-Tail Control, 3-Third version), developed for the supersonic threat.
8 BT-3A version added surface-to-surface capability by adding an end-burning sustainer motor which doubled the range of the missile. Nuclear warhead was added "for achieving the full effectiveness of a surface-to-surface mission;" General Dynamics, "Evolution of a Missile Family," January 1976, p. 6.
9 ACDA, FY 1980 ACIS, p. 272.
10 HASC, FY 1980 DOE, p. 84; HASC, FY 1981 DOD, Part 4, Book 2, p. 2310.

8
STANDARD-2 Missile

STANDARD-2 (SM-2) (RIM-67B)[1]

Figure 8.13 STANDARD-2 (RIM-67B) missile being fired from U.S.S. *Wainwright* (CG-28).

DESCRIPTION: Navy shipboard supersonic surface-to-air and surface-to-surface missile, with four versions, two of which will have a nuclear warhead (and designated SM-2(N)).

CONTRACTORS: General Dynamics Pomona, CA (prime/guidance)
Aerojet & Hercules (propulsion)
Atlantic Research (booster)
Motorola (fuze)

SPECIFICATIONS[2]: (SM-2ER only)

Length: 26 ft 7 in (319 in)

Diameter: 1.1 ft (13.5 in)

Stages: 2

Weight at Launch: 2996 lb

Propulsion: solid propellant rocket motor with booster

Speed: Mach 2.5

Guidance: command/inertial; mid-course guidance updates, semi-active radar terminal homing on illuminated target,[3] inertial reference system, ECM

Range: 65 mi

DUAL CAPABLE: yes[4]

NUCLEAR WARHEADS: one W81; low Kt (see W81)

DEPLOYMENT: STANDARD missiles are deployed on cruisers and destroyers equipped with AEGIS, TERRIER, or TARTAR fire control systems.[5] The SM-2ER was the replacement for the nuclear armed TERRIER missile,[6] but the medium-range STANDARD (SM-2MR) has also been designated as nuclear capable.[7]

Launch Platform: CG-16 and CG-28 class cruisers, CGNs and DDG-37 class ships[8]

Number Planned: 2044 SM-2 planned for procurement[9]

HISTORY:
IOC: 1968 (SM-1); 1978 (SM-2); 1982 (SM-2ER)

1968 SM-1 program started

Jun 1972 SM-2 missile program development begins

Nov 1972 first flight

1980 full production begins on SM-2ER missile

FY 1981 nuclear SM-2 funded in caretaker status[10]

FY 1982 integration of designed warhead and fuze into SM-2ER[11]

Nuclear Weapons Databook, Volume I **275**

8
STANDARD-2 Missile

Figure 8.14 **STANDARD-2MR** missiles on shipboard launcher.

FY 1983	target detecting device, nuclear safety studies, and warfare doctrine funded[12]
FY 1984	flight test of electronic fuzing and arming package of nuclear armed SM-2ER[13]

TARGETING:
Types: aircraft, nuclear armed antiship cruise missiles, surface ships[14]

Selection Capability: unknown

Retargeting: unknown

Accuracy/CEP: unknown

COST:[15]
Unit Cost: $643,000 (flyaway) (FY 1978)

FY	Number Procured	Total Appropriation ($ million)
1979 & prior		462.7
1980	55	50.8
1981	275	144.8
1982	375	222.3
1983	375	302.8
1984	450	347.3

COMMENTS: SM-2N is being proposed by the Navy as an "antiballistic" missile system to protect Navy ships against high-altitude, nuclear armed cruise missiles (e.g., the Soviet AS-4 cruise missile).[16] In FY 1983 Congressional hearings on the defense budget, it was revealed that the nuclear armed SM-2 program was being expanded to include SM-2MR (medium range) as well as ER (extended range) missiles.[17] Due to its small size (half that of current nuclear capable TERRIER), SM-2 can be carried by TARTAR-equipped ships (VIRGINIA class cruisers), giving them a nuclear AAW capability for the first time. This will result in an increase in the number of ships carrying nuclear weapons.[18]

---

1 The nuclear version of the STANDARD missile is designated SM-2(N).
2 *The World's Missile Systems*, 6th Ed., p. 228.
3 HASC, FY 1981 DOD, Part 4, Book 2, p. 2315.
4 The nuclear warhead will be "an integral nuclear ordnance section configured for use with both extended-range and medium-range versions of the STANDARD missile, completely alternate and interchangeable with the conventional ordnance section"; HASC, FY 1983 DOD, Part 3, pp. 112-113.
5 ACDA, FY 1981 ACIS, p. 367.
6 AW&ST, 9 March 1981, p. 137.
7 HASC, FY 1983 DOD, Part 3, p. 113.
8 ACDA, FY 1981 ACIS, p. 379.
9 *U.S. Missile Data Book, 1980*, 4th Ed., p. 2-104.
10 JCS, FY 1981, p. 48.
11 HASC, FY 1982 DOD, Part 2, p. 66.
12 HASC, FY 1983 DOD, Part 3, p. 113.
13 HASC, FY 1982 DOD, Part 2, p. 66.
14 HASC, FY 1980 DOE, p. 141; HASC, FY 1982 DOD, Part 2, p. 66.
15 ACDA, FY 1981 ACIS, p. 384.
16 Wilson, George C., "Floating Antimissile System Sails Smoothly," *Washington Post*, 8 February 1981, p. A2; also, ACDA, op. cit., p. 382; HASC, FY 1982 DOD, Part 2, p. 66.
17 HASC, FY 1983 DOD, Part 3, p. 113.
18 ACDA, FY 1980 ACIS, p. 189.

W81

Figure 8.15 **STANDARD-2 (RIM-67B)** missile on launcher. The W81 warhead will be fitted to the STANDARD missile.

FUNCTION: Warhead for the STANDARD-2 (SM-2) surface-to-air and surface-to-surface system.

WARHEAD MODIFICATIONS: none

SPECIFICATIONS:
Yield: low Kt fission[1]

Weight: unknown

Dimensions: unknown

Materials: reportedly uses oralloy as fissile material; IHE (PBX-9502)[2]

SAFEGUARDS AND ARMING FEATURES: CAT F PAL

DEVELOPMENT:
Laboratory: LANL[3]

History:
IOC: FY 1987[4]
1975 and prior: DOE/DOD Phase 1 and Phase 2 studies completed[5]
1977: Phase 2 engineering development initiated[6]
1978[7]: Lab assignment (Phase 3) (through 1983)
1979-1980: OSD defers development funds for W81[8]
FY 1980: W81 funded in Phase 3 at $2.016 million[9]
FY 1987: initial deployment (Phase 5)

Production Period: 1984[10]

DEPLOYMENT:
Number Planned: approximately 350 (1983)[11]

Delivery System: dual capable STANDARD SM-2 on shipboard launchers and eventually in vertical launching system

Service: Navy

Allied User: none

Location: SM-2 is used on a variety of surface combatants with the TERRIER, TARTAR, and AEGIS air defense systems.

8
W81

COMMENTS: W81 is modification of the B61 "primary" already in the stockpile and tested.[12] The nuclear warhead section is "completely alternate and interchangeable with the conventional ordnance section."[13] The development funds for SM-2(N) were deferred by OSD in 1979 and 1980. Program funding for 1981 was zero with the program maintained in a caretaker status.[14] The FY 1983 DOE budget request included funds to initiate construction of production facilities for the W81 warhead.[15]

---

1 An enhanced radiation warhead for the W81 was under consideration but was rejected by the Secretary of Defense in 1978. It was, however, switched back to a fission weapon in 1979 mostly due to availability; the exact yield had still not been determined as of March 1982; HASC, FY 1983 DOD, Part 5, pp. 693-694.
2 HASC, FY 1982 DOE, p. 217; AW&ST, 22 March 1982, p. 19.
3 ACDA, FY 1981 ACIS, p. 380.
4 AW&ST, 6 December 1982, p. 17.
5 HASC, FY 1983 DOD, Part 3, p. 113.
6 Ibid.
7 HASC, FY 1980 DOE, p. 137; SASC, FY 1980 DOE, p. 164.
8 SASC, FY 1980 DOD, Part 6, p. 2350.
9 Ibid.
10 HAC, FY 1982 DOE, EWDA, Part 7, p. 167; SAC, FY 1981 EWDA, p. 818.
11 This number was estimated before the program was expanded beyond the SM-2ER version.
12 ACDA, FY 1980 ACIS, p. 186; HASC, FY 1983 DOD, Part 5, p. 693.
13 HASC, FY 1983 DOD, Part 3, p. 113.
14 HASC, FY 1981 DOD, Part 4, Book 2, p. 2315.
15 DOE Justification, FY 1983, p. 50.

9

Army Nuclear Weapons

Chapter Nine
Army Nuclear Weapons

The Army[1] uses a wide variety of nuclear weapon systems—medium range PERSHING 1a and short-range LANCE surface-to-surface missiles, NIKE-HERCULES surface-to-air missiles, 155mm and 8-inch (203mm) artillery, and atomic demolition munitions (nuclear land mines). The HONEST JOHN surface-to-surface rocket, although withdrawn from active U.S. use, is nuclear armed with some NATO allies. Army nuclear weapons are deployed with U.S. combat units throughout the United States, Europe, in South Korea, and among allied military forces. They vary in range from manually emplaced land mines to 460 miles, and in yield from sub (0.01) to 400 kilotons.

The PERSHING 1a is the longest range and highest yield Army nuclear weapon currently deployed. One hundred and eighty launchers, with more than 300 missiles, all armed with W50 nuclear warheads, are deployed in West Germany with the U.S. Army and the West German Air Force. First deployed in 1962 to replace the SERGEANT missile, the PERSHING represented an increase in range, accuracy, and reliability. A replacement system for the PERSHING 1a, the PERSHING II, is now planned for initial deployment in December 1983. Originally designed as a more accurate and reliable missile with similar range to the PERSHING 1a, the PERSHING II was subsequently made long-range (1800 km) with a small yield warhead (W85). The deployment of the new PERSHING II missile has become very controversial. It is uncertain whether the full complement of missiles will eventually be deployed and whether the current West German PERSHING 1a will be replaced. Another PERSHING missile, a short-range and accurate "fallback" system designated PERSHING 1b, is also under development for deployment with the West German Air Force, or for U.S. deployment in Europe should arms control negotiations successfully eliminate the long-range missiles.

More widely deployed than PERSHING is the short-range corps support LANCE 125 km range missile, first deployed in 1972, and equipped with the W70 nuclear warhead (as well as a conventional warhead). Approximately 100 LANCE launchers and 945 nuclear warheads are currently operational in the U.S. Army, and nuclear armed with U.S. warheads in the Belgian, British, Dutch, Italian, and West German armies. LANCE replaced HONEST JOHN in all of these countries, more than doubling the range and accuracy over the older missile, and providing greater mobility and reliability. A new warhead for the LANCE, an enhanced radiation version of the W70 (Mod 3) produced in 1981-1983, is being stored in the U.S. and awaits shipment to Europe. The HONEST JOHN short-range free-flight rocket, first deployed in 1954, remains deployed with W31 nuclear warheads in the Greek and Turkish armies. No plans are currently known for the replacement of HONEST JOHN in the above forces with the LANCE, but they will be obsolete in the late 1980s and impossible to support. A nuclear armed LANCE replacement is under development, called the Corps Support Weapon System, as part of the Army-Air Force Joint Tactical Missile System program to investigate and develop new medium-range battlefield weapons with greater accuracy and flexibility.

The NIKE-HERCULES is the only land-based nuclear armed surface-to-air missile in the U.S. military. Approximately 500 warheads and launchers are deployed in Europe. Nuclear warheads are supplied to the following five countries for their NIKE-HERCULES: Belgium, Greece, Italy, the Netherlands, and West Germany. South Korean NIKE-HERCULES may also be supplied with nuclear warheads. The missile was originally deployed as a strategic system to defend the United States against massed bomber attacks, but has since been withdrawn from that role and is now only deployed in Europe and South Korea to defend the rear area and key installations against tactical aircraft. The NIKE-HERCULES is a dual capable system, armed with conventional warheads or the W31 nuclear warhead. It can also be used in the surface-to-surface role. First deployed in 1958, the NIKE-HERCULES has a number of operational limitations associated with slow rate of fire and guidance and is gradually being withdrawn from use. Some nuclear warheads have already been withdrawn as NIKE-HERCULES batteries are being converted to conventional warheads only. The development of the more accurate and versatile conventionally armed PATRIOT surface-to-air missile, starting in 1984, will completely replace the NIKE-

---

[1] Marine Corps use of nuclear artillery and atomic demolition munitions is indicated throughout this chapter and discussed in more detail in Chapter Four.

9
Army Nuclear Weapons

HERCULES, initially in U.S. forces and eventually in other NATO forces.

Nuclear artillery is widely deployed within the U.S. Army and Marine Corps and with the following seven NATO armies: Belgium, Greece, Italy, the Netherlands, Turkey, the United Kingdom, and West Germany. Some 5000 nuclear artillery warheads are deployed. Two nuclear projectiles—M422 203mm projectile with W33 nuclear warhead and M454 155mm projectile with W48 warhead—are currently deployed in Europe. One new projectile—M753 203mm projectile with W79 enhanced radiation warhead—is in production, but is being stored in the United States.

Numerous nuclear capable artillery guns are currently operational. The most common are the M109 self-propelled 155mm howitzer and the M110 self-propelled 203mm guns. Originally fielded in 1956, the W33/M422 nuclear projectile is the oldest nuclear warhead deployed. Its safety, reliability, and usefulness have been widely questioned. However, since the replacement for the W33 was finally limited to an enhanced radiation design in 1982, it is unlikely that it will be deployed to Europe. A new 155mm nuclear projectile—the W82/M785—is also under development, with an enhanced radiation yield. It is planned for initial deployment in 1987.

The smallest nuclear weapons are atomic demolition munitions (ADMs) (nuclear land mines). Two types are currently deployed with both the U.S. Army and Marine Corps. A number of NATO units, including at least British, Dutch, and West German units, are also trained to emplace and fire ADMs. The Medium Atomic Demolition Munition (MADM) is the more commonly available. First deployed in 1965, it weighs about 400 pounds, and is assigned to ADM engineer units within combat divisions and corps. The Special Atomic Demolition Munition (SADM) is less common than the MADM, smaller (some 150 pounds), and more portable, designed for use by special forces and commando units for emplacement and detonation behind enemy lines. Although replacements for the ADMs have been discussed occasionally, there is currently no plan to replace them with new nuclear systems.[2]

---

2 DOD, FY 1983 RDA, p. VII-14.

9
HONEST JOHN Missile

Missiles and Rockets
HONEST JOHN (MGR-1B)

Figure 9.1 HONEST JOHN (MGR-1B) missile.

DESCRIPTION:	Short-range, free-flight (unguided), mobile, nuclear capable surface-to-surface, solid propellant, ballistic rocket used by the Greek and Turkish Armies, and the Army National Guard.
CONTRACTORS:	McDonnell-Douglas (prime/missile) Hercules (rocket motor) Thiokol (stabilization rockets) Emerson Electric (missile)
SPECIFICATIONS:	(MGR-1B)
Length:	24 ft 9.5 in (7.6 m) (297 in)
Diameter:	30 in (762 mm)
Stages:	1
Weight at Launch:	4332 lb[1]
Propulsion:	solid propellant rocket motor
Speed:	Mach 1.5
Guidance:	spin stabilization with two pairs of spin rockets which are ignited automatically as the rocket leaves the launcher[2]
Range:	minimum range approximately 5000-6200 meters; maximum range of B version, 38 km[3]
DUAL CAPABLE:	yes
NUCLEAR WARHEADS:	one W31 in M47 and M48 nuclear warhead sections; 1-20 Kt range (see W31)
DEPLOYMENT:	
Launch Platform:	M386 self-propelled truck launcher, M289 self-propelled launcher and M33 towed launcher
Number Deployed:	about 200 nuclear versions remaining (1983); as many as six reload missiles per launcher estimated deployed
Location:	Greece, Turkey; probably some stored in U.S.; possible nuclear warheads in South Korea[4]
HISTORY:	
IOC:	1954
1951	firing tests of HONEST JOHN began
1906	MGR-1B enters service
TARGETING:	
Types:	tactical targets, headquarters command post, masses of armor, enemy short range nuclear weapons

9
HONEST JOHN Missile

Selection Capability: impact, low air and high air options with capability for selection of height of burst to 2000 meter maximum[5]

Accuracy/CEP: one nm; low level winds have a considerable effect on both the range and accuracy,[6] large vertical probable error[7]

COMMENTS: HONEST JOHN has been replaced by LANCE missiles in all NATO armies except in Greece and Turkey.[8] MGR-1B modification incorporated reduction in length and improvement in performance. The HONEST JOHN must be warmed with special electric blankets for 24-48 hours prior to firing; this enables it to attain a predetermined temperature (77° F) for proper and even propellant burn.[9]

---

1 With nuclear warhead; weight at launch of conventional HONEST JOHN is 4719 lb; USACGSC, *Selected Readings in Tactics*, RB 100-2, Vol VI, June 1977, p. 1-66; *The World's Missile Systems*, 6th Ed., p. 282; USA, *Field Artillery Battalion, Honest John*, FM 6-61, April 1966, p. 4.
2 FM 6-61, *op. cit.*, p. 5.
3 USA, *Field Artillery Rocket Honest John with Launcher M289*, FM 6-60, December 1974, p. 12; *Army Magazine*, October 1978, p. 146; FM 6-61, *op. cit.*, p. 5.
4 Last active U.S. HONEST JOHN battalion in South Korea retired in 1979 with missiles and equipment turned over to South Korean forces. Presumably nuclear warheads remain in South Korea under U.S. control; HAC, FY 1980 DOD, Part 1, p. 740; HASC, FY 1981 DOD, Part 1, p. 931.
5 USA, *Field Artillery Honest John Gunnery*, FM 6-40-1, June 1972, pp. 4-38, 4-39, 6-1.
6 FM 6-40-1, *op. cit.*, p. 1-2.
7 FM 6-61, *op. cit.*, p. 14.
8 JCS, FY 1982, p. 76; DOD, FY 1982 Annual Report, p. 127.
9 FM 6-40-1, *op. cit.*, p. 5-1; FM 6-60, *op. cit.*, p. 18; FM 6-61, *op. cit.*, p. 5.

9
LANCE Missile

LANCE (MGM-52C)

Figure 9.2 LANCE (MGM-52C) missile on mobile launcher.

DESCRIPTION:	Short-range, Army all-weather, highly mobile, guided nuclear-capable surface-to-surface ballistic missile.
CONTRACTORS:	LTV Corporation Warren, MI (prime/missile) Farm Machinery Corporation (launcher) Rocketdyne (motor) LTV/American Bosch Arma Corp/Systron-Doner Corp. (guidance)
SPECIFICATIONS:	(MGM-52C)
Length:	248 in
Diameter:	22.2 in
Stages:	1
Weight at Launch:	2900 lb (nuclear missile);[1] 2834 lb[2]
Propulsion:	storable prepackaged liquid P8E-6 motor, 46,450 lb thrust (maximum)[3]
Speed:	Mach 3
Guidance:	inertial with mid-course correction made by distance measuring equipment from ground station via radio link; AN/DJW-48 missile guidance set; directional control, automatic meteorological compensation guidance system (DC-AUTOMET) (weight: 36 lb)
Range:	5-125 km[4]
Ceiling:	1350 m (minimum range), 45,700 m (maximum range)[5]
DUAL CAPABLE:	yes
NUCLEAR WARHEADS:	one W70 in M234 nuclear warhead section;[6] five different warhead sections for missile, two tactical (one nuclear, one nonnuclear), two trainers and one practice; 1-100 Kt range (fission mods); circa 1 Kt (ER mod) (see W70)
DEPLOYMENT:	
Launch Platform:	M752 self-propelled launcher tracked launch platform, with 55 km/h speed and amphibious capability, carries one missile; the M740 launcher zero length is a towed launcher with the basic launch fixture of the M752 for use in special operations such as airborne missions
Number Deployed:	approximately 945 nuclear missiles (1983); 2133 missiles procured,[7] 1450 missiles in Army inventory[8]

9
LANCE Missile

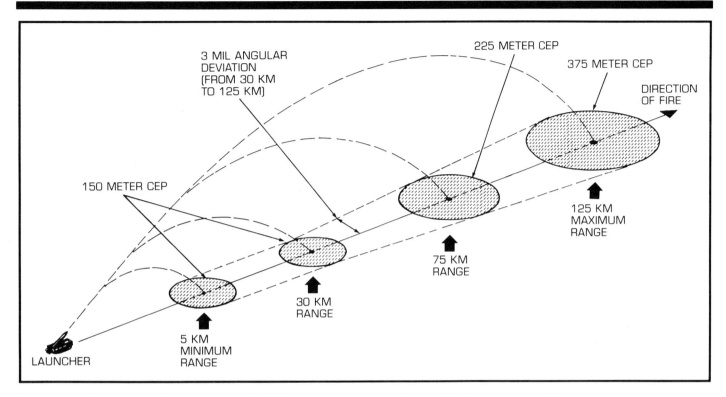

Figure 9.3 The accuracy of a nuclear armed **LANCE** missile is portrayed, showing diminished accuracy at longer ranges.

Location: six battalions with 6 launchers each deployed with U.S. Army in West Germany; 2 battalions in the U.S. at Fort Sill, OK;[9] with British, Belgian, Dutch, Italian, and West German armies[10]

HISTORY:
IOC: 1972[11]

Nov 1962 basic LANCE program begins

1965 first test firing of basic LANCE

Mar 1967 development of improved LANCE begins

Mar 1969 first test firing of improved LANCE

Sep 1971 production begins

Dec 1972 first firing of conventional LANCE warhead

1976 LANCE begins deployment with NATO armies

1979 adaption kits for installation of Mods 3 and 4 on LANCE procured[12]

1980 production of LANCE missile completed

TARGETING:
Types: command posts, logistical elements, troop concentrations, transportation elements, chokepoints, missile units, forward airfields, fixed air defense sites, critical terrain features (defiles, bridgeheads, main supply routes).[13] Because of its large CEP, LANCE cannot be effectively used for targets on or near the FEBA.[14]

Selection Capability: five heights of burst: ground (impact), air low, air low/ground backup, air high, or air high/ground backup[15]

Nuclear Weapons Databook, Volume I **285**

9
LANCE Missile

Retargeting: short reaction time (15 minutes) due to storable propellant, simplified prefire checkout;[16] rate of fire is 4 missiles per hour, eight missiles per day[17]

Accuracy/CEP: (see Figure 9.3)

COST:
Unit Cost: $142,000 (FY 1981) (flyaway);
$516,000 (FY 1981) (program);
$215,300 (FY 1978)

FY	Number Procured	Total Appropriation ($ million)
1977 & prior	1574[18]	1026.8
1978	360	76.9
1979	?	78.2
1980	-	4.1
1981	-	11.4

COMMENTS: LANCE has replaced HONEST JOHN and SERGEANT missiles in the U.S. Army and NATO Armies on a less than one-for-one basis. The FMC M688 auxiliary vehicle carries a load hoist and two reserve missiles. Both the M752 and M688 are mounted on the M667 basic vehicle. LANCE missile is composed of three main subsections: the missile main assemblage, the warhead section, and four control surfaces (fins). The M29 large control surfaces are used with the M234 nuclear warhead and provide aerodynamic stability to the missile by maintaining axial spin during flight. The launcher has an on-carriage traverse limitation of 285 mils with the nuclear warhead. Reliability of system is 90 percent of properly checked ready to fire missiles, 95 percent of properly fired rounds.[19]

---

1 Weight of nuclear missile; weight of nonnuclear missile is 3400 lb; USACGSC, *Selected Readings in Tactics*, RB-100-2, Vol VI, June 1977, p. 1-66.
2 *The World's Missile Systems*, 6th Ed., p. 284.
3 USA, *System Description for Lance Guided Missile System*, TM 9-1425-485-10-1, February 1972, CONFIDENTIAL (DECLASSIFIED), p. 1-46.
4 TM 9-1425-485-10-1, *op. cit.*, pp. 1-11, 1-45; in USA, *Field Artillery Battalion, Lance*, FM 6-42, w/Ch. 1, 29 December 1980, p. 2-1, the range is given as 8-75 km in non-nuclear configuration and 8-115 km in the nuclear configuration at sea level. If the launcher altitude is greater than 1000 meters, the maximum ranges are extended to 80 km and 133 km respectively; *Military Balance* lists range as 70-110 km, *The World's Missile Systems*, 6th Ed., p. 284, lists range as 3-70 mi.
5 TM 9-1425-485-10-1, *op. cit.*, pp. 1-61 - 1-63.
6 Warhead section is stored and shipped in the M511 container. The container is environmentally sealed, RF shielded, and pressure and humidity controlled. It has a small door to provide access to the CDS and PAL device without opening the container. FM 6-42, *op. cit.*, p. 2-13.
7 *U.S. Missiles Data Book*, 1980, 4th ED., p. 2-36; *Army Magazine*, October 1982, p. 324.
8 *Army Magazine*, October 1978, p. 144.
9 One of these battalions is earmarked for non-European contingencies; ACDA, FY 1980 ACIS, p. 153.
10 ACDA, FY 1982 ACIS, p. 244.
11 "Lance," *Armies & Weapons*, No. 42, April 1978, pp. 55-62.
12 ACDA, FY 1982 ACIS, p. 245.
13 FM 6-42, *op. cit.*, p. V; TM 9-1425-485-10-1, *op. cit.*, p. 1-7.
14 FM 6-42, *op. cit.*, p. 4-5.
15 USA, *Field Artillery Lance Missile Gunnery*, FM 6-40-4, 15 June 1979, p. 2-4, and B-1. Examples of HOBs in feet given in the manual are 232 ft for Air Low and 840 ft for Air High.
16 TM 9-1425-485-10-1, *op. cit.*, pp. 1-7, 2-7.
17 TM 9-1425-485-10-1, *op. cit.*, p. 2-3.
18 *U.S. Missile Data Book*, 1980, 4th Ed., p. 2-38.
19 TM 9-1425-485-10-1, *op. cit.*, p. 2-4.

NIKE-HERCULES (MiM-14)

Figure 9.4 **NIKE-HERCULES (MiM-14)** missile.

Figure 9.5 **NIKE-HERCULES** missile launch.

DESCRIPTION: Medium-range, fixed, guided, surface-to-air nuclear-capable missile used by the Army in West Germany and widely deployed in the following NATO countries: Belgium, Greece, Italy, Netherlands, and West Germany.

CONTRACTORS: Western Electric Company Burlington, NC (prime/guidance)
Thiokol (propellant)
Hercules, Inc. (boosters)
McDonnell-Douglas (missile airframe)

SPECIFICATIONS:

Length: 41 ft 6 in (498 in)[1]

Diameter: 31.5 in[2]

Stages: 2

Weight at Launch: 10,400 lb

Propulsion: solid propellant with four boosters, solid sustainer

Speed: Mach 3.3

Guidance: radio command

Range: 75-100 mi;[3] 120-160 km;[4] 140 km surface-to-surface range[5]

DUAL CAPABLE: yes

NUCLEAR WARHEADS: one W31, 1 Kt range (see W31)

DEPLOYMENT:

Launch Platform: fixed launchers at fixed locations

Number Deployed: 500 nuclear warheads (1983), circa 200 launchers

Location: Greece, Italy, West Germany (U.S., Belgian, Dutch, and German NIKE-HERCULES are deployed in West Germany)[6]

9

NIKE-HERCULES Missile

HISTORY:
IOC: 1958

TARGETING:
Types: aircraft, secondary surface-to-surface missions[7]

Selection Capability: Target is acquired by acquisition radars and then tracked by target tracking radars which issue command, guidance, and detonation instructions to the missile's computer and warhead.

COMMENTS: Conventional warheads can be used, but the system was designed to break up formations of attacking bombers with nuclear warheads. Nuclear warheads are being reduced in Europe.[8] Concurrent with PATRIOT deployment, NIKE-HERCULES will be phased out by FY 1985.[9] Launchers in nuclear role have already been reduced in the German and Greek forces.[10] U.S. NIKE-HERCULES Battalion has 576 men.[11]

---

1 *The World's Missile Systems*, 6th Ed., p. 204; USA, *ADA Employment - Nike Hercules*, FM 44-95, April 1968, p. 4-1 - 4-8.
2 Ibid.
3 Ibid.
4 USA, *ADA Reference Handbook*, FM 44-1-2, 30 June 1978, p. 3-4.
5 USA, *Fire Support in Combined Arms Operations*, FM 6-20, 30 September 1977, p. F-4.
6 Bases in Florida, Alaska, and other locations overseas have been deactivated; training unit is located at Ft. Bliss, TX.
7 FM 44-95, *op. cit.*, p. 2-15.
8 Walter Pincus, *Washington Post*, 1 November 1981, p. 1.
9 HAC, FY 1983 DOD, Part 4, p. 8.
10 JCS, FY 1981, p. 46.
11 FM 44-1-2, *op. cit.*, p. 3-4.

PERSHING 1a (MGM-31A/B)[1]

Figure 9.6 PERSHING 1a (MGM-31A/B) missile launch.

DESCRIPTION: Medium-range, two-stage guided surface-to-surface, mobile, nuclear ballistic missile used by the Army and the West German Air Force.

CONTRACTORS:
Martin Marietta Corporation
Orlando, FL
(prime/missile)
Thiokol Chemical Corporation
(powerplant)
Oregon Metallurgical
(jet vanes)
Singer
(hydraulic actuator)
Gulton
(static inverter)
Intercontinental
(motor case)
H.I. Thompson
(nozzle)
Bendix
(guidance)
Ford Motor Company
(transporter)
Sperry Rand
(fusing and arming)

SPECIFICATIONS:

Length: 34 ft 6 in (10.39 m)

Diameter: 39.4 in;[2] 1016 mm

Stages: 2

Weight at Launch: 10,273 lb[3]

Propulsion: solid propellant rocket motors

Speed: Mach 8

Guidance: inertially guided from RV separation to target

Range: up to 460 mi (740 km); 160-720 km;[4] 115-460 mi;[5] 185-740 km[6]

DUAL CAPABLE: no

NUCLEAR WARHEADS: one W50 warhead, in three warhead sections, with three yields: 60, 200, 400 Kt (see W50)

DEPLOYMENT: battalion personnel strength is 1368 men

Launch Platform: M757 truck/TEL (5 ton, 8x8) (originally deployed on M474 tracked vehicles), air transportable in C-130, C-141, and C-5

Number Deployed: some 800+ missiles procured; 180+ launchers deployed (108 U.S. launchers in West Germany and 72 West German launchers); approximately one reload per launcher available; approximately 13 missiles are returned to the U.S. annually from West Germany and fired for training[7]

9
PERSHING 1a Missile

Figure 9.7 Launch sequence of **PERSHING 1a**.

Location: U.S. Army and West German Air Force battalions/squadrons at Schwaebisch Gmuend (US); Neckars Ulm (US); Neu Ulm (US); Landsberg Lech (FRG); Geilenkirchen/Tevren (FRG); each battalion also includes a permanent "combat alert site" where launchers are on alert.

HISTORY:

IOC:	1962
Jan 1958	program initiated to replace REDSTONE missile
Jan 1960	testing of PERSHING began
Jun 1962	first PERSHING battalion activated at Fort Sill, Oklahoma
Apr 1964	first battalion deployed to West Germany
1965	PERSHING assigned a limited QRA mission in Europe
1970	changeover from P1 to P1a for U.S. Army units completed and assumption of full QRA mission
1971	changeover from P1 to P1a for West German units completed
1974	missile is outfitted with new digital guidance and high reliability components
1976	sequential launch adaptor, which permits three missiles to be launched before shifting power and air supply cables, and an automatic reference system, is deployed allowing launch from non-presurveyed sites[8]
Dec 1979	NATO endorses deployment of 108 PERSHING IIs to replace U.S. PERSHING 1s in West Germany

9
PERSHING 1a Missile

Dec 1983 — PERSHING IIs planned for initial deployment in West Germany

TARGETING:

Types: nuclear delivery units, command and control posts, airfields, command headquarters

Accuracy/CEP: 0.2 nm (450 m); 0.5 nm; 82 ft;[9] 400 m at max range [10]

COST: $3.117 m (FY 1978)

FY	Number Procured	Total Appropriation ($ million)
1979	-	78.1
1980	-	79.0
1981	-	11.8

COMMENTS: PERSHING 1a upgrade replaced tracked vehicles with wheeled vehicles, added new support equipment increasing rate of fire, improved erector-launcher and systems reliability. The major innovation was the incorporation of the ability to fire from unsurveyed firing positions. Missile was unchanged. PERSHING 1a is planned to be replaced by the PERSHING II by 1985 in U.S. Forces. The PERSHING missile is the only U.S. delivery system currently dedicated solely to the tactical delivery of nuclear weapons.[11]

---

1 Much of the descriptive information was provided to the authors by the Pershing Program Office, USAMICOM, Redstone Arsenal, AL.
2 *The World's Missile Systems*, 6th Ed., p. 290.
3 Ibid.
4 *Military Balance*, 1980-1981, p. 88.
5 *Missiles of the World*, p. 88.
6 USACGSC, *Selected Readings in Tactics*, RB 100-2, Vol VI, June 1977, p. 1-66.
7 DOD, FY 1980 RDA, p. VII-8.
8 *Army Magazine*, October 1977, p. 164; October 1978, p. 142.
9 *The World's Missile Systems*, 6th Ed., p. 290.
10 "Pershing II: First Step in NATO Theatre Nuclear Force Modernization?" *International Defense Review*, August 1979.
11 DOD, FY 1981 RDA, p. 11-23.

9

PERSHING II Missile

PERSHING II

PERSHING II (PII) is a new land mobile surface-to-surface ballistic missile being developed as the follow-on replacement for the medium-range PERSHING 1a (P1a). PII is designed to provide a significant increase in range, greater accuracy, and reduction in yield over the PERSHING 1a missile. With its longer range, it will be capable of striking targets on Soviet territory from its bases in West Germany. Greatly increased accuracy is achieved by using a maneuvering RV equipped with terminal guidance radar.

The major innovation of PERSHING II is its all-weather Radar Area Correlation Guidance (RADAG) which takes radar "pictures" of the target area, compares them with digital information stored in the RV's computer, and makes course adjustments until the pictures correspond and the warhead hits the target.

The PERSHING II missile originated as a short-range design with higher reliability and accuracy than the PERSHING 1a.[1] In 1977, the extended-range missile was adopted with an alternative reduced-range missile (designated PERSHING RR or PERSHING 1b) being maintained as an option.[2] The PERSHING II currently has two configurations: (1) a two-stage missile with a second stage propulsion section and (2) a single-stage missile consisting only of the first stage.

The Air Force has also proposed a Medium Range Ballistic Missile (MRBM)[3] as a competitor or follow-on to the GLCM and PERSHING. The MRBM under consideration was a longer range, road mobile, MIRVed missile. A feasibility study concerning the MRBM was completed by the Air Force in March 1979. A number of systems were considered as an MRBM candidate, including a two-stage MINUTEMAN III with modifications to guidance and reentry systems. The MRBM, however, was shelved in favor of the PERSHING II.

Planned deployment of PII includes a brigade of three U.S. battalions, containing four firing batteries (9 launchers each), which each have three firing platoons. P1a's are currently deployed overseas with U.S. Army and West German Air Force units in West Germany. A total of 108 PII launchers will initially be deployed in West Germany, replacing U.S. launchers on a one-for-one basis.

PERSHING II operations will include one platoon (3 launchers) of each battalion on "quick reaction alert" (QRA) at all times. During wartime, batteries would disperse into wooded areas and launchers would be set up requiring only a clear space of six feet diameter above the missile to fire.[4] After a missile is fired, the unit would quickly relocate to another area and set up again to refire.[5] "Combat alert sites" (QRA sites) are not considered to be survivable and missiles must reach "covert field firing positions" to avoid detection and target acquisition.[6]

In the FY 1982 budget process the W86 earth penetration warhead (EPW) for the PERSHING II was cancelled by DOD. Because the PII had become a long-range system designed to meet NATO requirements, the need for an EPW was no longer thought justified. The short-range system first envisioned five years earlier had different targets for which the EPW was thought important.[7]

---

1 Martin Marietta began promoting a PERSHING II two-stage 400 nm missile and single-stage variant which would allow two missiles per launcher in 1977; see "Pershing, Pershing 1a, Pershing II: Evolution of a Total Weapon System," November 1977.
2 HAC, FY 1983 DOD, Part 4, p. 427.
3 HAC, FY 1980 DOD, Part 2, pp. 442-444.
4 HAC, FY 1983 DOD, Part 4, p. 402.
5 HASC, FY 1983 DOD, Part 3, p. 764.
6 SASC, FY 1981 DOD, Part 5, p. 2827.
7 HASC, FY 1981 DOD, Part 4, Book 2, p. 2309.

PERSHING II[1]

Figure 9.8 **PERSHING II** missile launch.

DESCRIPTION: Two-stage, solid propellant, medium-range, highly accurate, low yield, ballistic missile for use in Europe.

CONTRACTORS: Martin Marietta Aerospace
Orlando, FL
(prime)
Goodyear Aerospace Corporation
Akron, OH
(guidance)
Singer Co., Kearfott Div.
Little Falls, NJ
(inertial measuring system)
Bendix Corporation
Teteboro, NJ
(computers and power supplies)
Hercules, Inc.
Salt Lake City, UT
(propulsion)

SPECIFICATIONS:

Length:	413.5 in
Diameter:	39.4 in[2]
Stages:	2
Weight at Launch:	15,873 lb;[3] 10,143 lb;[4] 16,400 lb[5]
Propulsion:	solid
Speed:	Mach 8

9

PERSHING II Missile

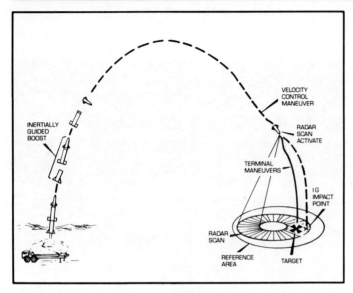

Figure 9.9 Flight sequence of **PERSHING II** missile, showing radar scanning to increase accuracy.

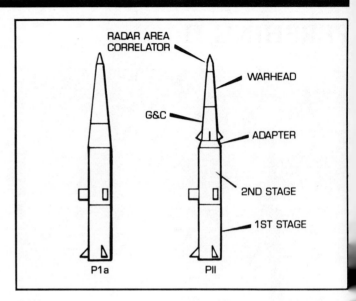

Figure 9.10 Comparison of **PERSHING 1a** and **PERSHING II** missiles.

Guidance:	boost guided phase inertial equipment; all-weather radar area correlation (RADAG) terminal guidance.[6] Radar guidance compares a prestored map of the target area with return signals, generating course changes in the RV.
Throwweight/ Payload:	1470 lb;[7] 650 lb,[8] payload comparable to P1a[9]
Range:	1300 km,[10] 1500-1800 km[11]
DUAL CAPABLE:	no[12]
NUCLEAR WARHEADS:	one W85, 5-50 Kt range (see W85)
DEPLOYMENT:	
Launch Platform:	PERSHING II will use the P1a erector-launcher with upgraded ground support equipment.[13]
Number Planned:	approximately 384 missiles;[14] PII will reportedly replace U.S. P1a missiles on a one-for-one basis[15]
Location:	PII Battalions (12 batteries) will be located at Neu Ulm, Neckars Ulm, and Schwaebisch Gmuend, West Germany; one training battalion (4 and 2/3 batteries) will be located at Fort Sill, OK
HISTORY:	
IOC:	December 1983[16]
Jan 1972	development of PERSHING follow on begins
Apr 1974	Martin Marietta awarded contract to develop more accurate version of PERSHING 1a
FY 1975	PII system requirements document approved and PII Advanced Development initiated[17]
Nov 1977	first flight of new guidance aboard modified P1a
Aug 1978	PII program is directed to work toward the extended-range variant by the Secretary of Defense[18]

294 Nuclear Weapons Databook, Volume I

9

PERSHING II Missile

Dec 1978	PII program approved to enter FSED and a planned 57-month Engineering Development program[19]	end 1986	PII replacement of U.S. P1a's completed[25]
Feb 1979	engineering development begins	**TARGETING:** Types:	hardened and soft missile sites, airfields, naval bases, nuclear, biological and chemical storage sites, command and control centers, headquarters, rail yards, road networks/choke points, ammunition and petroleum storage facilities, troop concentrations and facilities, dams/locks, masonry bridges, and tunnels[26]
Aug 1979	Secretary of Defense directs Army to plan for August 1983 IOC[20]		
Dec 1979	NATO endorses deployment of 108 PIIs and readjusts IOC to match GLCM IOC		
Feb 1980	President assigns PII DX-status, highest national priority for development[21] President approves Program of Cooperation with West Germany to develop P1a follow-on[22]	Retargeting:	rapid retargeting capability; immediate retargeting with target data available, can generate new target data immediately[27]
Apr 1982	first planned flight test is cancelled	Accuracy/CEP:	45 m; 20-45 m CEP objective; RADAG correlates returns from an initial 350 square nm area surrounding the target with a prestored reference map of the target area, obtains several such correlations during terminal descent, and updates the inertial position of the RV[28]
Jun 1982	full production contract awarded in accelerated concurrent development and production program		
Jul 1982	first flight test of PII is unsuccessful[23]		
Dec 1983	initial deployment in West Germany[24]		

9

PERSHING II Missile

COST:

Unit Cost: $5,475,824 (FY 1983) (unit)[29]

Program Cost:
- Development: $625.7 m
- Procurement: $927.3 m
- Operating/Support: $2781.3 m (FY 1980)[30]
- Development: $691.6 m
- Procurement: $2120.8 m (FY 1983);[31]
- $2737.6 m (Dec 1982)

FY	Number Procured	Total Appropriation ($ million)
1977 & prior	-	63.3[32]
1980 & prior	-	255.7
1981 & prior	-	408.0[33]
1981	-	151.4
		(146.0 requested)
1982	21	372.2[34]
1983	(91)	110.9[35]
1984	95	457.4
1985	104	447.3
1986	77	?

COMMENTS: PII will use same ground support equipment as P1a system. Option for a short-range, single-stage version of PERSHING II (known as PERSHING II RR (reduced-range) or PERSHING 1b to meet requirements to replace West German P1a's is being maintained.[36] The short-range PERSHING II is also considered for deployment to U.S. Army units if the long-range missile is rejected.[37]

---

1 Much of the descriptive information was provided to the authors by the Pershing Program Office, USAMICOM, Redstone Arsenal, AL.
2 *The World's Missile Systems*, 6th Ed., p. 290.
3 *Ibid.*
4 *U.S. Missile Data Book*, 1980, 4th Ed., p. 2-65.
5 Pershing Program Office, Redstone Arsenal, AL.
6 ACDA, FY 1979 ACIS, p. 115.
7 *The World's Missile Systems*, 6th Ed., p. 290.
8 *U.S. Missile Data Book*, 1980, 4th Ed., p. 2-65.
9 ACDA, FY 1981 ACIS, p. 235.
10 HAC, FY 1982 DOD, Part 3, p. 522.
11 DOD, FY 1982 Annual Report, p. 66, lists the unclassified PII range as 1000 km. The true range is classified.
12 DOD, FY 1981 RDA, p. VII-7.
13 16-2/3 "battery sets" will be procured comprising 9 launchers, 9 missiles, 4 platoon control centers and other ground support equipment per battery set; HASC, FY 1983 DOD, Part 3, p. 762.
14 Of the approximately 384 missiles planned, approximately 228 will be for operational purposes (basic load and reloads/spares), 24 will be maintenance missiles and the remainder will be for 10 years of weapons testing; HASC, FY 1983 DOD, Part 3, p. 762, 764.
15 HASC, FY 1982 DOD, Part 2, p. 237.
16 SASC, FY 1982 DOD, Part 7, p. 3803; IOC is defined as 9 launchers and 13 missiles; HAC, FY 1983 DOD, Part 4, p. 431.
17 HAC, FY 1980 DOD, Part 2, p. 863; ACDA, FY 1979 ACIS, p. 116.
18 *Ibid.*
19 *Ibid.*
20 HASC, FY 1982 DOD, Part 2, p. 237.
21 HASC, FY 1982 DOD, Part 2, p. 238; HAC, FY 1983 DOD, Part 4, p. 408, 444; SASC/SAC, Joint FY 1981 Mil Con, p. 387.
22 *Ibid.*
23 Walter Pincus, *Washington Post*, 23 July 1982, p. 8.
24 DOD, FY 1983 RDA, p. VII-13.
25 *Ibid.*
26 ACDA, FY 1979 ACIS, p. 116.
27 SASC, FY 1980 DOD, Part 6, p. 3469; new radar reference scenes for targets which have not been preplanned will be generated in the field by the battalion Reference Scene Generation Facility, utilizing a Defense Mapping Agency data base; HAC, FY 1982 DOD, Part 3, p. 522.
28 ACDA, FY 1979 ACIS, p. 115.
29 DOD, Procurement Programs (P-1), 8 February 1982, p. A-6.
30 HAC, FY 1980 DOD, Part 2, p. 864; in escalated dollars, through 1996.
31 HAC, FY 1983 DOD, Part 3, p. 773.
32 ACDA, FY 1979 ACIS, p. 117.
33 *Army Weapons Systems*, 80, n.d., p. 19.
34 39 missiles were requested, but only 21 were authorized due to cost overruns; HAC, FY 1983 DOD, Part 4, p. 398.
35 HASC, FY 1983 DOD, Part 3, p. 762; $619.9 m was requested for 91 missiles, but only $110.9 was appropriated; DOD, FY 1984 Annual Report, p. 233.
36 HASC, FY 1982 DOD, Part 2, p. 238.
37 HAC, FY 1983 DOD, Part 4, p. 443.

W85

FUNCTION:	Air burst/surface burst warhead being developed for the PERSHING II long-range theater ballistic missile system.	**DEVELOPMENT:** Laboratory:	LANL

FUNCTION: Air burst/surface burst warhead being developed for the PERSHING II long-range theater ballistic missile system.

WARHEAD MODIFICATIONS: none

SPECIFICATIONS:

Yield: selectable;[1] low Kt; 5-50 Kt range; 10-20 Kt[2]

Weight: less than 1600 lb[3]

Dimensions:
 Length: 41.7 in[4]
 Diameter: 12.4 in[5]

Materials: oralloy warhead;[6] contains IHE[7]

SAFEGUARDS AND ARMING FEATURES: CAT F PAL; airburst and surface burst; launch requires PAL, warhead intent (safety), and missile ignition enabling coded signals

DEVELOPMENT:

Laboratory: LANL

History:
 1979 Lab assignment (Phase 3) (through FY 1983)[8]
 1983 initial deployment (Phase 5)

Production Period: 1983-[9]

DEPLOYMENT: 108 PERSHING II launchers are scheduled to be deployed in West Germany beginning in 1983; some 384 missiles will be deployed.

Service: Army

Allied User: none[10]

Location: West Germany

COMMENTS: W85 will be an adaptation of the already developed B61 Mod 4 bomb.[11] Also considered as possible candidates for PII were an adaptation of the W70 Mod 1 or Mod 2,[12] and W80.[13] Warhead section is a welded aluminum monocoque structure overwrapped with a rubber modified silica phenolic heatshield. Warhead and electrical connections including the safe and arm fuze (SAF) system are mounted inside the warhead section.

---

1 ACDA, FY 1982 ACIS, p. 203; SASC, FY 1982 DOD, Part 7, p. 3888.
2 *Aerospace Daily*, 3 July 1980, p. 19; *International Defense Review*, August 1979.
3 Weight of reentry vehicle, AW&ST, 2 August 1982, p. 20.
4 Information provided to authors by Pershing Program Office, USAMICOM, Redstone Arsenal, AL.
5 *Ibid.*
6 HASC, FY 1982 DOE, p. 103.
7 HASC, FY 1982 DOE, p. 217.
8 DOE Budget Justification, FY 1983, p. 51; remained in Phase 3 during FY 1980; SASC, FY 1981 EWDA, p. 818.
9 Funds for production of W85 are included in FY 1983 DOE Budget.
10 West Germany, presently equipped with the PERSHING 1 system, has not yet decided on a replacement system.
11 HAC, FY 1980 DOD, Part 2, p. 863; USANCA "Materiel and Safety Significant Activities (1 January 1978-31 December 1978)," n.d., p. 2; "Pershing II: First Step in NATO Theatre Nuclear Force Modernization," *International Defense Review*, August 1979.
12 ACDA, FY 1979 ACIS, p. 115.
13 SASC, FY 1979 DOE, p. 59.

9

Corps Support Weapon System

Corps Support Weapon System

The replacement for the LANCE missile, the Corps Support Weapon System (CSWS), often designated LANCE II or Improved LANCE, is currently under development and planned for the early 1990s.[1] The new missile is envisioned as an all-weather, dual capable, air transportable replacement with an improved CEP and rate of fire.[2] The CSWS would be deployed at the Corps level with the mission of interdicting enemy surface-to-air missile systems and second echelon ground forces at a range of 120-140 miles, with precision conventional, nuclear (enhanced radiation), and chemical warheads.[3] The CSWS will have the capability of striking targets three times further, five times more accurately, and with a higher rate of fire than the present LANCE.[4] An anti-armor capability, by deploying terminally guided submunitions and advanced target acquisition and guidance systems, is also planned.

The CSWS is part of the larger Army-Air Force Joint Tactical Missile System (JTACMS) program to develop standoff weapons to attack moving rear echelon targets deep behind enemy lines. The JTACMS will use Assault Breaker technology, developed under the Defense Advanced Research Projects Agency (DARPA) program, to develop a surface-to-surface weapon system for conventional and nuclear "deep battlefield interdiction."[5] Two delivery modes are under investigation within JTACMS: air-launched (the Air Force's Conventional Standoff Weapon) and ground-launched (the Army's CSWS). The JTACMS program originates with the Assault Breaker program started in 1978 to develop new standoff weapons for second echelon armor strikes. Assault Breaker (and now JTACMS) examined alternatives such as derivatives of Multiple Launch Rocket System (MLRS), PATRIOT, and LANCE as capable of carrying new warheads and being compatible with new guidance and target acquisition systems. The LANCE

Figure 9.11 U.S. Army Assault Breaker prototype missile, similar in design to the nuclear armed **Corps Support Weapon System**.

---

1 DOD, FY 1983 RDA, p. VII-12.
2 DOD, FY 1981 RDA, p. VII-6.
3 HASC, FY 1982 DOE, p. 45; AW&ST, 1 November 1982, p. 77.

4 Information provided by Vought Corporation.
5 DOD, FY 1984 Annual Report, p. 132.

Corps Support Weapon System

replacement, originally designated the Nuclear Corps Support Missile System, was an original part of the Assault Breaker concept.

The merging of the Army and Air Force programs in FY 1982 led to a slowing down of the CSWS program pending a clearer definition of requirements and operational concepts.[6] The joint development program incorporates the Air Force and Army development programs and takes advantage of common guidance, propulsion, conventional warheads, and electronic components.[7] Procurement plans, according to one report, are now for some 5500 missiles for the Army, with an accelerated IOC of 1986-1987.[8]

Competitor missiles for CSWS include the T22, a LANCE missile variant being developed by Vought and first tested in August 1979; the T16, a PATRIOT missile variant being developed by Martin Marietta with a nuclear warhead and surface-to-surface capability;[9] and the T19, a "generic" missile capable of both Army and Air Force use. A nuclear warhead for the CSWS is being developed. According to one report, it would range from 10-40 Kt and would incorporate enhanced radiation features.[10]

COST:

FY	Number Procured	Total Appropriation ($ million)
1981	-	7.6
1982	-	11.8
1983	-	6.1
1984	-	50.2
1985	-	126.5

---

6 *Aerospace Daily*, 25 August 1982, p. 305.
7 HAC, FY 1982 DOD, Part 9, p. 478.
8 AW&ST, 1 November 1982, p. 77.

9 SASC, FY 1980 DOD, Part 6, p. 3444.
10 AW&ST, 1 November 1982, p. 77.

9 Nuclear Artillery

Nuclear Artillery

Nuclear artillery is one of the most widely dispersed and numerous of weapons. From 1953, when the mammoth 280mm cannon was first sent to West Germany, until today, six nuclear artillery warhead types have been produced, and virtually every large artillery gun has become capable of firing nuclear rounds. Nuclear artillery is now in the midst of a major modernization, with guns and projectiles being upgraded and replaced.

The U.S. Army has over 3500 nuclear capable artillery guns deployed: 748 M114 155mm guns, 2200 M109 155mm guns, and 1046 M110 8-inch (203mm) guns.[1] Nuclear artillery is also used in the Marine Corps. The seven NATO allies with nuclear artillery—Belgium, Greece, Italy, the Netherlands, Turkey, United Kingdom, and West Germany—all use U.S.-designed artillery, particularly the standard M109 and M110 guns, but also deploy a number of obsolete guns (see Table 9.1).

Approximately 5000 nuclear artillery shells are estimated to be deployed, and of these, most are in Europe. Nuclear artillery is low yield, with the explosive capacity of the warheads varying from about 0.1 kiloton (for the W48) to about 12 Kt (for the W33). Three warhead types are currently deployed—the 1-12 Kt W33 8-inch fission warhead, the 0.1 kiloton W48 155mm fission warhead, and the 1-2 Kt W79 8-inch enhanced radiation warhead (see Chapter Three). The projectiles are always fired as air bursts[2] with accuracies (for W48 and W33) of 40, 100, and 172 meters CEP at "short, medium, and long range."[3]

Both 155mm and 203mm guns are being increasingly adapted in military formation, with the replacement of both 105mm and 175mm non-nuclear capable guns. In addition, there has been an increase in the number of guns in combat units both as a measure to generally increase firepower, and as an increase following the retirement of HONEST JOHN rockets from divisions.[4]

Figure 9.12 The first live nuclear artillery test, **Shot Grable**, on 25 May, 1953. A 280mm nuclear artillery shell with an explosive yield of 15 Kt was exploded in an airburst over the Nevada Test Site.

---

1 HASC, FY 1979 DOD, Part 3, Book 1, p. 773.
2 *Military Applications of Nuclear Technology*, Part 1, p. 13; Part 2, p. 35.
3 *Ibid.*, Part 2, p. 43.
4 JCS, FY 1981, p. 46.

9
Nuclear Artillery

Two new nuclear capable 155mm guns are beginning to enter the U.S. and NATO armies: a new long-range towed howitzer (M198) equipping light Army units and the Marines, and a European-designed and produced gun—FH-70/SP-70—available in towed and self-propelled designs. The SP-70 is the self-propelled design which features a high rate of fire, an automatic loader, and improved survivability and mobility. Older M1, M44, and M109 guns will be withdrawn from Italian, British, and West German units as newer FH-70/SP-70 guns are deployed in the mid-1980s. U.S. production lines for M109A2 and M198 will be open in FY 1984 and beyond, primarily for reserve forces.[5]

The 155mm artillery unit can contain 18-24 guns, organized with three batteries of either six or eight guns. An eight inch artillery unit is organized as either a 12, 18, or 24 howitzer battalion with two or three batteries, and four to six sections of four guns, the basic firing element. Self propelled artillery is assigned to mechanized and armor units for support, while towed artillery is assigned to infantry and other light units. Nuclear artillery warheads are maintained available to the firing element, stored in special two-and-one-half ton trucks, each with a one-and-one-half ton trailer.[6]

The range and design of nuclear artillery is beginning to reach practical limits, with restraints in noise, target acquisition, accuracy, and reliability.[7] The newest guns and barrels have ranges of 30 km, with older M109 and M110 guns being converted with new longer tubes and muzzle brakes (and designated A1 and A2 versions). The new tubes—33 vs. 20 caliber in the M109 and 37 vs. 25 caliber in the M110—increase muzzle velocity and range from 18 to 30 km for 155mm guns and from 14 to 30 km for 8-inch guns. Nuclear artillery projectiles are also becoming more sophisticated (and expensive), incorporating timing and memory assemblies, fuze subcomponents, power supplies, electronic programmers, target sensors, and rocket motors in the shell (see Future Artillery).

Future artillery guns are being examined to provide "important capabilities for responding rapidly and accurately to fire missions" and "reduced emplacement" times.[8] The new weapon is envisioned to have short recoil cycle time, burst rate-of-fire, automatic ammunition handling, loading and resupply, and automated position location and weapons alignment. During 1978, the Defense Nuclear Agency initiated a study to determine the feasibility of Long Range Cannon Artillery, an 8-inch, 80 km, nuclear capable shell for the mid-1980s. A Division Support Weapon System is also being developed by FMC. It will be a 155mm, 45 caliber, automatic loading gun.

Table 9.1
Nuclear Artillery Guns

Type	Country in Use
155mm	
M-2	Italy, Turkey
M59	Turkey
M115 (M1)	Italy (M1A1), Turkey, United States, West Germany
M44 SP	Belgium, Italy, Turkey
M109 SP	Belgium, Greece, Italy, Netherlands (M109A1), Turkey, United Kingdom, United States, West Germany (M109G)
M198	United States
FH-70/SP-70	Italy, United Kingdom, West Germany
8-inch	
M115	Belgium, Greece, Italy, Netherlands, Turkey
M55 SP	Italy
M110 SP	Belgium, Greece, Netherlands, Turkey, United Kingdom, United States, West Germany

Source: DOD, "Rationalization/Standardization Within NATO," Fifth Report, 28 January 1978; Seventh Report, January 1981.

---

5 DOD, FY 1984 Annual Report, p. 133.
6 U.S. Army, "3X8 Direct Support Field Artillery Battalion," TC 6-50-1, January 1981, p. 9-2.
7 HASC, FY 1979 DOD, Part 3, Book 1, p. 772.
8 Information provided by the Cannon Artillery Weapons System Project Manager.

9

M109 155mm Gun

M109 155mm Gun[1]

Figure 9.13 M109 155mm self-propelled gun.

DESCRIPTION:	Heavy, self-propelled 155mm artillery gun used by the U.S. Army, Marine Corps, NATO, and allied armies
MODIFICATIONS:	M109A1(SP), M109A2(SP), M109A3(SP), M109G(SP)
CONTRACTORS:	Bowen-McLaughlin-York York, PA (final assembly) Detroit Diesel (engine) Cadillac Division, General Motors Corp. (development)

SPECIFICATIONS:[2]

Max Range:	14,600 m (initial); 18,100 m (A1); 24,000 m (A2); 30,000 m (A3 and upgraded A2)
Weight:	52,461 lb (initial); 53,060 lb (A1); 54,700 lb (A2)
Weight of HE Round:	95-104 lb
Rate of Fire:	4/min (first 3 min); 1/min thereafter
Crew:	6
Prime Mover:	armored tracked vehicle, max speed 35 mph (roads), amphibious to 42 inches of water (with kit); air transportable (C-5A)

9
M109 155mm Gun

NUCLEAR WARHEADS: W48, 0.1 Kt yield (see W48); compatible with W82

DEPLOYMENT:

Number Deployed: 1608 (Army) (1978);[3] 2100 (Army);[4] 2200 (Army) (1983)[5]

Location: United States, Europe, South Korea, Japan (Okinawa)

Allied User: West Germany (M109G) (gun modified by Rheinmetall); Italy (M109G), Netherlands, Belgium, South Korea

HISTORY:

IOC: 1969

1953 development of M109 began

COST:[6] M109(A2)

FY	Number Procured	Total Appropriation ($ million)
1980 & prior	?	474.5
1981	0	27.3
1982	0	?

COMMENTS: Basic M109 carried 20 caliber (156 in) gun; A1 and later versions upgraded to 33 caliber (238 in) gun; M109A3 is the name for the retrofitted M109A1; M109A2 improves gun mount design, hydraulic components, safety features, and ammunition stowage (36 rounds),[7] in full scale production; new production M109A1s and later versions contain crew safety and ammunition stowage improvements.

---

1 USACGSC, *Selected Readings in Tactics*, RB 100-2, Vol. VI, June 1977, pp. 1-68, 1-64; *Army Magazine*, October 1977, October 1981.
2 USA, *155mm Howitzer, M109, M109A1, Self Propelled*, FM 6-8, June 1974.
3 HASC, FY 1979 DOD, Part 3, Book 1, p. 773.
4 *Army Magazine*, October 1978, p. 137.
5 HASC, FY 1979 DOD, Part 3, Book 1, p. 773.
6 USA, *Army Weapons Systems*, 80, n.d.
7 *Military Review*, May 1980, p. 83.

9
M198 155mm Gun

M198 155mm Gun[1]

Figure 9.14 M198 155mm towed gun.

DESCRIPTION:	Medium, towed 155mm artillery gun used by light units of the Army and Marine Corps
MODIFICATIONS:	none
CONTRACTORS:	Rock Island Arsenal IL (gun mounts and final assembly)
	Condec Corp. Greenwich, CT (carriages)
	Watervliet Arsenal NY (cannon)
	Numax Electronics Hauppauge, NY (fire control)

SPECIFICATIONS:	
Max Range:	18,000 m;[2] 30,000 m (rocket assisted projectiles)
Weight:	15,795 lb
Weight of HE Round:	95-104 lb
Rate of Fire:	4/min (first 3 min); 20/hr thereafter
Prime Mover:	M813 5-ton truck, speed 34 mph (roads); helicopter and air transportable; fordable to 30 inches of water
Crew:	11

9

M198 155mm Gun

NUCLEAR WARHEADS:	W48, 0.1 Kt yield (see W48); compatible with W82
DEPLOYMENT:	
Number Deployed:	468 (Army) (1983)[3]
Location:	United States, Japan (Okinawa), South Korea
Allied User:	none known
HISTORY:	
IOC:	1979
1968	development of M198 begins

COST:[4]

FY	Number Procured	Total Appropriation ($ million)
1980 & prior	0	138.0
1981	0	44.9
1982	-	?

COMMENTS: M198 is used by United States airborne and light infantry units, replacing the M114.

---

[1] USACGSC, *Selected Readings in Tactics*, RB 100-2, Vol. VI, June 1977, pp. 1-63, 1-64; *Army Magazine*, October 1977, October 1981.
[2] With standard ammunition.
[3] HASC, FY 1979 DOD, Part 3, Book 1, p. 773.
[4] USA, *Army Weapons Systems*, 80, n.d.

9

M110 8-inch (203mm) Gun

M110 8-inch (203mm) Gun[1]

Figure 9.15 M110 8-inch (203mm) self-propelled howitzer.

DESCRIPTION:	Heavy, self-propelled 8-inch (203mm) artillery howitzer used by the Army, Marine Corps, NATO, and allied armies.
MODIFICATIONS:	M110A1(SP), M110A2(SP)
CONTRACTORS:	Bowen-McLaughlin-York York, PA (final assembly) Detroit Diesel (engine) Pacific Car and Foundry Renton, WA (NBC protection system)

SPECIFICATIONS:

Max Range:	16,800 m (initial); 20,600 m (A1); 29,000 m (A2)
Weight:	58,500 lb (initial); 60,100 lb (A1); 62,500 lb (A2)
Weight of HE Round:	228 lb
Rate of Fire:	1 every 2 min
Crew:	13
Prime Mover:	tracked vehicle; speed 9 mph (cross country); 34 mph (roads); fordable to 42 inches of water; air transportable (C-5A)
NUCLEAR WARHEADS:	W33, sub-12 Kt (see W33); compatible with W79 (see W79)

9

M110 8-inch (203mm) Gun

DEPLOYMENT:

Number Deployed: 720 (Army) (1978); 1046 (Army) (1983)[2]

Location: United States, Europe, South Korea, Japan (Okinawa)

Allied User: Belgium, Italy, Netherlands, United Kingdom, West Germany

HISTORY:

IOC: 1961

1983: in full scale production (A2)

COST:[3] M110(A2)

FY	Number Procured	Total Appropriation ($ million)
1980 & prior	?	252.3
1981	0	11.3
1982	0	?

COMMENTS: A1 version carried 25 caliber gun, no muzzle brake; A2 gun incorporated muzzle brake, longer (37 caliber) cannon; conversion of M110A1s to A2 configuration by field installation of muzzle brakes is in progress. M110 is deployed with 12-24 guns per battalion.

---

1 USACGSC, *Selected Readings in Tactics*, RB 100-2, Vol. VI, June 1977, p. 1-65; *Army Magazine*, October 1977, October 1981.
2 HASC, FY 1983 DOD, Part 3, Book 1, p. 773.
3 USA, *Army Weapons Systems*, 80, n.d.

9

Future Nuclear Artillery

Future Nuclear Artillery

The W82 155mm nuclear artillery projectile is currently under development, with a planned deployment date of late-1986. The development program went largely unfunded during most of the Carter Administration, but has been brought to full funding in the Reagan budgets. The new projectile will have an enhanced radiation capability.[1] The W82 thus will be the third neutron weapon in the stockpile, joining the W70-3 and W79 already in production.

Development of a new 155mm nuclear projectile began in 1969 when the Army argued that a modernized artillery shell was needed to replace the W48. Development of the W74 began that year and continued until 1973 when the Joint Committee on Atomic Energy terminated the program due to excessive cost and the use of obsolete (non "neutron bomb") technology. Army requests to reinitiate the 155mm development were denied by Congress in 1975 and 1976, but in 1977 a new analysis of modernization requirements led to Congressional authorization and appropriation of funds to reinitiate a research program. The new projectile (W82), with an improved fission yield component design, has the technical capability for conversion to enhanced radiation yield in the field. This projectile continued in development until 1979 when Secretary of Defense Brown directed a 67 percent cut in the fiscal year 1981 budget request, and then zeroed out fiscal year 1982 funds.[2] The 155mm nuclear artillery shell presently in research and development, however, has again risen in cost.[3] That factor, along with the political implications of the enhanced radiation yield, make the weapon controversial.

Much controversy within the nuclear weapons program has been created over whether the new warhead should be fission yield, enhanced radiation yield, or both. Although technically the warhead is capable of accepting an ER option,[4] it has been referred to in some official reports as having a fission yield.[5] According to some reports,[6] the proposal to build the warhead as an ER warhead had been dropped, but it appears that planning and development continued with the intent of at least having an ER capability.[7]

Compared to the W74, which was cancelled by Congress in 1973, the features of the W82 are not impressive. W74 was designed for two yields, larger than the 0.1 kiloton W48.[8] The W82 yield also exceeds that of the W48. The W74 reportedly had a CEP of 20, 60, and 110 meters at short, medium and long ranges, comparable to the W82.[9] The W74, when cancelled by Congress, cost $452,000 each,[10] and the W82 cost is now estimated at close to $3 million each. The huge cost can be largely attributed to the more expensive enhanced radiation design, with its tritium requirements.

Table 9.2
Comparison of Old and New 155mm Nuclear Artillery Shells

	W48	W82
Users:	Army, Marine Corps, NATO	Army, Marine Corps, NATO
Yield:	0.1 kiloton	less than 2 kilotons
Type:	pure fission	enhanced radiation
Range:	16 km	30 km
Weight:	128 lb	95 lb
Materials:	plutonium	plutonium and tritium
Cost:	less than $200,000 each	circa $2.5-3.0 million each
Development Began:	Aug 1957	1969
First Deployed:	1963	1984
Number Deployed:	3000	1000+ planned
Locations:	US, Europe, South Korea	US

---

1 HASC, FY 1983 DOD, Part 5, p. 693.
2 HASC, FY 1982 DOD, Part 2, p. 30; HASC, FY 1982 DOD, Part 1, p. 45.
3 Some estimates put the cost per projectile at $3 million.
4 JCS, FY 1981, p. 47.
5 JCS, FY 1982, p. 78; HASC, FY 1981 DOD, Part 4, Book 2, p. 2305.
6 Walter Pincus, *Washington Post*, 27 October 1981, p. A10.
7 Walter Pincus, *Washington Post*, 11 December 1981, p. A9.
8 *Military Applications of Nuclear Technology*, Part 1, pp. 10, 12.
9 Ibid.
10 *Military Applications of Nucler Technology*, Part 3, p. 101.

W82

FUNCTION: Warhead for the XM-785 Artillery Fired Atomic Projectile (AFAP) for 155mm artillery, to replace the current W48/M454 AFAP.

WARHEAD MODIFICATIONS: Warhead is built in components, allowing easy conversion from fission yield to enhanced radiation yield.

SPECIFICATIONS:
Yield: less than 2 Kt;[1] exceeds that of the 0.1 Kt W48;[2] capable of accepting an ER option convertible in the field[3]

Weight: circa 95 lb, ballistically similar to M549 rocket assist projectile (RAP) conventional round[4]

Dimensions:
 Length: 34.3 in
 Diameter: 6 in

Materials: if fission yield,[5] probably with plutonium, utilizing large amounts of tritium for ER version

SAFEGUARDS AND ARMING FEATURES: Category D PAL, nonviolent command disable in M617 storage container,[6] improved radar fuzing for more accurate height of burst[7]

DEVELOPMENT:
Laboratory: LLNL

History:
 IOC: 1986
 May 1976: feasibility study completed (Phase 2)[8]
 Sep 1977: Lab assignment (Phase 3) (through FY 1983)[9]
 Dec 1980: production engineering (Phase 4) deferred[10]
 1986: initial deployment

Production Period: 1986-

DEPLOYMENT:
Number Planned: 1000 initially (1983); as many as 2500 overall

Delivery System: dual capable M198 and M109/A1/A2/A3 155mm howitzers; older 155mm howitzers

Service: Army and Marine Corps[11]

Allied User: NATO artillery units; the projectile is "compatible with the new family of howitzers being developed by the NATO allies"[12]

Location: United States, Greece, Italy, West Germany, Turkey, South Korea

9
W82

COMMENTS: Reagan FY 1982 DOD budget requested $44.4 million for the W82.[13] The FY 1982 supplemental request for DOE also requested $7.5 million to initiate construction on W82 production facilities.[14] W82 eliminates "current projectile deficiencies in range, vulnerability, fuzing, and yield."[15] It includes a rocket assist module which extends the range over that of the M-454 AFAP. The range will be up to 30 km (in the new M-198 howitzer) and 24 km in the M-109A1 howitzer, compared to 16 km for the present projectile.[16] It is ballistically similar to a conventional round, precluding the need for a special spotting round.[17] The new projectile is more accurate than the W48, with same accuracy as the current conventional round.[18] Chamberlain Manufacturing Co. received an Army contract on 21 March 1980 for $6 million for development of XM-785 components. Training rounds include XM820 "Type X" and XM841 "Type W."

---

1 Walter Pincus, *Washington Post*, 27 October 1981, p. A10.
2 ACDA, FY 1982 ACIS, p. 280.
3 JCS, FY 1981, p. 47.
4 HASC, FY 1981 DOD, Part 4, Book 2, p. 2305.
5 JCS, FY 1982, p. 78; HASC, FY 1981 DOD, Part 4, Book 2, p. 2305; ACDA, FY 1981 ACIS, p. 278.
6 ACDA, FY 1979 ACIS, p. 136; ACDA, FY 1980 ACIS, p. 151.
7 ACDA, FY 1971 ACIS, p. 136; ACDA, FY 1980 ACIS, p. 155.
8 ACDA, FY 1979 ACIS, p. 137.
9 Development engineering requested by DOE; ACDA, FY 1982 ACIS, p. 277; ACDA, FY 1979 ACIS, p. 136. The W82 remained in Phase 3 with no production funds in FY 1979 and FY 1980; SAC, FY 1981 EWDA, p. 818.

10 HASC, FY 1982 DOD, Part 1, p. 45.
11 ACDA, FY 1979 ACIS, p. 136.
12 JCS, FY 1982, p. 78.
13 ACDA, FY 1982 ACIS, p. 278.
14 DOE, FY 1982 Supplemental Request to the Congress, Atomic Energy Defense Activiti[es], March 1982, p. 4.
15 JCS, FY 1982, p. 78.
16 DOD, FY 1981 RDA, p. VII-6.
17 ACDA, FY 1982 ACIS, p. 278.
18 HASC, FY 1981 DOD, Part 4, Book 2, p. 2305.

Atomic Demolition Munitions and Earth Penetration Weapons

Atomic Demolition Munitions and Earth Penetration Weapons

The Army (and Marine Corps) currently have two atomic demolitions munitions (ADMs) deployed: the Medium ADM (MADM) and the Special ADM (SADM) (see Chapter Three for technical description). The MADM is a 1-15 Kt nuclear land mine weighing some 400 pounds, first deployed in 1965. The SADM is a sub-1 Kt nuclear land mine weighing some 150 pounds, first deployed in 1964. MADM is emplaced by engineer teams and carried by jeeps and helicopters. SADM is man portable and emplaced by special forces and commandos teams behind enemy lines. Approximately 600 ADMs are estimated to be deployed, mostly in Europe, South Korea, Guam, and the United States. ADM teams are earmarked to provide ADM support in Allied corps sectors in Central Europe and Italy.[1] A number of NATO engineer units (British, Dutch, and West German) are also trained to emplace ADMs.

ADMs are emplaced in chambers in the ground, on bridges, or in tunnels and dams, and are detonated by timer or remote command. They would be used primarily to disrupt the movement of enemy forces and to make them concentrate in a mass to bypass obstacles (and thus create other targets for nuclear weapons). SADMs would be emplaced behind enemy lines, particularly at airfields, command posts, transportation, communications and industrial terminals, and petroleum supplies.

Work on the next generation of earth penetration weapons (EPW)/demolition munition nuclear weapons has been ongoing since the 1970s. High explosive tests that simulate the effects of low yield buried nuclear munitions on structural targets have been conducted to determine future ADM/EPW requirements. A Tactical Earth Penetrator Warhead compatible with Army missiles was proposed in 1976 and was eventually designed as the W86 EPW for the PERSHING II missile.

The W86 earth penetration warhead was a small diameter,[2] single yield[3] design. Laboratory (Phase 3) work began in 1979[4] with the idea of supplementing ADMs by providing a remote delivery capability to create barriers and a means of attacking hard, point, and subsurface targets with maximum damage and minimum fallout.[5] The W86 warhead was designed to dive about nine stories underground before exploding, to

Figure 9.16 Medium Atomic Demolition Munition (MADM) mock-up.

destroy point targets that require earthshock or cratering as the primary damage mechanism.[6]

During the PERSHING II DSARC II deliberations (February 1979), the Army conducted a study examining EPW needs and effectiveness and, given its potential benefits, recommended to continue its development.[7] The EPW program was then cancelled in January 1981 because the carrying missile had changed from its original battlefield mission in 1976 to a long-range missile with different targets.[8] Cancellation was also due to budget constraints.[9] Development of the warhead was then completed and the technology was put on the shelf pending a future requirement.[10]

Another weapon, the Shallow Burst Munition—a nuclear device used at shallow depths which would provide air blast kills with reductions in thermal and nuclear radiation—was examined during 1976 but was rejected in Phase 1.

A replacement for the current MADM system has also been considered in the form of a modified B61 bomb, designated the Nuclear Cratering Explosive (NCE).[11] The number of ADMs will be gradually reduced as improved conventional capabilities are achieved. Currently there are no plans to replace them with new nuclear systems.[12] It appears that, at least for the present, there are no plans to produce a new ADM or EPW system.

---

1 SASC, FY 1983 DOD, Part 7, p. 4386.
2 HASC, FY 1982 DOE, p. 218.
3 ACDA, FY 1981 ACIS, p. 235.
4 Remained in Phase 3 during FY 1980; SAC, FY 1981 EWDA, p. 818.
5 HAC, FY 1980 DOD, Part 2, p. 863.
6 ACDA, FY 1979 ACIS, p. 115; USA, "Equipping the Army of the Eighties, A Statement to the Congress on the FY 1981 ARMY RDTE and Procurement Appropriations," n.d., p. 33.
7 HAC, FY 1982 DOD, Part 2, p. 576.
8 DOD, FY 1982 RDA, p. VII-8; HASC, FY 1981 DOD, Part 4, Book 2, p. 2309.
9 HASC, FY 1981 DOD, Part 4, Book 2, p. 2309.
10 HAC, FY 1982 EWDA, Part 5, p. 34; DOD, FY 1982 RDA, p. VII-8.
11 USANCA "Materiel and Safety Division Activities, (January 1976 through December 1976)," n.d.; (July 1976 through June 1977), n.d.
12 DOD, FY 1983 RDA, p. VII-14.

GLOSSARY

Glossary

Terms

Glossary of Terms

ABM System:
A system to counter strategic ballistic missiles in flight, and consisting of: 1) ABM interceptor missiles; 2) ABM launchers; 3) ABM radars, which are radars constructed and deployed for an ABM role, or of a type tested in an ABM mode.

Air-Breathing of Missile:
A missile with an engine requiring the intake of air for combustion of its fuel, as in a ramjet or turbojet. (To be contrasted with the rocket-powered missile, which carries its own oxidizer and can operate beyond the atmosphere.)

Airburst:
An explosion of a nuclear warhead above the surface as distinguished from an explosion on contact with the surface or after penetration. Also, the explosion of a nuclear weapon in the air, at height greater than the maximum radius of the fireball.

Air Defense:
Defensive measures designed to destroy attacking enemy aircraft or missiles in the earth's envelope of atmosphere, or to nullify or reduce the effectiveness of such attack.

Air-Launched Cruise Missile (ALCM):
A cruise missile transported aloft by a carrier aircraft and launched from that aircraft in flight.

Air-to-Air Missile (AAM):
A missile launched from an aircraft at a target above the surface.

Air-to-Surface Missile (ASM):
A missile launched from an aircraft to impact on a surface target.

Anti-Ballistic Missile (ABM):
A defense missile used to intercept and destroy or otherwise neutralize an attacking ballistic missile in the upper reaches of the atmosphere and beyond (endoatmosphere and exoatmosphere).

Anti-Submarine Warfare (ASW):
Operations conducted with the intention of denying the enemy the effective use of submarines.

Arming:
As applied to weapons and ammunition, the changing from a safe condition to a state of readiness for initiation.

Atomic Bomb:
An explosive projectile (usually, a gravity bomb) whose warhead contains nuclear-fissionable radioactive materials as the explosive charge, producing nuclear fusion or fission effects to destroy a target. More narrowly, a fission bomb (see), as distinguished from fusion, or hydrogen bomb.

Atomic Demolition Munition (ADM):
Nuclear device designed to be detonated on or below the surface, or under water, to block, deny and/or canalize enemy forces.

Avionics:
The application of electronics to aviation and astronautics.

Ballistic Missile:
Any missile designed to follow the trajectory that results when it is acted upon predominantly by gravity and aerodynamic drag after thrust is terminated. Ballistic missiles typically operate outside the atmosphere for a substantial portion of their flight path and are unpowered during most of the flight.

Glossary

Terms

Ballistic Missile Defense (BMD): Measures for defending against an attack by ballistic missiles; for example, a system composed of antiballistic missiles and radar and control equipment designed to intercept and destroy attacking ballistic missiles before they reach their targets.

Ballistic Trajectory: The trajectory traced after the propulsive force is terminated and the body is acted upon only by gravity and aerodynamic drag.

Beam Rider/Riding:
1. A missile guided by a radar, radio, or laser beam.
2. A missile guided by an electronic beam.

Blast: The brief and rapid movement of air vapor or fluid away from a center of outward pressure, as in an explosion or in the combustion of rocket fuel; the pressure accompanying this movement. This term is commonly used for "explosion," but the two terms may be distinguished.

Bomber (Light, Medium, Heavy):
1. Light: A bomber designed for a tactical operating radius of under 1000 nautical miles at design gross weight and design bomb load.
2. Medium: A bomber designed for a tactical operating radius of between 1000 and 2500 nautical miles at design gross weight and design bomb load.
3. Heavy: A bomber designed for a tactical operating radius over 2500 nautical miles at design gross weight and design bomb load.

Booster: An auxiliary or initial propulsion system which travels with a missile or aircraft and which may or may not be separated from the parent craft when its impulse has been delivered. A booster system may contain or consist of one or more units.

Bus: The projectile of a missile, with multiple reentry vehicles (MRVs), including the RVs, guidance system, propellant, and thrust device for altering the ballistic flight path so that RVs can be ejected sequentially toward respective targets. Also known as post-boost vehicle (PBV).

Circular Error Probable (CEP): A measure of the delivery accuracy of a weapon system. It is the radius of a circle around a target of such size that a weapon aimed at the center has a 50% probability of falling within the circle.

Collateral Damage: Physical harm inflicted by intent or otherwise on persons and property as a result of attack (specifically, nuclear attack) on a primary military target.

Combat Radius: The maximum distance which an operational aircraft characteristically armed for a combat mission can fly unrefueled from its starting point and return safely, allowing for fuel expenditure involved in combat action typical of the mission profile.

Command Disable System (CDS): A device integrated in a storage container to disable a nuclear warhead by destroying critical components. Cannot be activated until a code is inserted.

Counterforce: The employment of strategic air and missile forces in an effort to destroy, or render impotent, military capabilities of an enemy force.

Countervalue: The employment of strategic air or missile forces to attack selected enemy population centers, industries, and resources and installations which constitute the social fabric of the nation.

Glossary

Terms

Cruise Missile:
: A guided missile which uses aerodynamic lift to offset gravity and propulsion to counteract drag. A cruise missile's flight path remains within the Earth's atmosphere.

Cruise Missile Carrier:
: An aircraft equipped for launching a cruise missile.

Decoy:
: A model, electromagnetic reflector, or other device accompanying a nuclear weapon delivery vehicle in order to mislead enemy defensive systems so as to increase the probability of penetration and weapon delivery.

Delivery System:
: An aerospace vehicle considered as a whole, with all associated components, and integral with launchers and other installations employed in transporting, launching, targeting, guiding, and delivering on target its nuclear weapon(s).

Dual-Capable Weapons:
: Weapons, weapons systems, or vehicles capable of selective equipage with nuclear or non-nuclear munitions.

Electromagnetic Pulse (EMP):
: The electromagnetic radiation from a nuclear explosion, caused by Compton-recoil electrons and photoelectrons from photons scattered in the materials of the nuclear device, in a surrounding medium. The resulting electric and magnetic fields may couple with military systems to produce damaging current and voltage source.

Electronic Countermeasures (ECM):
: Electronic warfare involving actions taken to prevent or reduce the effectiveness of enemy equipment and tactics employing or affected by electromagnetic radiations, and to exploit the enemy's use of such radiations.

Endoatmosphere:
: From sea level to about 40 nautical miles altitude.

Enhanced Radiation (ER):
: The effects of and the technology employed in that class of controlled-effects nuclear weapons designed to intensify nuclear radiation in the target area by attenuating blast and heat.

Equivalent Megatonnage (EMT):
: A measure used to compare the destructive potential of differing combinations of nuclear warhead yield against relatively soft countervalue targets. EMT is computed from the expression: $EMT = NY^x$, where N = number of actual warheads of yield Y; Y = yield of the actual warheads in megatons; and x = scaling.

Exoatmosphere:
: Higher than about 40 nautical miles above sea level.

Externally Observable Differences:
: Externally observable design features used to distinguish between those heavy bombers of current types (and air-launched cruise missiles) which are capable of performing a particular SALT-limited function and those which are not. These differences need not be functionally related but must be a physical design feature which is externally observable.

Fallout:
: The precipitation to earth of radioactive particulate matter from a nuclear cloud; also applied to the particulate matter itself.

Fighter-Bomber:
: Tactical aircraft configured for ground attack and interdiction as well as for air combat. As dual-capable systems, fighter-bombers (such as F-111s) constitute a non-central system with potential for strategic missions.

Fission:
: The process whereby the nucleus of a particular heavy element splits into (generally) two nuclei of lighter elements, with the release of substantial amounts of energy.

Glossary

Terms

Fission Weapon: Nuclear warhead whose material is uranium or plutonium which is brought to a critical mass under pressure from a chemical explosive detonation to create an explosion that produces blast, thermal radiation, and nuclear radiation. The complete fission of one pound of fissionable material would have a yield equivalent to 8000 tons of TNT. Commonly known as atomic bomb.

Forward Based Systems: A term introduced by the U.S.S.R. to refer to those U.S. nuclear systems based in third countries or on aircraft carriers and capable of delivering a nuclear strike against the territory of the U.S.S.R.

Fractionation: The division of the payload of a missile into several warheads. The use of a MIRV payload is an example of fractionation.

Functionally Related Observable Differences (FROD): The means by which SALT II provides for distinguishing between those aircraft which are capable of performing certain SALT-limited functions and those which are not. FRODs are differences in the observable features of airplanes which specifically determine whether or not these airplanes can perform the mission of a heavy bomber, or whether or not they can perform the mission of a bomber equipped for cruise missiles of a range in excess of 600 km, or whether or not they can perform the mission of a bomber equipped for Air-to-Surface Ballistic Missiles (ASBMs).

Fusion: The process accompanied by the release of tremendous amounts of energy, whereby the nuclei of light elements combine to form the nucleus of a heavier element.

Fusion Weapon: Nuclear warhead containing fusion materials (e.g., deuterium and tritium) which are brought to critical density and temperature conditions by use of a primary fission reaction (thermonuclear) in order to initiate and sustain a rapid fusion process, which in turn creates an explosion that produces blast, thermal radiation, and nuclear radiation. The complete fusion of one pound of fusion material is equivalent to 36,000 tons of TNT. Commonly known as hydrogen bomb.

Ground Alert: That status in which aircraft on the ground/deck are fully serviced and armed, with combat crews in readiness to take off within a specified short period of time (usually 15 minutes) after receipt of a mission order.

Guidance: The entire process by which target intelligence information received by the guided missile is used to effect proper flight control to cause timely direction changes for effective target interception.

Guided Missile: An unmanned vehicle moving above the surface of the earth, whose trajectory or flight path is capable of being altered by an external or internal mechanism.

Half-Life: The time required for the activity of a given radioactive species to decrease to half of its initial value due to radioactive decay. The half-life is a characteristic property of each radioactive species and is independent of its amount or condition. The half-life of tritium is 12.3 years.

Hard Target: Any weapon site, command and control facility, production center, blast shelter or other strategic target which has been hardened for protection against the effects of nuclear attack.

Glossary

Terms

Height of Burst (HOB):	1. The vertical distance from the earth's surface or target to the point of burst. 2. For nuclear weapons, the optimum height of burst for a particular target (or area) is that at which it is estimated a weapon of a specific energy yield will produce a certain desired effect over the maximum possible area.
High Explosive (HE):	Generally applied to the bursting charges for bombs, projectiles, grenades, mines, and demolition charges.
Homing:	The technique of tracking along a position line toward the point of origin of a radio, radar or other navigation aid.
Homing Overlay Experiment (HOE):	The HOE is designed to demonstrate the ability of optics to acquire targets in flight; isolate RVs from accompanying chaff, penetration aids, and booster fragments; and guide the missile to intercept with a goal of a miss distance small enough to permit RV destruction by other than nuclear means. HOE would demonstrate the capability and illustrate the advantages of ex-oatmosphere, non-nuclear intercept at relatively long ranges.
Howitzer:	A cannon which combines certain characteristics of guns and mortars. The howitzer delivers projectiles with medium velocities, by either low or high trajectories.
Hydrogen Bomb:	A nuclear weapon in which part of the explosive energy is obtained from nuclear fusion (or thermonuclear) reaction.
Inertial Confinement:	A concept for attaining the density and temperature condition that will produce nuclear fusion by use of lasers or other high power sources to compress and heat small pellets containing fusionable fuel. The energy released is in the form of fast neutrons, X-rays, charged particles, and debris, and can be used in much the same way as the energy output of any other fusion (or fission) process.
Inertial Guidance:	A guidance system designed to project a missile over a predetermined path, wherein the path of the missile is adjusted after launch by devices wholly within the missile and independent of outside information. The system measures and converts accelerations experienced to distance traveled in a certain direction.
Initial Operational Capability (IOC):	The date when the first combat missile unit is equipped and trained, and logistic support established to permit performance of combat missions in the field. An initial operational capability date is associated with each new missile system as a target date for delivery of combat equipment, repair parts, maintenance equipment, and publications, plus supply of trained personnel.
Intercontinental Ballistic Missile (ICBM):	A land-based fixed or mobile rocket-propelled vehicle capable of delivering a warhead to intercontinental ranges. Once they are outside the atmosphere, ICBMs fly to a target on an elliptical trajectory. An ICBM consists of a booster, one or more reentry vehicles, possibly penetration aids, and, in the case of a MIRVed missile, a post-boost vehicle.
Intermediate Range Ballistic Missile (IRBM):	A ballistic missile, with a range capability from about 1500 to 3000 nautical miles.

Glossary

Terms

Kiloton (Kt): A unit of measure of a nuclear weapon's yield, equivalent to the explosive energy of one thousand tons of TNT. Thirteen kilotons was the approximate yield of the atomic bomb detonated at Hiroshima.

Kiloton Weapon: A nuclear weapon, the yield of which is measured in terms of thousands of tons of trinitrotoluene (TNT) explosive equivalents, producing yields from 1 to 999 kilotons.

Launch Weight: The weight of the fully loaded missile itself at the time of launch. This would include the aggregate post-boost vehicle (PBV) and the payload.

Laydown: Weapons employment from an aircraft where a delayed fuzing and arming of the warhead permits low level delivery and safe escape.

Mach Number: The ratio of the velocity of a body to that of sound in the surrounding medium.

Maneuverable Reentry Vehicle (MaRV): A reentry vehicle capable of performing preplanned flight maneuvers during the reentry phase.

Maximum Range: The greatest distance a weapon can fire without consideration of dispersion, or the greatest distance a weapon system can fly.

Medium-Range Ballistic Missile (MRBM): A ballistic missile with a range capability from about 600 to 1500 nautical miles.

Megaton (Mt): A unit of measurement for nuclear yield equivalent to the energy released from one million tons of TNT.

Midcourse Guidance: The guidance applied to a missile between termination of the launching phase and the start of the terminal phase of flight.

"Mod" Designator Number: Modifications made to the major assembly design of a weapon system. Mod-0 is the first version of a weapon design, with subsequent modifications of the weapon design numbered consecutively.

Multiple Independently-Targetable Reentry Vehicle (MIRV): Multiple reentry vehicles carried by a ballistic missile, each of which can be directed to a separate and arbitrarily located target. A MIRVed missile employs a post-boost vehicle (PBV) or other warhead dispensing mechanism. The dispensing and targeting mechanism maneuvers to achieve successive desired positions and velocities to dispense each RV on a trajectory to attack the desired target. Alternately, the RVs might themselves maneuver toward their targets after they reenter the atmosphere.

Multiple Reentry Vehicle (MRV): The reentry vehicle of a ballistic missile equipped with multiple warheads where the missile does not have the capability of independently targeting the reentry vehicles—as distinct from a missile equipped for MIRVs.

Nuclear Radiation: Particulate and electromagnetic radiation emitted from atomic nuclei in various nuclear processes. The important nuclear radiations, from the weapons effects standpoint, are alpha and beta particles, gamma rays, and neutrons.

Nuclear Weapon: A device in which the explosion results from the energy released by nuclear reactions involving atomic nuclei; either fission, fusion, or both.

Nuclear Yield: The energy released in the detonation of a nuclear weapon, measured in terms of kilotons or megatons of trinitotluene explosive (TNT) required to produce the same energy release. Yields are categorized as: Very Low—less than 1 kiloton; Low—1 kiloton to 10 kilotons; Medium—over 10 kilotons to 50 kilotons; High—over 50 kilotons to 500 kilotons; Very High—over 500 kilotons.

Optimum Height: The height of an explosive which will produce the maximum effect against a given target.

Glossary

Terms

Over Pressure:
: The pressure resulting from the blast wave of an explosion. It is referred to as "positive" when it exceeds atmospheric pressure and "negative" during the passage of the wave, when resulting pressures are less than atmospheric pressure.

Payload:
: Weapons and penetration aids carried by a delivery vehicle. In the case of a ballistic missile, the RV(s) and antiballistic missile penetration aids placed on ballistic trajectories by the main propulsion stages or the PBV; in the case of a bomber, those bombs, missiles, or penaids carried internally or attached to the wing or fuselage.

Penetration Aids (Active and Passive):
: Devices employed by offensive weapon systems, such as ballistic missiles and bombers, to increase the probability of penetrating enemy defenses. They are frequently designed to simulate or to mask an aircraft or ballistic missile warhead in order to mislead enemy radar and/or divert defensive antiaircraft or antimissile fire.

Permissive Action Link (PAL):
: A coded switch which serves as a mechanical supplement to the administrative controls exercised over nuclear weapons employment. When installed, they make weapon-enabling, or access to the warhead itself, dependent upon possession of the code.

Personnel Reliability Program (PRP):
: Program in which individuals who have responsibilities in the nuclear release process are kept under scrutiny to determine if behavior affects the conduct of the work.

Post-Boost Vehicle (PBV):
: That part of a missile which carries the reentry and thrust devices for altering the ballistic flight path so that the reentry vehicles can be dispensed sequentially toward different targets (MIRVs). Ballistic missiles with single RVs also might use a PBV to increase the accuracy of the RV by placing it more precisely into the desired trajectory.

Projectile:
: An object projected by an applied exterior force and continuing in motion, as an artillery shell.

Propellant:
: That which provides the energy required for propelling a projectile. Specifically, an explosive charge for propelling a bullet, shell or the like; also a fuel, either solid or liquid, for propelling a rocket or missile.

Radar:
: *Radio Detection And Ranging* equipment that determines the distance and usually the direction of objects by transmission and return of electromagnetic energy.

Radar Cross-Section (RCS):
: The image produced by radar signals reflected from a given target surface. Because the size of the image is a function not only of the target's size, but of structural shape and the refractory characteristics of its materials as well, radar cross-section is an important design characteristic for air and space vehicles.

Radius of Action:
: The maximum distance a ship, aircraft, or vehicle can travel away from its base along a given course with normal combat load and return without refueling, allowing for all safety and operating factors.

Ramjet:
: A jet propulsion engine containing neither compression nor turbine, which depends for its operation on the air compression accomplished by the forward motion of the engine.

Glossary

Terms

Range:
1. The distance between any given point and an object or target.
2. Extent or distance limiting the operation or action of something, such as the range of an aircraft, ship, or gun.
3. The distance which can be covered over a hard surface by a ground vehicle, with its rated payload, using the fuel in its tank and in cans normally carried as part of the ground vehicle equipment.
4. Area equipped for practice in shooting at targets.

Reduced Blast/Enhanced Radiation Weapon (RB/ER): A nuclear weapon designed to produce significantly more and/or higher energy output(s) of neutron, X-ray, gamma rays, or a combination thereof than a normal weapon of the same total yield.

Reentry Vehicle (RV): That portion of a ballistic missile which carries the nuclear warhead. It is called a reentry vehicle because it reenters the earth's atmosphere in the terminal portion of the missile trajectory.

Residual Radiation: Nuclear radiation caused by fallout, radioactive material dispersed artificially, or irradiation which results from a nuclear explosion and persists longer than one minute after burst.

SAFEGUARD: A ballistic missile defense primarily designed to protect U.S. land-based retaliatory forces against direct attack, and protect the U.S. against a possible accidental launch or small attack. The principal subsystems were the SPRINT and SPARTAN missiles, Missile Site Radar, Perimeter Acquisition Radar, and the Data Processing System.

Short-Range Attack Missile (SRAM): An air-to-surface missile with a range under 600 miles (and generally under 100 miles) carried by U.S. B-52 and FB-111 bomber aircraft as penetration aids for suppression of enemy air defenses.

Short-Range Ballistic Missile (SRBM): Land-based, rocket-propelled vehicle capable of delivering a warhead through space to a target at ranges up to about 600 nautical miles. The U.S. PERSHING and LANCE, and Soviet SCUD, are tactical missile systems classified as SRBMs.

Silo: Hardened, underground facility for a fixed-site ballistic missile and its crew, designed to provide pre-launch protection and to serve as a launching platform. High-yield, precision nuclear weapons are required to destroy a silo construction.

Stockpile: Nuclear storage. Also, the total number of nuclear weapons which a nation maintains in storage at all locations and potentially available for deployment.

Strategic Forces: Nuclear weapons and delivery systems designed for nuclear attack against strategic targets or for active defense against such an attack: bombers, missile systems, and strategic interceptors. Commonly refers to offensive weapons in the U.S. and the U.S.S.R. that can deliver a nuclear strike on each other or a third party.

Strategic Offense: Forces and measures existing to mount a nuclear attack against enemy strategic targets, designed to destroy the enemy's war-making capacity.

Sub-Kiloton Weapon: A nuclear weapon producing a yield below one kiloton.

Glossary

Terms

Submarine-Launched Ballistic Missile (SLBM): A ballistic missile carried in and launched from a submarine, which affords mobility and concealment for a missile force. The SALT II Treaty includes the following definition: "submarine-launched ballistic missile (SLBM) launchers are launchers of ballistic missiles installed on any nuclear-powered submarine or launchers of modern ballistic missiles installed on any submarine, regardless of its type." [Article II (2)]

Surface-to-Air Missile (SAM): A surface-launched missile designed to operate against a target above the surface.

Surface-to-Surface Missile (SSM): A surface-launched missile designed to operate against a target on the surface.

Tactical Nuclear (Forces, Weapons): The use of nuclear weapons by land, sea, or air forces against opposing forces. Also supporting installations or facilities, in support of operations, which contribute to the accomplishment of a military mission of limited scope, or in support of the military commander's scheme of maneuver, usually limited to the area of military operations.

Terminal Guidance: The guidance applied to a missile between mid-course guidance and its arrival in the vicinity of the target.

Terrain Contour Matching (TERCOM): Guidance system, presently employed in cruise missiles, which correlates preprogrammed contour map data with the terrain being overflown, in order to take periodic fixes and adjust the flight path accordingly. TERCOM improves the accuracy provided by inertial guidance alone.

Theater: The geographical area outside the continental United States for which a commander of a unified or specified command has been assigned military responsibility.

Thermonuclear Weapon: A weapon in which very high temperatures are used to bring about the fusion of light nuclei, such as those of hydrogen isotopes (e.g., deuterium and tritium), with the accompanying release of energy. The high temperatures required are obtained by means of an atomic (fusion) explosion.

Throw-Weight: Ballistic missile throw-weight is the useful weight which is placed on a trajectory toward the target by the boost or main propulsion stages of the missile. For the purposes of SALT II, throw-weight is defined as the sum of the weight of:
- the RV or RVs;
- any PBV or similar device for releasing or targeting one or more RVs; and
- any antiballistic missile penetration aids, including their release devices.

Transporter-Erector-Launcher (TEL): The vehicle designed to move a land-based mobile missile within its shelter and to break through the overhead cover, raise the missile into firing position, and serve as a platform for the launch of the missile.

Triad: The tripartite U.S. strategic deterrent force, which consists of land-based ICBMs, submarine-launched ballistic missiles, and strategic bombers. The capabilities and characteristics of each system complement the others. Disproportionate reliance on any one system is avoided, so that the ends of deterrence and stability are served, and the risks of technological surprise are reduced.

Turbojet Engine: A jet engine whose air is supplied by a turbine driven compressor, the turbine being activated by exhaust gases.

Glossary

Terms

Warhead: The part of a missile, projectile, torpedo, rocket, or other munition which contains either the nuclear or the thermonuclear system, high explosive system, chemical or biological agents, or inert materials, intended to inflict damage.

Yield: The energy released in an explosion. The energy released in the detonation of a nuclear weapon is generally measured in terms of the kilotons (Kt) or megatons (Mt) of TNT required to produce the same energy release.

Glossary

Abbreviations and Acronyms

Glossary of Abbreviations and Acronyms

AAM	Air-to-Air Missile	ANGB	Air National Guard Base
AB	Airbase	AOE	Ammunition Ship
ABM	Anti-Ballistic Missile	AP	Airport
ABRES	Advanced Ballistic Reentry Systems	AMaRV	Advanced Maneuvering Reentry Vehicle
ABRV	Advanced Ballistic Reentry Vehicle	AS	Submarine Tender (Ship)
		ASALM	Advanced Strategic Air-Launched Missile
ACDA	Arms Control and Disarmament Agency	ASAT	Anti-Satellite
ACIS	Arms Control Impact Statement	ASBM	Air-to-Surface Ballistic Missile
		ASM	Air-to-Surface Missile
ACMT	Advanced Cruise Missile Technology	ASMS	Advanced Strategic Missile System
AD	Destroyer Tender (Ship)	ASROC	Anti-Submarine Rocket
ADCOM	Aerospace Defense Command	ASW	Anti-Submarine Warfare
ADM	Atomic Demolition Munition	ASWSOW	Anti-Submarine Warfare Stand-Off Weapon
AE	Ammunition Ship		
AEC	Atomic Energy Commission	ATB	Advanced Technology Bomber
AFAP	Artillery-Fired Atomic Projectile	ATBM	Anti-Tactical Ballistic Missile
		ATP	Advanced Technology Program
AFB	Air Force Base		
AFM	Air Force Manual	AW&ST	*Aviation Week & Space Technology*
AFR	Air Force Regulation		
AGM	Air-to-Surface Missile	AWACS	Airborne Warning and Control System
AIR	Air-to-Air Missile		
AIRS	Advanced Inertial Reference Sphere	BB	Battleship
		BDM	Bomber Defense Missile
ALBM	Air-Launched Ballistic Missile	BMD	Ballistic Missile Defense
ALC	Air Logistics Center	CANTRAC	Catalog of Navy Training Activities
ALCM	Air-Launched Cruise Missile		
AMAC	Airborne Monitoring and Control System	CDS	Command Disable System
		CEP	Circular Error Probable
AMSA	Advanced Manned Strategic Aircraft	CG	Guided Missile Cruiser
		CGN	Nuclear Powered Guided Missile Cruiser
ANG	Air National Guard		

324 Nuclear Weapons Databook, Volume I

Glossary

Abbreviations and Acronyms

cm	Centimeter	FY	Fiscal Year
CMCA	Cruise Missile Carrier Aircraft	GAO	General Accounting Office
CMI	Cruise Missile Integration	GLCM	Ground-Launched Cruise Missile
CMP	Counter Military Potential	HAC	House Appropriations Committee
CONUS	Continental United States	HASC	House Armed Services Committee
CSB	Closely Spaced Basing	HE	High Explosive
CSWS	Corps Support Weapon System	HEU	Highly Enriched Uranium
CV	Aircraft Carrier	HOB	Height Of Burst
DARPA	Defense Advanced Research Projects Agency	HOE	Homing Overlay Experiment
DD	Destroyer	IAP	Improved Accuracy Program
DDG	Guided Missile Destroyer	IAP	International Airport
DOD	Department of Defense	ICBM	Intercontinental Ballistic Missile
DOE	Department of Energy	IHE	Insensitive High Explosives
DSARC	Defense Systems Acquisition Review Council	in	Inch
DU	Depleted Uranium	IOC	Initial Operational Capability
ECM	Electronic Counter Measure	IR	Infrared
EMP	Electro-Magnetic Pulse	IRBM	Intermediate-Range Ballistic Missile
EMT	Equivalent Megatonnage	JCMPO	Joint Cruise Missile Program Office
EPW	Earth Penetrator Weapon/Warhead	JCS	Joint Chiefs of Staff
ER	Enhanced Radiation ("Neutron Bomb")	JTACMS	Joint Tactical Missile System
ERB	Extended-Range Bomb	kg	Kilogram
ERDA	Energy Research and Development Administration	km	Kilometer
EWDA	Energy and Water Development Appropriations	Kt	Kiloton
FBM	Fleet Ballistic Missile	LANL	Los Alamos National Laboratory
FBS	Forward Based Systems	LLNL	Lawrence Livermore National Laboratory
FEBA	Forward Edge of the Battle Area	LoADS	Low Altitude Defense System
FF	Frigate	LRCA	Long-Range Combat Aircraft
FFG	Guided Missile Frigate	LRTNF	Long-Range Theater Nuclear Forces
FM	Field Manual	m	Meter, million
FRG	Federal Republic of Germany	MADM	Medium Atomic Demolition Munition
FROD	Functionally Related Observable Difference	MAPS	Multiple Aim Point System
ft	Feet	MaRV	Maneuvering Re-entry Vehicle
FUFO	Full Fuzing Option		

Glossary

Abbreviations and Acronyms

MCAS	Marine Corps Air Station	RB/ER	Reduced Blast/Enhanced Radiation
mi	Statute Mile	RCS	Radar Cross Section
MIRV	Multiple Independently Targeted Re-entry Vehicle	RDA	Research, Development, and Acquisition
MLRS	Multiple Launch Rocket System	RDT&E	Research, Development, Test, and Evaluation
MPM	Multipurpose Missile	RR	Reduced-Range
MPS	Multiple Protective Shelter	RV	Re-entry Vehicle
MRASM	Medium-Range Air-to-Surface Missile	SAC	Senate Appropriations Committee
MRBM	Medium-Range Ballistic Missile	SAC	Strategic Air Command
MRV	Multiple Re-entry Vehicle	SACEUR	Supreme Allied Command Europe
Mt	Megaton	SADM	Special Atomic Demolition Munition
MT	Metric Ton		
MX	Missile Experimental	SALT	Strategic Arms Limitation Treaty
NAS	Naval Air Station	SAM	Surface-to-Air Missile
NATO	North Atlantic Treaty Organization	SASC	Senate Armed Services Committe
nm	Nautical Mile		
NORAD	North American Aerospace Defense Command	SICBM	Small Intercontinental Ballistic Missile
OAS	Offensive Avionics System	SLBM	Submarine-Launched Ballistic Missile
OSD	Office of the Secretary of Defense	SLCM	Sea-Launched Cruise Missile
P1a	PERSHING 1a Missile	SLEP	Service Life Extension Program
PII	PERSHING II Missile	SNDV	Strategic Nuclear Delivery Vehicle
PAA	Primary Airvehicle Authorized		
PAL	Permissive Action Link	SNM	Special Nuclear Materials
PBV	Post-Boost Vehicle	SRAM	Short-Range Attack Missile
PGRV	Precision Guided Re-entry Vehicle	SRBM	Short-Range Ballistic Missile
		SSBN	Nuclear Powered Ballistic Missile Submarine
POC	Program of Cooperation		
POL	Petroleum, Oil, and Lubricants	SSM	Surface-to-Surface Missile
		SSN	Nuclear-Powered Attack Submarine
PRP	Personnel Reliability Program		
QRA	Quick Reaction Alert	STP	Systems Technology Program
RADAG	Radar Area Correlation Guidance	SUAWACS	Soviet Union AWACS
		SUBROC	Submarine Rocket
RAF	Royal Air Force	SUM	Shallow Underwater Mobile
RAP	Rocked Assisted Projectile	TAC	Tactical Air Command

Glossary

Abbreviations and Acronyms

TASM	TOMAHAWK Anti-Ship Missile	UE	Unit Equipment
TEL	Transporter-Erector-Launcher	UGM	Underwater-to-Surface Missile
TERCOM	Terrain Contour Matching	USA	United States Army
TLAM/C	TOMAHAWK Land-Attack Missile/Conventional	USAF	United States Air Force
		USN	United States Navy
TLAM/N	TOMAHAWK Land-Attack Missile/Nuclear	VHSIC	Very High Speed Integrated Circuits
TNF	Theater Nuclear Forces	VLA	Vertical Launch ASROC
TNW	Theater Nuclear War	VLS	Vertical Launching System
TY	Then Year	W	Warhead

INDEX

Index

A-1 (SKYRAIDER), 210
A-4 (SKYHAWK), 5, 49, 66, 94, 199, 200, 202, 204–206, 210, 212, 257
A-6 (INTRUDER), 5, 42, 49, 66, 92, 198–200, 202, 203, 207–208, 244, 251, 252, 253
A-7 (CORSAIR II), 5, 42, 49, 66, 92, 199, 200, 203, 209–210, 225, 244, 251–253
Advanced Ballistic Reentry Systems (ABRES), 108, 109
Advanced Ballistic Reentry Vehicle (ABRV), 76, 105, 108, 121, 125, 126, 127. See also Missile Reentry Vehicle
Advanced Cruise Missile, 16, 17, 80, 172, 191–192
Advanced Cruise Missile Technology (ACMT) program, 16, 18, 173–175
Advanced Fighter Technology Integration, 204
Advanced Maneuverable Reentry Vehicle (AMaRV), 109–110. See also Missile Reentry Vehicle
Advanced Manned Strategic Aircraft (AMSA), 156
Advanced Strategic Air Launched Missile (ASALM), 17, 18, 39, 71, 80, 155, 193–197, 245, 264, 272. See also Counter SUAWACS
  Lethal Neutralization System, 193, 195
  to replace SRAM, 18, 193, 194
Advanced Strategic Missile System (ASMS), 108, 109
Advanced Tactical Air-Delivered Weapon, 16, 18, 202
Advanced Tactical Fighter, 203–204, 219, 222
Advanced Technology Bomber (ATB) "Steath," 17, 106, 153, 159, 162, 172, 173, 194
AEGIS anti-air defense system, 257, 272, 275, 277. See also *Ticonderoga* class cruisers
Aerojet Strategic Propulsion Co., 111, 113, 124, 136, 168, 188
Aeronca, Inc., 161
Aerospace Defense Command (ADCOM), 214
Airborne Launch Control System (ALCS), 119
Air burst/ground burst, 42, 43, 58, 63, 126, 297
Aircraft, retired, 58
Aircraft carriers (CV), 82, 245, 246, 251–260
Aircraft Monitoring and Control system (AMAC), 198
AiResearch Manufacturing Co., 161, 176
Air Force, 17, 41, 49, 58, 59, 62, 66, 71, 75, 80, 113, 116, 121, 163, 215, 216, 219, 220, 230, 232, 299
  strategic role of, 84–86
  training locations, 85
  warheads used by, 39
  weapons location, 85
Air Launched Cruise Missiles (ALCM), 3, 12, 16, 17, 18, 38, 58, 79, 80, 86, 101, 102, 156, 159, 173–178, 191–193
  subcontractors, 175
Air Logistics Centers (ALC), 85
Air National Guard, 86, 209, 210, 214, 222, 230, 231
Air-to-air missiles, 41, 168, 194. See also GENIE, PHOENIX

Air-to-surface missiles. See B-52, FB-111 Bombers, Short Range Attack Missiles
Alameda Naval Air Station, CA, 94
Alamogordo, NM, Trinity site, 31
Allied countries, 94–97, 202, 280. See also Europe, NATO
Aluminum Co. of America, 176
Ammunition ship (AE), 92
Amphibious Assault Ships (LPH), 89, 244
Amphibious Transport Docks (LPD), 89
Anadyte-Kropp, 176
Andersen AFB, Guam, 86
Anti-air missiles, 4, 5, 52, 89, 273, 275, 277, 287. See also NIKE-HERCULES
Anti-Ballistic Missiles (ABM), 129, 163, 164, 276. See also Low Altitude Defense System, SAFEGUARD, SENTRY
Anti-ship cruise missile. See HARPOON, TOMAHAWK
Anti-submarine aircraft, 5, 236–237, 238–240. See also P-3, ORION, S-3 VIKING
Anti-submarine helicopters, 5, 92. See also SH-60F
Anti-Submarine Rocket (ASROC), 3, 5, 8, 51, 92, 94, 189, 244, 246, 257, 258, 260, 264, 265–268, 271
Anti-Submarine Warfare (ASW), 3, 8, 63, 265, 271
  air-delivered weapons, 16, 18, 39, 51, 64, 254, 266
  standoff weapons (ASWSOW), 244, 246, 265, 271
  SUBROC, 61, 94, 244, 246, 247, 249, 269–270
  surface-delivered weapons, 16, 18, 39, 51, 245
ARMADILLO carrier and launcher, 132, 133
Armed Services and Appropriations Committee, Conference Report, 16
Army, 4, 45, 53, 54, 60, 72, 73, 78, 280–311
  artillery used by, 88
  elimination of non-nuclear weapons in, 88
  National Guard, 282
  nuclear weapons: deployed by, 280–311
    locations of, 87
    training centers, 88
    units, 90, 281
    warheads, 39
  strategic role of, 86–89
  Technical Proficiency Inspection, 83
Artillery Fired Atomic Projectile (AFAP), 47, 54, 77, 309, 310. See also Nuclear artillery
Asia, SAC bases in, 11, 97
Assault breaker program, 298
ASTOR torpedo, 264, 265
Atlantic City AP, NJ, 88
Atlantic Command, 83
Atlantic Research Corp., 173
Atlas D, 12, 102

Nuclear Weapons Databook, Volume I   **331**

Index

Atomic Demolition Munitions (ADM), 2, 3, 5, 11, 89, 90, 91, 96, 280, 281, 311
    deployed in West Germany, 89
    warheads, 12
    See also Medium Atomic Demolition Munitions, Special Atomic Demolition Munitions
Atomic Energy Commission (AEC), 6, 14
Attack submarines (SSN), 5, 92, 244–248, 269. See also HARPOON, Submarine Rockets, TOMAHAWK
AV-8A/B (HARRIER), 91, 199, 202, 203, 206, 208, 211–212
AVCO Corp., 113, 124, 161

B-1 bomber (A/B), 42, 66, 80, 106, 156–161, 172, 174, 175, 194, 198, 200
B-52 bomber (G/H), xiv, 4, 17, 42, 49, 58, 66, 67, 80, 101, 102, 148–151, 154–157, 162, 174–176, 178, 189, 194, 198, 200
B-58 (HUSTLER), 156
B.F. Goodrich Co., 161
Bagotville, Canada, 214
Ballistic Missile Defense (BMD), 128, 129, 131, 163–169. See also Anti-Ballistic Missiles
Ballistic Missile Reentry Vehicles, 106–109, 292, 295. See also Missile reentry vehicles
Ballistic missile submarines, 69. See also POLARIS, POSEIDON, TRIDENT
Bangor, WA TRIDENT base, 94, 104
Barksdale AFB, LA, 85, 176
Baseline Terminal Defense Systems (BTDS), 18, 164
Battleships, 185–186, 244, 246, 255
Belgium, 45, 47, 54, 66, 73, 78, 94, 95, 222, 228, 229, 280, 281, 285, 287, 300, 303, 307
Bell Aerospace Textron, 116
Bell Telephone Labs, 136
Bendix Corp., 161, 289, 293
Bethe, Hans, 25, 29
Bitburg AB, West Germany, 218
Blytheville AFB, AR, 176
Boeing, 113, 118, 124, 132, 133, 148, 150, 154, 155, 161, 165, 172–175
    Boeing/Hughes/LTV, 193
    Boeing/Northrop, 162
Bomber Defense Missile (BDM), 193
Boosted fission weapons, 27. See also Fission
Brown, Harold, 246, 308
Brunswick, Corp., 161
Buchel, West Germany, 229
Burke class destroyers, 186, 244, 246, 255, 256, 261, 262

C-5 transport, 289
C-130 transport, 289
C-141 transport, 289
California class cruisers, 185, 186, 260
Calmendro, 109, 125, 127
Canada, 41, 94, 95, 199, 213
Canberra, 11
Cannon AFB, NM, 233

Capability Inspection, Air Force, 83
Carl Vinson class aircraft carriers, 251, 252
Carswell AFB, TX, 86, 176
Carter, Jimmy, 13, 159, 177, 248, 308
Castle AFB, CA, 88
Central Command, 83
Certification Inspection, Air Force, 83
CF-18, 223
CF-101, 93, 163, 168, 199, 202, 213–214, 224
Chain reaction, fission, 22–23, 25
Chamberlain Manufacturing Co., 78, 310
Chanute AFB, IL, 85
Charles Stark Draper Lab., 124
Cherokee Shot, thermonuclear test, 34
Clark AB, Phillipines, 216
Cleveland Pneumatic Co., 161
Closely Spaced Basing/Dense Pack (CSB), 120, 125, 129, 131
Coded switch system, 160
Cohen, S.T., 28
Collins, John, 100, 112, 115, 119
Comisco, Italy, GLCM base, 180
Command data buffer, 114, 118
Command disable, 30, 49, 54, 64, 72, 77, 160, 182, 200
Commission on Strategic Forces (Scowcroft Commission), 120, 129, 132, 133
Comox, Canada, deployment of CF-101B, 214
Computer Development Corp., 165
Congress, funding for warheads, 16
Consolidated Control Corp., 176
Consolidated Guidance, stockpile recommendations, 83
Continuous Patrol Aircraft (CPA), 130–131
Contractors, 111, 113. See also individual contractor names
Conventional standoff weapon, 298
CORPORAL, 11
Corps Support Command, 88
Corps Support Weapon System (CSWS), 16, 18, 280, 298
Costs of weapons, 70, 74, 122, 135, 140, 151, 159, 177, 180, 187, 190, 195, 206, 208, 210, 212, 214, 218, 221, 225, 235, 237, 240, 250, 254, 259, 261, 268, 276, 286, 291, 296, 299, 303, 305, 307
Counter SUAWACS, 193, 195
Critical mass, 24–25. See also Fission
Cruise Missile Carrier Aircraft (CMCA), 156
Cruise missiles, xiv, 2, 16, 172–195. See also Advanced Cruise Missile, Air Launched Cruise Missile, Ground Launched Cruise Missile, Sea Launched Cruise Missile, TOMAHAWK
Cruisers (CG/CCN), 51, 92, 245, 246, 255, 257–260, 267, 276
Crane Co., 161
"Custodial units" in allied countries, 38, 94
Cutler Hammer, 161
Cyclotol, 41, 42, 45, 59

Damascus AK, silo accident, 112
Davis-Monthan AFB, AZ, 87, 112
DAVY CROCKETT, 33, 34, 60
Defense Advanced Research Projects Agency (DARPA), 29, 191, 192

Index

Defensive forces. *See* Anti-Ballistic Missiles, Ballistic Missile Defense, Strategic Defense System
Deep Basing (DB), 130
DEA, Inc., 176
Delco Electronics, 111, 154, 194
Denmark, 45, 222
Department of Defense (DOD), 6, 15, 89, 110, 193, 296
    DARPA Directed Energy Program, 29, 191–192
    Joint Cruise Missile Project Office, 172
    and nuclear weapons stockpile, 38
    statements on nuclear weapons, 13, 272
    stockpile memorandum, 83
Department of Energy (DOE):
    PANTEX plant, Amarillo, TX, 85, 87, 93
    rate of nuclear arms production, 13, 266
    relationship to DOD, 14
    report on POSEIDON warheads, 70
    responsibility for nuclear weapons, 14
Depth charges, xiv, 2, 61, 63, 246
Derivative Fighter Aircraft, 19
Desired Ground Zeros (DGZ), 112
Destroyer Tender (AD), 92
Destroyers (DD/DDG), 51, 92, 245, 267
Detroit Diesel, 209
Deuterium, 22, 26, 27, 28, 35
Division Support Weapon System, 301
Dock Landing Ships (LSD), 89
D-T boosted fission, 56, 65, 68, 69, 72
D-T reaction, 27
Dyess AFB, TX, 159
Dynamics Research Corp., 124

Eagle Picher Industries, 176
Earth Penetration Warhead (EPW), 292, 311
Economics Technology Association, 124
Eglin AFB, FL, 218
Eisenhower, Dwight, D., and ballistic missile development, 12
Electric Boat, 104
Electrical safety on warheads, 30
Electromagnetic pulse (EMP), 29
Electronics Space Systems Corp., 165
Ellington AFB, TX, 88
Ellsworth AFB, SD, 176
Elmendorf AFB, AK, 88, 216, 218
Emergency Rocket Communications System (ERCS), 115
Emerson Electric, 282
Enewetak Atoll, 26, 34
Enhanced radiation weapons (ER), 14–15, 28–29, 308
    warheads for, 19, 55, 57, 72, 73, 278, 300
    yield of, 29
Environmental Sensing Device (ESD), 30
Ertech Western, 124
Europe, 27, 42, 45, 50, 64, 84
    deployment of US missiles in, xiv, 19, 83, 172, 173, 180, 183, 203, 221, 280, 288, 293, 300, 307, 311
    and SAC bases, 11, 87
    See also Allied countries, NATO

Explosive Technology, 176
Extended Range Bomb (ERB), 202

F-4 (PHANTOM II), 4, 5, 18, 41, 42, 49, 66, 86, 92, 94, 97, 163, 168, 199, 200, 202, 203, 215–216, 219, 221, 222, 225, 231, 251, 252, 253
F-14, 202, 203, 216, 251, 252
F-15 (EAGLE), 4, 5, 18, 19, 41, 49, 66, 157, 163, 168, 199, 203, 216–219, 222, 231
F-16 (FIGHTING FALCON), 5, 18, 19, 49, 66, 85, 86, 199, 200, 202, 204, 220–222, 229, 231
    F-16A, 157, 204
    F-16E, 203
F-18/A-18 (HORNET), 5, 92, 97, 199, 202, 203, 210, 214, 223–225, 252
F-89J, 168
F-100 (SUPERSABRE), 5, 42, 94, 199, 202, 226–227
F-101, 41, 86
F-102, 231
F-104 (STARFIGHTER), 5, 42, 94, 199, 202, 228–229, 242
F-106 (DELTA DART), 4, 18, 41, 86, 163, 168, 199, 202, 216, 219, 222, 230–231
F-111, 5, 19, 59, 66, 156, 199, 200, 202, 203, 219, 222, 232–233
FB-111, 4, 17, 49, 66, 67, 71, 80, 85, 102, 125, 152–154, 155, 156, 194, 199, 200
F.E. Warren AFB, WY, 87, 129
Fairay Corp., 222
Fairchild AFB, WA, 176
FALCON missile, 33, 60
Farm Machinery Corp., 284
FAT MAN, 6, 25, 26, 32, 33, 34
Ferrulmatics Inc., 78
FH-70/SP-70 artillery gun, 301
First strike capabilities, xiv
Fissile materials, 14, 23, 25, 30, 102
    availability of, 6
    core, 24, 27
    as design factor in warheads, 13
    for MX missile, 125, 127
Fission, nuclear, 22–26, 51, 53, 61, 63
Fleet attack SSN, 247
Florennes, Belgium, 180
FMC/Northrop Ordance, 173
Fokker Corp., 222
Force loadings, 4
Ford, Gerald, 77
Ford Motor Co., 289
Fort Bliss, TX, 288
Fort Sill, OK, 285, 290, 294
Fractional crit, 25
Free flight rocket, 4
Fresno Air Terminal, CA, 88
Frigates (FF), 51, 244, 245, 246, 255, 256, 262, 266, 267
Fuels, thermonuclear, 22–29
Fugro National, 124
Fusion weapons, 6, 22, 24, 26–32

Index

G & H Technology, 176
Gadget nuclear explosive device, 31, 32
Gaeta/Naples, Italy, 94
Garrett Corp., 161, 191
Garwin, Richard, 28
Geilenkirchen/Tevren, West Germany, 290
Gelb, Leslie, 133
General Dynamics, 111, 124, 133, 136, 142, 161, 165, 186, 191, 220, 222, 232
General Electric, 110, 111, 124, 136, 142, 161, 162, 165, 175, 203, 215, 228
General Precision, 154
GENIE rocket, 18, 86, 88, 168–169, 199, 213, 216, 218, 230, 231
Getler, Michael, 68, 80, 141
Ghedi-Torre, Italy, 229
Goodyear Aerospace Corp., 161, 265, 270, 293
Goodyear Tire & Rubber Co., 161
Grand Forks AFB, ND, 86, 164, 167, 176
Gray, Colin, 112, 119
Great Falls IAP, MT, 88
Greece, 39, 45, 47, 54, 66, 78, 94, 95, 199, 210, 216, 228, 280, 281, 282, 283, 287, 288, 300, 301, 309
Griffis AFB, NY, 86, 87, 88, 176
Griffiths, David R., 153
Ground Launched Cruise Missiles (GLCM), 15, 18, 80, 172, 173, 179–181, 183, 191, 292, 295
Groton, CT submarine base, 269
Groves, Major General Leslie, 31
Grummen Aerospace, 207, 208
GTE Sylvania, 113, 124, 166, 173
Guam, 86, 94, 311
Gulton Industries, 176, 289
Gun assembly technique, 26
Guns. See Nuclear artillery

H.I. Thompson, 289
Halloran, Richard, 146, 191
Hanford reactor, 11
HARPOON cruise missile, 19, 173, 188–190, 208, 235, 237, 245, 246, 247, 249, 250, 255–258, 261–267
Harris Corp., 161
Harry Diamond Lab, 78
HAWK improved missile, 95
Hector Field Air National Guard Base, ND, 88
Helicopter Combat Support Squadrons (HC), 92
Helicopter Mine Countermeasures Squadrons (HM), 92
Henningson, Durham and Richardson, 124
Hercules, Inc., 113, 124, 136, 142, 161, 282, 287, 298
Hi Shear Corp., 176
Hill AFB, UT, 221
Hiroshima, xiv, 6, 26
Holloman AFB, NM, 218
Holy Loch, United Kingdom, 94, 135
  missile accident in, 70
Homestead AFB, FL, 216
HONEST JOHN missile, 3, 4, 5, 8, 11–13, 19, 90, 91, 94, 95, 280, 282–283, 300

HOUND DOG missile, 43, 101
Howitzer. See Nuclear artillery
Hughes Aircraft, 136, 165, 168, 194, 205, 230
Hughes-Treitler Manufacturing Co., 161
Hydrogen weapons. See Fusion weapons

IBM, 111, 148, 161, 165, 188
Implosion, 6, 11, 26, 33
Improved nuclear torpedo, 245
Indian Head, MD, 85
Indian Ocean, 251
Indian Springs, NV, testing site for GENIE, 168
Insensitive High Explosives (IHEs), 31, 49, 109, 182, 200, 277, 297
Insertable nuclear component, 190
Instruments Systems Corp., 171
Intercontinental Ballistic Missiles (ICBM), 2, 11, 100, 101, 109, 113, 132–133, 164. See also MINUTEMAN, MX, Small ICBMs, TITAN
  advanced mobile, 18, 66
  basing of, 87, 132
  to replace MINUTEMAN, 85, 86, 102
  to replace MX, 102
Intermediate Range Ballistic Missiles (IRBMs), 12
Interstate Electronics Corp., 136
Irvin Industries, 176
Italy, 45, 47, 53, 54, 60, 66, 73, 94, 95, 228, 229, 241, 280, 281, 285, 287, 300, 301, 303, 307, 309, 311
ITT Labs, 136, 148

Jacksonville IAP, FL, 88
Jacobs, Major Roger, 90
Japan, 45, 302, 305, 307
Joint Chiefs of Staff, stockpile recommendations, 28, 83
Joint Planning Assessment Memorandum, 83
Joint Strategic Capabilities Plan, 83
Joint Strategic Planning Document, 83
Joint Tactical Missile System (JTACMS), 280, 298
JUPITER missile, 12

K.I. Sawyer AFB, MI, 86, 88
KC-135 tanker aircraft, 153
Kamen Aerospace Corp., 161
Kadena AB, Japan, 218
Kaplan, Fred, 28, 46, 73, 78
Kelsey Hayes Co., 161
Kevlar-29 parachute, 67
Killian Report, 104
Kings Bay Submarine Base, GA, 94, 135
Kirtland AFB, NM, 29, 79, 85
Kistiakowsky, George, 73, 78
Klein Brogel, Belgium, 229
Kollsman Instrument Co., 173, 176
Kunsan AB, South Korea, 221

LaMaddalena Submarine Base, Italy, 94
Landsberg Lech, West Germany, 290
Langley AFB, VA, 88, 218

Index

LANCE, 5, 19, 73, 88, 90, 91, 94, 95, 285, 286. See Corps Support Weapons Systems (CSWS)
Land-Based Missile System, 100–103. See also MINUTEMANs, MX missile, TITAN
Lawrence Livermore National Laboratory (LLNL), 29, 52, 54, 61, 62, 68, 69, 72, 77, 126, 182, 200, 277, 297, 309
Laydown detonation, 43, 49, 58, 63, 66
Leae Sieglee, 176, 188
Leitenberg, Milton, 2
Lethal Neutralization System, 18, 193, 195
Lithium, 7, 27
Lithium-6 deuteride, 27, 28, 42, 49, 58, 59, 62, 75
LITTLE BOY, 5, 6, 26, 32, 33
LITTLE JOHN, 53
Little Rock AFB, AR, 87, 112
Litton Industries, 154, 173, 176, 194
Lockheed, 104, 110, 135, 142, 145, 154, 165, 173, 191, 228
Logicon, Inc., 124
Long Beach Naval Base, CA, 94
Long range cannon artillery, 301
Long Range Combat Aircraft (LRCA), 156–157
Loring AFB, ME, 86
Los Alamos National Laboratory (LANL), 41, 43, 45, 51, 56–60, 71, 74, 75, 80, 125
Los Angeles class submarines, 186, 247, 249–250, 271
Low Altitude Defense System (LoAD), 18, 131, 164, 166, 167
Low or medium angle loft, 42, 49, 63
LTV Corp., 284
LX-09, 60
LX-10, 69
LX-14, 70

MACE missile, 43
Mach 3 supersonic bomber, 156
Malmstrom AFB, MT, 87, 119
M.A.N. Corp., 173
Maneuvering Reentry Vehicle (MaRV), 16, 18, 145
March AFB, CA, 86
Marconi-Elliot Corp., 220
Mare Island Naval Shipyard, CA, 134
Marine Corps, 47, 49, 53, 54, 60, 63, 66, 78, 244, 281, 300, 301, 302, 304, 306, 308, 309
  Air stations:
    MCAS Cherry Point, NC, 206, 208, 212
    MCAS El Toro, CA, 206, 208, 212, 224
    MCAS Iwakuni, Japan, 206, 208
  CH-46 and CH-53 helicopters, 90
  nuclear capable aircraft, 90
  Nuclear Weapons Acceptance Inspection (NWAI), 83
  role of, 89–91, 163
  units on Navy ships, 89, 90
MARK III Implosion bomb, 24, 32–33
MARK IV Implosion bomb, 32, 33
Martin Marietta Corp., 111, 124, 133, 161, 165, 166, 173, 193–194, 195, 272, 289, 292, 293, 294, 299
Materials, nuclear fuel. See Cyclotol, Deuterium, Lithium, Orraloy, Plutonium, Thorium, Tritium, Uranium

Mather AFB, CA, 86
Mayport Naval Base, FL, 94
McChord AFB, WA, 88, 218
McConnell AFB, KS, 87, 112
McDonnell Douglas Corp., 161, 165, 166, 168, 173, 176, 188, 193, 194, 203, 205, 213, 215, 217, 233, 282, 287
McPhee, John, 32, 60
Mediterranean, US aircraft in, 208, 210, 251
Medium Atomic Demolition Munition (MADM), 5, 52–53, 91, 96, 281, 311
Medium Range Ballistic Missile (MRBM), 292
Megatonage in current nuclear stockpile, 5
Menasco, Inc., 161
Mercury, NV, 122
Microcom Corp., 176
Mike device, 26, 34
Minot AFB, ND, 86, 87, 88
MINUTEMAN I missile, 3, 8, 102, 115, 118
MINUTEMAN II missile, 3, 4, 8, 87, 101, 111, 113–115, 116, 132
MINUTEMAN III missile, 3, 4, 8, 17, 34, 38, 39, 86, 87, 101, 112, 113, 114, 116–119, 120, 167, 292
MIDGETMAN. See Small ICBM
Missile launchers, 265
Missile reentry vehicles:
  Mk-1, 107
  Mk-3 (MIRV), 69, 103–105, 107
  Mk-4 (MIRV), 74, 103–105, 107
  Mk-5, 107, 126
  Mk-6, 59, 107
  Mk-11, 62, 107, 113, 115
  Mk-11C, 62, 103, 107, 115, 156
  Mk-12, 11, 68, 100, 107, 117, 119
  Mk-12A, 68, 75, 76, 100, 107, 117, 118, 119, 124, 127
  Mk-17, 107
  Mk-19, 107
  Mk-20, 107
  Mk-21, (ABRV), 76, 103, 105, 108, 109, 121, 125, 126, 127
  Mk-80, 108
  Mk-81, 108
  Mk-500 EVADER, 108, 110, 142, 143
  Mk-600, 108, 110
MIT Labs, 136, 142, 165
Moody, AFB, GA, 216
Motorola Corp., 78
Mountain Home AFB, ID, 233
Munster warhead, 109, 125
Multiple Independently Targeted Reentry Vehicle (MIRV), 68, 69, 70, 74, 100, 116, 121, 136, 137, 139, 142
Multiple Launch Rocket System (MLRS), 298
Multiple Protective Shelter (MPS), 128, 130, 147
Multiple Reentry Vehicle (MRV), 197. See also Missile reentry vehicles
Multipurpose Missile (MPM), 193
MX/PEACEKEEPER missile, 102, 107, 112, 120–125, 166, 176
  with ABRV, 121, 125, 126, 127

Index

basing, 128–132
costs of, 122
contractors, 125
chronology of development, 125
and Soviet missile silos, xv
stress testing of, 18
validation test of, 123
warhead and RV system, 109, 125

Nagasaki, 6, 26
National Guard, 45. See also Air National Guard, Army National Guard
NATO (North Atlantic Treaty Organization), 53, 54, 60, 63, 66, 73, 83, 87, 95, 96, 172, 180, 199, 202, 220, 221, 280, 281, 286, 290, 295, 300, 302, 306, 308, 309. See also Allied countries, Europe.
Naval Air Development Center, Warminster, PA., 271
Naval Air Stations:
    Atsugi, Japan, 210
    Barbers Point, HI, 235
    Cecil Field, FL, 210, 235
    Jacksonville, FL, 235, 238, 249
    Lemoore, CA, 210, 224
    Miramar, CA, 216
    Moffett Field, CA, 235
    North Island, CA, 235, 238, 240
Naval Coastal Systems Center, Panama City, FL, 271
Naval Detachment, Army Ammunition Plant, McAlister, OK, 93
Naval Ocean Surveillance Center, San Diego, CA, 271
Naval Ordnance Station, Indianhead, MD, 93
Naval Surface Weapons Center, Silver Spring, MD, 173, 271
Naval Underwater Systems Center, Newport, RI, 271
Naval Weapons Stations:
    Charleston, SC, 93, 94, 135
    China Lake, CA, 271
    Concord, CA, 93
    Seal Beach, CA, 93
    Yorktown, VA, 93
Navy, 5, 49, 51, 53, 61, 63, 66, 69, 74, 80, 83, 91–94, 163, 206, 207, 209, 210, 215, 223, 234, 235, 236, 238, 239, 246, 247, 270
Nellis AFB, NV, 85
Neptune anti-submarine aircraft, 94
Netherlands, 45, 47, 54, 63, 73, 78, 94, 95, 199, 222, 229, 234, 235, 280, 281, 285, 287, 300, 301, 303, 307
Neutron bomb. See Enhanced Radiation weapons
New bombers, 156–160
Newport Naval Base, RI, 94
Newport News, VA, 134
New Strategic Air Launched Missile Warhead, 16, 193. See also Lethal Neutralization System
Niagra Falls IAP, NY, 88
NIKE-HERCULES surface-to-air missile, 4, 19, 89, 90, 95
Nitze, Paul, 70, 112, 115, 119, 137, 143
Norfolk Naval Base, VA, 94
North Africa, SAC bases in, 11

North American Aerospace Defense Command (NORAD), 94
North American Corp., 226
Norway, 45, 222
Novaya Zemlya, 34
Nuclear artillery:
    M1, 301
    M2, 301
    M44, 301
    M55, 47, 301
    M59, 301
    M109 155 mm, 54, 88, 90, 95, 281, 300–303, 309
    M110, 8" (203 mm), 54, 88, 90, 95, 300, 301, 306, 307
    M114, 300, 305
    M115, 47, 77, 301
    M198 155 mm, 54, 90, 91, 301, 304–305, 309, 310
    M422, 281
    M454, 281
    M785, 281
Nuclear capable tactical aircraft, 199–243
Nuclear certified unit, 82
Nuclear consent switch, 198
Nuclear Cratering Explosives (NCE), 311
Nuclear land mines. See Atomic Demolition Munitions
Nuclear Powered Ballistic Missile Submarine (SSBN), 140, 147
Nuclear warheads, 12, 38–80
    allocation by service branch, 83
    average age of in stockpile, 40
    development of, 11
    estimated production through 1995, 16
    inactive, 10–11
    safety and control features of, 30–31
    seven phases of, 14, 17
    used in Navy, 39
    variable yields, 38
    See also Warheads
Nuclear weapons:
    aircraft and bombs, 198–243
    Army, 280–311
    accidents, 12–13
    cruise missiles, 172–198
    defined, 22
    delivery systems, inactive, 10–11
    design, 27–28
    Navy, 92, 244–279
    reactor propulsion, 247, 251, 253
    research and miniaturization trends, 14
    safety considerations in design, 12
    by service branch, 82–97
    stockpiles, 3, 6–14, 39
    strategic, 2, 100–171
    and strategy of massive retaliation, 12
    tactical, 3
    technicians, cost of training, 84
    theater, 2–3
Nuclear Weapons Acceptance Inspection (NWAI), 83
Nuclear Weapons Deployment Plan, 83

Index

Nuclear Weapons Development Guidance, 83
Nuclear Weapons Stockpile Memorandum, 83
Nuclear Weapons Support Command, 86

Ohio class submarines (SSBN), 103, 140, 142
Oklahoma Aerotronics, 176
Oliver Hazard Perry class frigates, 261, 262, 263
One point safe, 65, 67, 77, 200, 201
Operation Ivy, 26
Oppenheimer, Robert, 32
Oregon Metallurgical, 289
Orraloy, 41, 47, 49, 58, 59, 65, 77, 79, 126, 182, 200, 277, 297
Osan AB, South Korea, 216
Otis AFB, MA 88

P-3 ORION, 5, 63, 92, 94, 188, 198, 234–235
Pacific region, 83, 84
   Air Force weapons in, 83, 86
   new weapons deployed in, 6
Palomares, Spain, accident, 13
PANTEX Plant, Amarillo, TX, 59, 70, 85, 87, 93
Parker Hannifin Inc., 161
PATRIOT missile, 45, 95, 298, 299
   replacing NIKE-HERCULES, 29, 280–281, 288
Patrol combatants, 255
PB-9502, 65, 79
PB-9505, 42, 65
PEACEKEEPER. *See* MX missile
Pearl Harbor Naval Base, HI, 93, 94, 269
Pease AFB, NH, 86, 153
Pegasus, 255
Penetrator naval depth bombs, 11
Per Udsen Co., 222
Permissive Action Link (PAL), 65, 72, 77, 80, 160, 182, 200, 277, 297, 309
   CAT-B PAL, 65
   CAT-D PAL, 65, 72, 77, 80, 200, 309
   CAT-F PAL, 65, 182, 277, 297
PERSHING 1a, 88, 89, 94, 95
PERSHING II, 57, 132, 133
PERSHING III, 18, 132, 133
Person, Brinkerhoff, Quade and Douglas, 124
Personnel Reliability Program (PRP), 83–84, 88
Phillipines, 50, 216
PHOENIX, air-to-air missile, 18, 41, 202, 245, 252
Physics International, 124
Pincus, Walter, 48, 73, 78, 112, 288, 308, 310
Pits, 26, 38
Pittsburgh Plate & Glass Inc., 161
Plattsburgh AFB, NY, 86, 153
Plutonium, 14, 22, 23, 24, 27, 31, 32, 52, 60, 62, 68, 69, 71, 72, 74, 75, 77, 79, 200, 308, 309
POLARIS, 12, 104, 136, 137
Polmar, Norman, 2, 227, 229, 231, 268, 270
Portsmouth Naval Shipyard, NH, 134
Portland IAP, OR, 88

Portugal, 210
POSEIDON C3 missile system, 3, 4, 8, 92, 102, 104, 136–137
   launch platform, 137
   on POSEIDON submarines, 93
   subcontractors, 136
   warheads, 17
POSEIDON submarines, 17, 74, 134–135, 137
   locations, 135
   warheads, 134, 135
Post Boost Vehicle (PBV), 120
Pratt and Whitney, 148, 152, 203, 205, 207, 217, 220, 226, 230, 232
Programs of Cooperation (POC), 94
Pyronetics Devices, 176

Quick Reaction Alert (QRA), 180, 181, 233, 290, 292

Radar Area Correlation Guidance (RADAG), 204, 205, 292
Radiation from fusion secondary, 27
Ralph M. Parsons Co., 124
Rapid Deployment Force, 83
Raytheon Co., 136, 142, 161, 173, 220
RCA, Princeton Lab, 136
Reagan Administration, 2, 59, 103, 112, 120, 147, 156, 159, 164, 175, 177, 186, 246, 251, 256, 308, 310
   Commission on Strategic Forces, 120, 129, 132, 133
   and MX deployment plans, 62, 120, 122, 125, 218
REDSTONE missile, 290
Reentry vehicles, 106–109
   development and chronology, 108
   maneuverable versus ballistic, 100
   on submarines, 103–105
   See also Missile reentry vehicles
REGULAS cruise missile, 172
Replenishment Oiler, 92
Rimini AB, Italy, 229
Robins AFB, GA, 86
Rocketdyne Division, 124, 284
Rockwell International, 113, 158, 165, 198
Rockwell/Lockheed, 162
Rolls Royce, 210, 211
Rosemont Corp, 176

S-3 VIKING, 5, 63, 93, 198, 236–237, 244, 251, 253
SABCA Corp., 222
SALT (Strategic Arms Limitation Talks), 109, 143
SAFEGUARD ABM system, 38, 163–165
San Diego Naval Base, CA, 94, 269
Sandia National Laboratories, 78, 124, 201, 202
Sandra Corporation, 31
Savannah River reactor, 11
SCAD program, 175
Schlesinger, James, 125
Science Applications, 124
Scoville, Herbert, Jr., 28
Sea-based missile systems, 103–105. *See also* POSEIDON, TRIDENT

Index

Sea-Launched Cruise Missiles (SLCM), 18–20, 100, 187, 191, 244–264. *See also* HARPOON, TOMAHAWK
Selectable yields, 31
Selfridge ANGB, MI, 88
Senate Appropriations Committee, 13, 15
Seneca Army depot, NY, 78, 87
SENTRY, 16, 17, 18
SENTRY missile, 164–167
SERGEANT missile, 12, 19, 280
Service Life Extension Program (SLEP), 251
Seymour Johnson AFB, NC, 86, 216
SH-60F (SEAHAWK), 199, 238, 239–240, 252
Shallow Burst Munition, 311
Shallow Underwater Missile system (SUM), 128, 247
   as alternative to MX, 147
   multiple protective shelter, 147
Shepherd AFB, TX, 85
Sherwin, Martin, 31
Short Range Attack Missiles, (SRAM), 17, 80, 100, 154–155
   and Advanced Strategic Air Launched Missile, 155, 191, 193
   contractors for, 154
   warheads, 154
Shot George, 27
   warheads, 154
Sierra Army Depot, CA, 87
Sierracin Corp., 161
Silo test model for MX, 129. *See also* MX missile
Simmons Precision Inc., 161
Singer, 289
Singer-Kearfott Division, 154, 161, 293
Single Integrated Operational Plan (SIOP), 82
SIPRI Yearbook, 2, 137
Small ICBMs (SICBM), 109, 125, 129, 132, 133, 216
SNARK cruise missile, 172
Soesterberg AB, Netherlands, 218
SofTech Inc., 124
South Korea, 45, 50, 53, 60, 66, 78, 94, 95, 280, 282, 283, 302, 305, 307, 308, 309, 311
Soviet Union, AWACS, 11, 193
   military and foreign policies, 2
Spangdahlem AB, West Germany, 216
SPARTAN missile test, Marshall Islands, 163
Special Atomic Demolition Munitions (SADM), 3, 5, 8, 33, 34, 60, 91, 281, 311
Sperry Corp., 161
Sperry Rand, 289
Sperry Systems, 136
Sperry Vickers Co., 161
Spruance class destroyers, 185, 186, 256, 257, 268
SSBN-X program, 147
Stainless Steel Products Co., 161
STANDARD 2, 189, 244, 246, 255, 256, 257, 258, 261, 262, 272–276, 277
STANDARD 2ER, 244, 255, 262, 265, 276
STANDARD/MR, 244, 263
"Stealth," (ATB), 153, 156, 159, 162, 172, 173, 191, 203

Sterrer Engineering and Manufacturing Co., 161
Stewart-Warner Electronics Division, 154
Strat-X, 104
Strategic ACLM Launcher (SAL), 156
Strategic Air Command (SAC), 11, 83, 85, 125, 150, 152, 153, 160, 172, 214
Strategic bomber forces, 105–106
   basing locations, 86
Strategic Defense System, 163–169. *See also* Anti-Ballistic Missiles, Ballistic Missile Defense
Strategic interception forces, basing of, 88
"Strategic Missile Systems 2000," 109
Strategic nuclear forces, 101
Strategic Reserve Force, 100
Strategic weapons development, 16–19
STRIKE EAGLE, 203, 204, 219
Sturgeon class submarines, 186, 247, 271
Subic Bay Naval Base, Phillipines, 94
Subcritical mass, 24–26
Submarine Launched Ballistic Missile (SLBM), 2, 100, 101, 144, 266. *See also* POSEIDON, TRIDENT
Submarine rocket (SUBROC), 3, 5, 8, 61, 92, 94, 244, 246, 247, 249, 269–270
   ANALOG fire control system, 270
Submarine tender (AS), 92
Subsonic Cruise Armed Decoy (SCAD), 172
Sunstrand Aviation Corp., 161, 176
Sunstrand Data Control, 161
Subsurface and delivered ASW Standoff weapons, 18
Super orraloy bomb, 34
Supercritical mass, 24, 26
Surface-to-air missiles, 5, 52, 89, 273, 275, 277, 287. *See also* NIKE-HERCULES, PATRIOT, STANDARD, TERRIER
Surface-to-Surface missiles (SSM), 4, 45, 46, 72, 75, 179, 273, 275, 282, 284, 289
System Development Corp., 165
System Operational Range, 178
Systems Science and Software, 124
Sylvania Electric Products Co., 136
Systron-Doner Corp., 284

T-16, 299
T-22, 299
Tactical Air Command (TAC), 86, 214
Tactical Air to Surface Munition (TASM), 202
Tactical Fighter Derivative program, 203
Tactical nuclear weapons, 3, 12, 19–20
Tactical warheads, mobilization trends, 19–20
Taegu AB, South Korea, 216
Taiwan, 45
TALOS surface-to-air missiles, 20, 272
Tamper, 25, 26, 28
Tank Landing Ship (LST), 89
TARTAR surface-to-air missile, 272, 275–277
TASC, 124
Taylor, Theodore, 34
Teal Dawn, 191, 192

Index

Technical Proficiency Inspection, 83
Teledyne, 173, 176, 188, 191
Teledyne Brown Engineering, 165, 166
Teledyne Ryan, 148
Teller, Edward, 127–129
Terminal Guided and Extended Range (TIGER) I & II, 202
Terrain Contour Matching (TERCOM), 179, 184, 186
TERRIER missile, 3, 4, 5, 8, 92, 95, 189, 244, 246, 251, 252, 255, 264, 267, 272, 273–277
Texas Instruments, 188
Theater warheads, 2, 4
Thermonuclear weapons, 26–28. See also Nuclear weapons
Third generation weapons, 29
THOR, 12
Thule, Greenland, nuclear accident in, 12
Ticonderoga class cruisers, 186, 244, 246, 256, 257–260
TITAN missile, 84, 86, 112
   I, 12, 102
   II, 3, 4, 8, 16, 34, 58, 59, 86, 87, 101, 111–112, 120
     locations, 112
     MK-6 reentry vehicle 100
     replacement of, 103
TOMAHAWK (SLCM), 79, 80, 172, 179, 181, 183–187, 189, 191, 244–247, 249, 250, 255, 256, 258, 261, 262, 264, 265–266, 270
TORNADO, 19, 94, 199, 202, 229, 241–242
Torrejon AB, Spain, 216
TRESTLE EMP simulator, 29
"Triad," 100, 101, 125
TRIDENT, 3, 4, 8, 16, 18, 74, 76, 100, 102, 104, 109, 126, 128
   TRIDENT I C4 missile system, 4, 18, 92, 104, 133, 137, 142–143, 144, 146, 169
   TRIDENT II D5 missile, 144–146
   TRIDENT submarine, 17, 138–141
Trinity test, 7, 24, 26, 31
Tritium, 14, 27, 308, 309
TRW, Inc., 124, 161, 165, 166
Tsipis, Kosta, 80, 178
Turkey, 45, 46, 65, 66, 78, 94, 95, 216, 226, 280, 281, 282, 283, 300

Ultra Systems Inc., 124
Underground testing of nuclear weapons, 14
Undersea Long Range Missile System (ULMS), 103, 104
Unidynamic, 173, 176
United Aircraft Products, Inc., 161
United Kingdom, 47, 54, 63, 66, 69, 73, 78, 94, 95, 104, 241, 280, 281, 285, 300, 307
   US weapon locations in:
     RAF Greenham Common, 180
     RAF Molesworth, 180
     RAF Upper Heyford, 233
United Nations, report on nuclear weapons, 51
United States Air Forces Europe (USAFE), 86
United States, 47, 49, 50, 58, 59, 60, 61, 62, 66, 68, 69
   missile locations in, 74, 76

SAC bomber bases, 71
   warheads for strategic defense, 83
United Technologies, 176, 194
Universal Match Corp., 154
University of Houston, 124
Uranium, 22, 23, 26, 27, 32, 38
USS Adams, with ASROC, 268
USS Agerholm, 264
USS Alabama, 140
USS America, 252
USS Ashtabula, 92
USS Bainbridge, 260
USS Belknap, 260
USS Ben Franklin, 103, 134
USS Bronstein, 262, 263, 268
USS Brooke, 262, 263, 267, 268
USS Constellation, 252
USS Coral Sea, 251, 252
USS Daniel Webster, 104
USS Decatur, 262
USS Dwight D. Eisenhower, 252
USS Enterprise, 252
USS Ethan Allen, 102
USS Farragut, 262
USS Florida, 140
USS Forrestal, 251, 252
USS Forrest Sherman, 262
USS Garcia, 262, 263, 268
USS George Washington, 104
USS Georgia, 140
USS Glover, 262, 263, 268
USS Guitarro, 185
USS Independence, 252
USS James Madison, 103, 104, 134, 137
USS John F. Kennedy, 252
USS Josephus Daniels, 260
USS Kidd, 255, 256, 262, 268
USS Kitty Hawk, 251, 252
USS Knox, 262, 263, 268
USS Lafayette, 103, 104, 134, 137
USS Leahy, 260
USS Lipscomb, 247
USS Long Beach, 186, 260
USS Merrill, 185
USS Michigan, 140
USS Midway, 251, 252
USS Missouri, 256
USS Narwal, 247
USS New Jersey, 246, 256
USS Nimitz, 253
USS Ohio, 104
USS Permit, 186, 247
USS Ranger, 252
USS Rhode Island (SSBN), 140
USS Roosevelt, 202
USS Sam Rayburn, 134
USS Saratoga, 251, 252

Index

USS *Sherman*, 268
USS *Ticonderoga*, 93
USS *Truxton*, 260
USS *Wainwright*, 275
USS *Will Rogers*, 104
USS *Wisconsin*, 256
UTC, Hamilton Standard Division, 161

Vandenburg AFB, CA, 85, 117
Variable yields, 38
Vertical Launch ASROC (VLA), 246, 255, 256, 261, 262, 265, 266, 267, 268
Vertical Launching Systems (VLS), 80, 185, 186, 193, 246, 250, 256, 258, 259, 264, 265, 266, 272, 277
Very High Speed Integrated Circuits (VHSIC), 203
Vickers Aerospace Co., 161
Virginia class cruisers, 185, 186, 206, 276
Vitro Labs, 136, 173
Volkel AB, Netherlands, 229
Vought Corp., 161, 209, 299

WALLEYE missile, 202
Warheads, active:
  W25, 3, 4, 7, 12, 18, 35, 39, 41, 86, 94, 95, 100, 168, 218, 231
  B28, 3, 4, 5, 7, 31, 42–44, 85, 86, 95, 159, 198, 199, 200, 205, 208, 210, 216, 227, 229, 242, 253
  W31, 3, 4, 5, 7, 12–13, 16, 39, 45–46, 94, 95, 280, 282, 287
  W33, 3, 5, 7, 12, 13, 19, 26, 38, 39, 47–48, 55, 78, 90, 281, 300, 306
  B43, 3, 4, 5, 49, 85, 86, 90, 92, 93, 94, 95, 149, 153, 198, 199, 200, 205, 208, 210, 221, 227, 229, 242, 244, 267, 271
  W44, 3, 4, 5, 8, 39, 51, 94, 244, 267, 271
  W45, 3, 5, 8, 12, 20, 39, 52–53, 94, 95, 244, 272, 273
  W48, 3, 5, 12, 19, 39, 54, 55, 90, 95, 281, 300, 302, 305, 308, 309, 310
  W50, 2, 3, 4, 5, 8, 12, 39, 56–57, 95, 289
  B53, 3, 4, 8, 39, 58, 86, 90, 199, 200
  W53/Mk-6, 3, 4, 5, 8, 38, 39, 59, 111, 149
  W54, 3, 12, 33, 34, 38, 39, 60, 96
  W55, 2, 3, 4, 5, 8, 39, 61, 244, 269, 271
  W56 Mk-11c, 3, 4, 8, 39, 62, 111
  B57, 3, 4, 5, 8, 39, 86, 92, 93, 94, 149, 198, 199, 205, 206, 210, 216, 221, 224, 227, 229, 233, 235, 237, 238, 240, 242, 244, 251, 253, 255, 266, 271
  B61, 3, 4, 5, 8, 16, 31, 38, 39, 85–86, 90, 91, 94, 95, 149, 153, 183, 198, 199, 202, 205, 207, 210, 216, 221, 224, 229, 233, 242, 244, 271, 253, 278, 297, 311
  W62/Mk-12, 3, 4, 8, 39, 68, 71, 100, 118
  W68/Mk-3, 3, 4, 8, 39, 69–70, 71
  W69, 3, 4, 9, 39, 71, 86, 149, 152, 153, 159
  W70, 3, 4, 5, 9, 15, 19, 39, 72–73, 95, 280, 284, 297
  W76/Mk-4, 3, 4, 9, 15, 16, 17, 38, 39, 74
  W78/Mk-1a, 3, 4, 9, 15, 38, 39, 68, 75–76, 111, 116, 118, 122, 124, 126–127, 133
  W79, 3, 5, 9, 16, 19, 38, 39, 48, 77–78, 90, 281, 300, 306
  W80, 3, 4, 9, 16, 17, 18, 19–20, 38, 39, 79–80, 149, 150, 159, 172, 175, 185, 191, 194, 244, 297
Warheads:
  allocation by service branch, 83
  average age of, 140
  development, 11
  inactive, 10–11
  safety and control features, 30–31
  deployed by Navy, 39
  retired: W23, 12; W19, 12; W29, 12; W40, 12; W52, 12; W66, 38; W71, 38; W71, 38; W72, 202; W74, 308; W8(?) 9, 292, 311
Warheads under development:
  B77, 9, 198, 201
  B83, 9, 16, 18, 39, 43, 49, 58, 149, 153, 159, 198–202, 216, 233
  W82, 9, 15, 16, 18, 19–39, 90, 208–210, 281
  W84, 9, 16, 18, 19, 38, 80, 172, 179, 182–183
  W85, 9, 16, 18, 39, 133, 294, 297
  W87, 9, 15, 16, 18, 38, 90, 120, 122, 133
Weak link/strong link, 30, 80, 200
Wellman Dynamics Corp., 176
West German Air Force, 280, 289, 290, 292
West Germany, 45, 47, 53, 54, 56, 57, 60, 63, 66, 78, 88, 89, 94, 95, 199, 207, 209, 211, 229, 241, 280, 281, 285, 287, 288, 289, 290–292, 296, 300, 301, 302
  American bases in:
    Bitburg AB, 218
    Buchel AB, 229
    Hahn AB, 221
    Memmingen AB, 229
    Neckars Ulm, 290
    Neu Ulm, 290
    Norvenich AB, 229
    Ramstein AB, 216
    Schwaebisch Gmuend AB, 290
    Wueschein, 180
Western Electric Corp., 136, 287
Westinghouse Electric, 124, 136, 149, 161, 173, 220
Whiteman AFB, MO, 87, 114, 115
Williams International, 173, 176, 191
Wilson, George, 178

X-ray laser, 29

Yield-to-weight ratio, 22, 23, 36–27
York, Herbert, 34

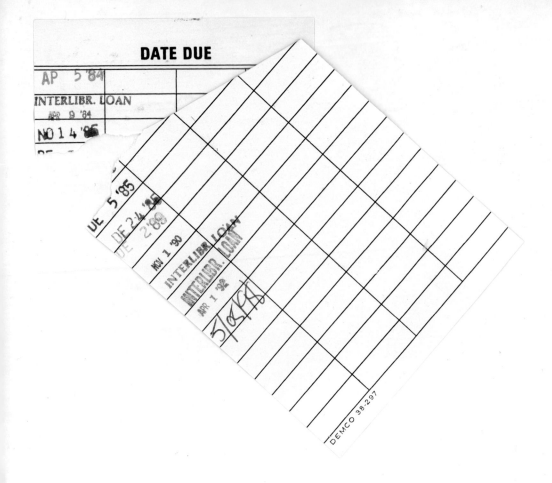